The Mollusca

VOLUME 3

Development

The Mollusca

Editor-in-Chief

KARL M. WILBUR

Department of Zoology
Duke University
Durham, North Carolina

The Mollusca

VOLUME 3
Development

Edited by

N. H. VERDONK
J. A. M. VAN DEN BIGGELAAR

Zoological Laboratory
State University of Utrecht
Utrecht
The Netherlands

A. S. TOMPA

Museum of Zoology
The University of Michigan
Ann Arbor, Michigan

1983

ACADEMIC PRESS
A Subsidiary of Harcourt Brace Jovanovich, Publishers
New York   London
Paris   San Diego   San Francisco   São Paulo   Sydney   Tokyo   Toronto

ACADEMIC PRESS, INC.
111 Fifth Avenue, New York, New York 10003

United Kingdom Edition published by
ACADEMIC PRESS, INC. (LONDON) LTD.
24/28 Oval Road, London NW1 7DX

Library of Congress Cataloging in Publication Data

Main entry under title:

The Mollusca.

    Includes index.
    Contents: v. 1. Metabolic biochemistry and
molecular biomechanics / edited by Peter W.
Hochachka -- v. 2. Environmental biochemistry
and physiology / edited by Peter W. Hochachka --
v. 3. Development / edited by N.H. Verdonk &
J.A.M. van den Biggelaar & A.S. Tompa -- v. 4-5.
Physiology / edited by A.S.M. Saleuddin & Karl M.
Wilbur.
    1. Mollusks--Collected works. I. Wilbur, Karl M.
QL402.M6 1983   594   82-24442
ISBN 0-12-751403-1 (v. 3)

PRINTED IN THE UNITED STATES OF AMERICA

83 84 85 86   9 8 7 6 5 4 3 2 1

To Christiaan P. Raven and Anthony C. Clement

Contents

1. Gametogenesis

M. R. DOHMEN

2. Meiotic Maturation and Fertilization

FRANK J. LONGO

3. Early Development and the Formation of the Germ Layers

N. H. VERDONK AND J. A. M. VAN DEN BIGGELAAR

4. Organogenesis

BEATRICE MOOR

5.   Origin of Spatial Organization

J. A. M. VAN DEN BIGGELAAR AND P. GUERRIER

6.   Morphogenetic Determination and Differentiation

N. H. VERDONK AND J. N. CATHER

7.   The Biochemistry of Molluscan Development

J. R. COLLIER

8. Physiological Ecology of Marine Molluscan Larvae

B. L. BAYNE

Contributors

Numbers in parentheses indicate the pages on which the authors' contributions begin.

B. L. Bayne (299), Natural Environmental Research Council, Institute for Marine Environmental Research, Plymouth PL1 3DH, United Kingdom

J. A. M. van den Biggelaar (91, 179), Zoological Laboratory, State University of Utrecht, 3508 TB Utrecht, The Netherlands

J. N. Cather (215), Department of Zoology, The University of Michigan, Ann Arbor, Michigan 48104

J. R. Collier (253), Biology Department, Brooklyn College of the City University of New York, Brooklyn, New York 11210

M. R. Dohmen (1), Zoological Laboratory, State University of Utrecht, 3508 TB Utrecht, The Netherlands

P. Guerrier (179), Station Biologique, 29.211 Roscoff, France

Frank J. Longo (49), Department of Anatomy, The University of Iowa, Iowa City, Iowa 52242

Beatrice Moor (123), CH-4000 Basel, Switzerland

N. H. Verdonk (91, 215), Zoological Laboratory, State University of Utrecht, 3508 TB Utrecht, The Netherlands

General Preface

This multivolume treatise, *The Mollusca,* had its origins in the mid 1960s with the publication of *Physiology of Mollusca,* a two-volume work edited by Wilbur and Yonge. In those volumes, 27 authors collaborated to summarize the status of the conventional topics of physiology as well as the related areas of biochemistry, reproduction and development, and ecology. Within the past two decades, there has been a remarkable expansion of molluscan research and a burgeoning of fields of investigation. During the same period several excellent books on molluscs have been published. However, those volumes do not individually or collectively provide an adequate perspective of our current knowledge of the phylum in all its phases. Clearly, there is need for a comprehensive treatise broader in concept and scope than had been previously produced, one that gives full treatment to all major fields of recent research. *The Mollusca* fulfills this objective.

The major fields covered are biochemistry, physiology, neurobiology, reproduction and development, evolution, ecology, medical aspects, and structure. In addition to these long-established subject areas, others that have emerged recently and expanded rapidly within the past decade are included.

The Mollusca is intended to serve a range of disciplines: biological, paleontological, and medical. As a source of information on the current status of molluscan research, it should prove useful to researchers of the Mollusca and other phyla, as well as to teachers and qualified graduate students.

Karl M. Wilbur

Preface

Developmental biology has become more and more an analysis of the mechanisms that underlie development with increasing emphasis on the cellular and the molecular approach. As a consequence it appears inevitable that the embryo is lost from sight. Even when developmental biology is studied at the supracellular level, the embryo proper may be overlooked as emphasis is laid on pattern formation in general. Nevertheless, the systematic study of the embryo proper is essential to the full comprehension of the biology of the respective taxa.

Fortunately, editors of monographs on the biology of special taxonomic groups often include a chapter or even a volume on development. Therefore, we welcomed the decision of Karl M. Wilbur to publish a special volume on development (and its companion volume on reproduction). We accepted his invitation to serve as its editors, the more so because no monograph on molluscan development has appeared since the book *Morphogenesis, The Analysis of Molluscan Development* by C. P. Raven appeared in 1958.

The extensive study of the embryonic development of molluscs and their wide usage as a model system for the resolution of problems in developmental biology mandated the participation of a number of distinguished scientists in the genesis of this volume. We have tried to cover all aspects of molluscan development starting with gametogenesis through the physiological ecology of the larvae. As far as possible, all groups of molluscs on which data are available are taken into account. Because the development of the cephalopods stands quite apart within the Phylum Mollusca, we had planned a separate chapter on this group. However, circumstances beyond our control have forced us to omit this chapter. Although we regret this very much, we hold the view that, because of the isolated position of the cephalopods, the remaining chapters represent an account of the development of spiralian molluscs that is complete and intelligible in itself.

The considerable progress in the experimental approach in recent decades finds concrete shape in the framework of this volume. Several chapters are devoted to an experimental analysis of the basic phenomena of molluscan development, and even in the more descriptive parts of the volume, a causal analysis often substantiates the presented data.

This volume, as a part of *The Mollusca,* is primarily meant to inform established investigators in various domains of molluscan reseach on the development of this group of animals. However, it will also help fellow scientists active in other fields of developmental biology to gain insight into the state of affairs of molluscan embryology.

We would like to express our thanks to our collaborators and most of all to the authors who first willingly accepted our invitation to write a chapter and later patiently coped with our comments, mostly intended to avoid duplication and overlap of chapters. We wish to thank Miss Angela de Lange for assistance with the final preparation of the manuscripts.

This volume is dedicated to two pioneers in the field of experimental embryology of molluscs: Christiaan P. Raven and Anthony C. Clement, who laid the fundamentals on which their students and many others are building further. We sincerely hope that this volume stimulates further research in molluscan development.

N. H. Verdonk

J. A. M. van den Biggelaar

A. S. Tompa

Contents of Other Volumes

Volume 2: Environmental Biochemistry and Physiology

Volume 4: Physiology, Part 1

Volume 5: Physiology, Part 2

Volume 6: Ecology

1

Gametogenesis

M. R. DOHMEN

Zoological Laboratory
State University of Utrecht
Utrecht, The Netherlands

I. Origin and Determination of Germ Cells

The origin of germ cells in molluscs has been completely elucidated in one species, the bivalve *Sphaerium striatinum* (Woods, 1931, 1932). In this species, the cell lineage of the germ cells can be accurately established because they contain a conspicuous marker consisting of a large aggregate of mitochondria. This mitochondrial cloud is already present in the oocyte and it segregates into the 4d micromere and subsequently into the primordial germ cells.

It is unknown whether this plasm determines the germ cell line. Cytoplasmic localizations that are thought to act as germ cell determinants have been observed in several groups of animals, notably in insects and amphibians (see Bounoure, 1939; Beams and Kessel, 1974; Nieuwkoop and Sutasurya, 1979, 1981; Ma-

1

howald et al., 1979). In two molluscan species, *Crepidula fornicata* and *Buccinum undatum*, plasms that are structurally similar to the germ plasms in insects and amphibians have been found in the polar lobe that is formed at first cleavage (Dohmen and Lok, 1975; Dohmen and Verdonk, 1979a,b). In *Buccinum*, the fate of these plasms is unknown. In *Crepidula*, there is some evidence that they segregate into the 4d micromere, but it is not known if they are incorporated into the primordial germ cells. These few data suggest that in a number of molluscs the germ cells may be determined in a preformistic way by means of the segregation of an ooplasmic determinant into the presumptive germ cell line.

In many molluscs, the germ cells can be recognized at an early stage by their characteristic morphology. They constitute small clusters of large cells that are carried to the place where they will multiply and form the gonads together with nongerminal cells (Woods, 1931; Brisson, 1973; Moor, 1977; and Chapter 4, this volume). The precocious morphological distinction of the germ cells probably means that they are determined before gonad formation. However, this determination does not exclude the possibility that some of the descendants of the primordial germ cells differentiate into nongerminal cells. For example, in *Sypharochiton* (Selwood, 1968), the primary oogonia divide four times, giving rise to 16 secondary oogonia. The outer oogonia in this cluster become follicle cells, the inner ones become oocytes. Hogg and Wijdenes (1979) argue that in the gastropod *Deroceras* the germinal epithelium consists of one cell type, the gonadal stem cell, that persists throughout adult life and from which both gametes and nongerminal cells develop.

In most species investigated thus far the germ cells appear to derive from mesodermal cells, either located in the pericardium or present as isolated cell masses in the body cavity (see Raven, 1966; Chapter 4, this volume). In a few pulmonates, however, an ectodermal origin has been postulated. Luchtel (1972a,b) and Hochpöchler (1979) reported that the gonads in these pulmonates develop from an ectodermal invagination or proliferation and found no evidence for the migration of mesodermal primordial germ cells into this gonad anlage. According to Moor (Chapter 4, this volume) it is more likely that these authors have overlooked the early stages in which the mesodermal primordial germ cells become incorporated into the tip of the ectodermal proliferation, as observed in other pulmonates (see also Griffond and Bride, 1981, for *Helix*).

Under experimental conditions, germ cells may derive from a normally somatic tissue, namely the spermoviduct in pulmonates. This is inferred from the observation that functional gonads can regenerate from implanted pieces of spermoviduct or after castration (Hogg and Wijdenes, 1979).

A. Sexualization of the Germ Cells

In molluscs the male and female gametes probably differentiate from sexually nondetermined protogonia. In gonochorists, the differentiation of primordial germ cells into male or female gametes is controlled by a neurohormone, the androgenic factor. Experiments in which isolated gonads were cultured under different conditions have shown that in the absence of androgenic factor, the primordial germ cells differentiate into female gametes. Differentiation into male gametes requires the presence of androgenic factor (see Le Gall and Streiff, 1975).

In simultaneous hermaphrodites, the presence of an androgenic factor has also been demonstrated, for example, in *Helix aspersa* (see Gomot, 1973). The question of how female gametes can develop in these hermaphrodites has not been solved. The spatial organization of the gonadal acini may play a role in the differentiation of germ cells. In a number of species the acini are divided into a male and a female compartment (Joosse and Reitz, 1969; Luchtel, 1972a,b; Sabelli and Scanabissi, 1980). These compartments may provide the microenvironmental conditions that control the differentiation of the germ cells. It is by no means certain, however, that the position of sexually undetermined protogonia in a certain compartment determines whether they will differentiate into male or female gametes. It is conceivable that the position in the acinus is not a determinative factor but rather a permissive factor in the sexualization of germ cells.

B. Atypical Sperm and Nutritive Eggs

An intriguing problem concerning germ cell determination is the origin of atypical sperm and nutritive eggs. Atypical sperm are abnormal varieties of sperm that are consistently produced by many gastropods (see Section II,A). Nutritive or food eggs are initially normal-looking eggs whose development is arrested or becomes abnormal at some point. They serve as food for normally developing larvae. Both atypical sperm and food eggs show species-specific characteristics and the percentage of atypical cells is constant. This suggests that these types of germ cells do not result from random defects during gametogenesis.

Reinke (1914) reports that atypical sperm of *Strombus* do not differentiate from spermatogonia but from accessory cells. Hence, their development is not comparable to spermatogenesis. However, most authors agree that atypical sperm originate from spermatogonia. Several detailed accounts of the development of atypical sperm have been published (e.g., Gould, 1917; Bulnheim, 1968; Kohnert, 1980). The spermatogonia are all alike initially, but they follow divergent paths of development at an early stage. In several species, the first differences are observed at the end of the proliferation period of the spermatogonia. In *Viviparus*, for example, atypical spermatogonia can be recognized by their well-developed ergastoplasm (Griffond, 1981).

Many atypical sperm grow much larger than normal sperm and store yolklike substances (see Bulnheim, 1968). This has led Ankel (1930) to suggest that oogenetic as well as spermatogenetic potencies are activated during atypical sperm development. This view does not hold for all species, as discussed by Bulnheim (1968).

It is completely unknown how atypical sperm are determined. Yamasaki (1966) suggests that in *Cipangopaludina* the distribution of Golgi bodies during the proliferation period of the spermatogonia determines atypical spermatogenesis.

Unlike atypical sperm, whose divergent development is evident at an early stage during spermatogenesis, nutritive eggs cannot be distinguished from normal eggs until after fertilization at the earliest. Portmann (1927) has suggested that nutritive eggs may result from fertilization by atypical sperm. This has not been convincingly demonstrated. Staiger (1951) argues that genetically determined incompatibility between a certain percentage of normal sperm and normal eggs might be responsible for the defective development of a certain number of eggs. This incompatibility is comparable to the self-sterility exhibited by some hermaphrodites. It may result in various defects in the process of fertilization, such as nonentrance of sperm into the egg, immobility of the male pronucleus, inhibition of amphimixis, and polyspermy. Even in nutritive eggs that show a seemingly normal development until rather advanced stages, the block of further development seems to be due to defective fertilization. In *Murex trunculus,* for example, nutritive eggs can proceed through a normal spiral cleavage until about the 60-cell stage, but this development is parthenogenetic. The male genome does not take part in it, as the male pronucleus remains immobilized and inactive at the sperm entrance point in the vegetal hemisphere (Staiger, 1951). It would be interesting to attempt to transplant the nucleus or part of the cytoplasm from an atypical egg to a normal egg and vice versa in order to see to what extent cytoplasmic or nuclear factors determine the defective development of a nutritive egg.

II. Spermatogenesis

Spermatogenesis starts with a spermatogonium mother cell dividing mitotically, giving rise to 32 primary spermatogonia. These cells grow and then divide, resulting in 64 secondary spermatogonia that differentiate into primary spermatocytes. Subsequently, two meiotic divisions give rise first to 128 secondary spermatocytes and then to 256 haploid spermatids. Such details are mentioned for the gastropod *Biomphalaria* (de Jong-Brink et al., 1977) and for *Helix aspersa* (Bloch and Hew, 1960) but it is not known whether they have general validity. The spermatids then enter into the process of spermiogenesis, that is, the differentiation into mature sperm. At all stages of spermatogenesis, starting with spermatogonia, cytoplasmic bridges are encountered between the cells constituting one clone. These bridges result from incomplete cytokinesis during the mitotic divisions that precede meiosis, and they may be retained until late spermatid stages, when the redundant cytoplasm is being discarded. The bridges are thought to ensure synchronous division and differentiation (Fawcett, 1961). Sperm cells are highly polar cells. This polarity is apparent at an early stage. In *Viviparus*, for example, the secondary spermatogonia show a polar organization in the distribution of organelles (Griffond, 1978). In species in which the developing germ cells are attached to a sertoli cell, a nutritive cell, or simply to the acinar epithelium, the apex of the cell corresponds with the attachment side. This suggests that the intercellular contact somehow determines the polarity of the germ cell. However, spermatogonia may also lie free in the lumen of the gonad, for example in *Crepidula* (Gould, 1917). It should be interesting to investigate the origin of polarity in such a system.

A. Sperm Types

In the monograph by Baccetti and Afzelius (1976), four main types of sperm are distinguished: primitive, modified, biflagellate, and aflagellate. In molluscs, both primitive and modified types of sperm occur. Primitive sperm consist of three distinct regions: head, middle piece, and tail. The middle piece consists of a few mitochondria surrounding the two centrioles. The tail is of the conventional nine plus two formula. In modified sperm, the borders between head, midpiece, and tail are indistinct. The mitochondria are modified; they often form long strands along the flagellum. The tail often contains an extra set of nine accessory elements, the dense fibers, outside the nine plus two filaments. Primitive sperm are produced by species that spawn their gametes into the water, where fertilization takes place. Some archaeogastropods and most bivalves, chitons, and scaphopods belong to this category. Modified sperm are produced by species with internal fertilization: most prosobranchs and all opisthobranchs and pulmonates.

In many meso- and neogastropods and some archaeogastropods, sperm dimorphism occurs, that is, one individual produces typical as well as atypical sperm. This phenomenon also occurs in insects, annelids, myriapods, and rotatoria (see Fain-Maurel, 1966). Recently, the term paraspermatozoon or paraspermatic cell has been introduced in order to avoid the connotation of abnormality in the term atypical sperm (Melone et al., 1980; Healy and Jamieson, 1981). Typical or eupyrene sperm have a normal chromosome complement and serve for the fertilization of eggs. Atypical or dyspyrene sperm have an abnormal chromosome complement and their function is largely unknown. Three classes of abnormalities in chromosome numbers can be distinguished: (1) hyperpyrene sperm contain more than the normal haploid set of chromosomes, (2) oligopyrene sperm contain less than a haploid set, and (3) apyrene sperm lack chromosomes altogether. Atypical sperm often possess multiple flagella, up to 3000 in the prosobranch *Opalia* (Bulnheim, 1962). Considerable differences between atypical and typical sperm have been observed in other aspects as well. Nishiwaki (1964) distinguishes eight basic types of atypical sperm in prosobranchs and Tochimoto (1967) adds still another type. A peculiar form of atypical sperm are the so-called spermatozeugmata, giant carriers of eupyrene sperm, for example, in *Epitonium* (Bulnheim, 1968). According to some authors, nurse cells can be regarded as homologous with atypical sperm (Buckland-Nicks and Chia, 1977). On the other hand, there are reports of dyspyrene sperm that can hardly be distinguished as such. In *Bithynia,* for instance, the oligopyrene sperm differs from eupyrene sperm only in having a smaller nucleus (Kohnert, 1980). The function of atypical sperm is unknown. Bulnheim (1968) concluded that their main function is probably nutritive. In this respect, they resemble another atypical germ cell in molluscs, the nutritive egg cell (see Section I,B).

B. Spermiogenesis

The allotted space does not permit a detailed comparative analysis of molluscan spermiogenesis and sperm structure. More information can be retrieved from the literature (Baccetti and Afzelius, 1976; Anderson and Personne, 1976; Roosen-Runge, 1977; Giese and Pearse, 1977, 1979; Fawcett and Bedford, 1979). The purpose of this section is to give a short description of the structures that are present in molluscan sperm and their origin.

1. Acrosomal Complex

The main parts of the acrosomal complex are the acrosomal vesicle and the subacrosomal substance, which is also called perforatorium, acrosomal rod, or axial rod. The acrosomal vesicle generally originates from proacrosomal granules derived from one or more Golgi complexes. In the prosobranch *Nerita* (Garreau de Loubresse, 1971) and the nudibranch *Spurilla* (Eckelbarger and Eyster, 1981), the Golgi complex does not seem to be involved in the formation of the acrosome. The proacrosomal granules fuse to form an acrosomal granule, which becomes visible in the early spermatid stage. This granule increases in size and develops into the acrosomal vesicle. The contents of the vesicle enable the sperm to penetrate the barriers that surround the egg (see Chapter 2, this volume).

The acrosome is usually situated at the apex of the sperm. However, in a number of species, such as the bivalves *Laternula* (Kubo, 1977), *Lyonsia* (Kubo and Ishikawa, 1978), and *Myodora* (Popham, 1979), the acrosome first forms in the anterior part of the spermatid, but then migrates posteriorly near the end of spermiogenesis. The posteriorly located acrosome does not react during fertilization (Kubo, 1977; Kubo et al., 1979). It is unknown which devices allow this type of sperm to penetrate through the egg envelopes. The same holds for a number of species in which the mature sperm do not contain an acrosome at all, as is the case in a number of bivalves (Franzén, 1955), the opisthobranch *Aplysia* (Thompson and Bebbington, 1969), and chitons (Pearse and Woollacott, 1979).

The subacrosomal substance consists of fibrous material between the acrosomal vesicle and the nucleus. It acts as the core of the acrosomal process. The accumulation of this material has been described in detail for the pulmonate snail *Euhadra hickonis* (Takaichi and Dan, 1977). In several species, the subacrosomal substance is organized into a more or less completely preformed acrosomal filament or axial rod (Baccetti and Afzelius, 1976; Popham, 1979). In other species, the subacrosomal substance polymerizes at the time of the acrosome reaction (Baccetti and Afzelius, 1976). In the archaeogastropod *Haliotis*, for instance, only about one-third of the length of the acrosome process preexists in the form of a bundle of actin filaments. The extension to its full length of 7 μm occurs during the acrosome reaction (Lewis et al., 1980). The acrosome process probably serves to bind the sperm to the egg (see Chapter 2, this volume).

2. Nucleus

During spermiogenesis the chromatin is condensed, the pores in the nuclear envelope disappear, the two laminae of the nuclear envelope adhere to each other, and the shape of the nucleus may change considerably. Yasuzumi (1974) distinguishes three basic patterns of chromatin condensation: granular, fibrillar, and lamellar. In molluscs, all three types have been observed. Successive stages of condensation and concomitant reduction of nuclear size are illustrated in Fig. 1. During this process, the chromosomes probably retain their individuality (see Baccetti and Afzelius, 1976). Basic nuclear proteins play a role in the condensation of sperm chromatin. During spermiogenesis, the somatic histones in the spermatids may be partly or wholly substituted by sperm-specific histones or protamine-like proteins (protamines are proteins that contain more than 40% arginine). According to de Jong-Brink et al. (1977), chromatoid bodies might serve as a store of basic nuclear proteins. In *Biomphalaria*, these bodies are already present in spermatogonia mother cells and persist until young spermatid stages. They disintegrate during the transition from somatic histones to sperm-specific histones. In some cases these transitions have been observed to coincide with changes in the conformation of the chromatin. In *Thais* (Walker and Mac-Gregor, 1968) and *Colus* (West, 1978a), the transition coincides with the formation of nuclear lamellae. Nuclear magnetic resonance studies on the complex of chromatin and H1-like histone from sperm of *Mytilus* and other marine invertebrates indicate that the highly contracted state of sperm chromatin is directly related to the high arginine content of the sperm-specific histone (Puigdoménech et al., 1976).

There is considerable variation in the types and relative amounts of basic

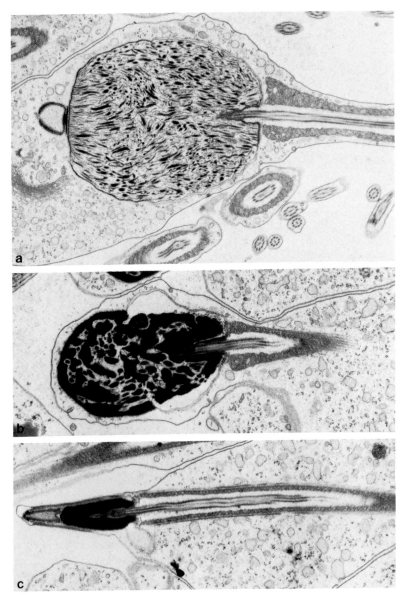

Fig. 1. Successive stages of chromatin condensation and concomitant reduction of nuclear size during spermiogenesis in the prosobranch *Bithynia tentaculata*. Magnifications of (**a**), (**b**), and (**c**) are identical.

nuclear proteins between sperm of different species. For instance, sperm of *Patella* do not contain any histone component, but only a protamine. Sperm of *Haliotis* and *Gibbula* contain both a protamine and an H2b-like histone (Subirana et al., 1973; Colom and Subirana, 1981). In a number of bivalve species, studied by Zalensky and Zalenskaya (1980) and Odintsova et al. (1981), the sperm contain both histones and protamine-like proteins. Bloch and Hew (1960) studied the biochemical changes in histone and the transition to protamine during sperm maturation in *Helix*.

During condensation, the chromatin strands often become oriented parallel to the anteroposterior axis of the cell. The development of the anteroposterior polarity of the nucleus has been described in detail for the pulmonate *Euhadra* (Takaichi, 1978). Polarization of the nucleus starts with the deposition of dense material against the nuclear envelope at the future anterior and posterior poles. Subsequently, a posterior invagination, the centriolar fossa, develops and the basal body of the flagellar axoneme moves into it.

The shape of the nucleus in mature sperm varies greatly between molluscan species. Even among the primitive sperm of the bivalves, the nucleus may vary from nearly spherical, ovoid, conical, and barrel-shaped forms, to pencil-shaped and long, tapering, curved forms (see Popham, 1979). In the modified sperm, column-shaped, helically coiled, and cylindrical nuclei are found (see Koike and Nishiwaki, 1980). In the cylindrical nuclei, the centriolar fossa extends to the very tip of the nucleus, thus forming an intranuclear canal (see Chapter 2, this volume). This type is found in *Littorina* (Buckland-Nicks and Chia, 1976), *Colus* (West, 1978a), and *Theodoxus* (Giusti and Selmi, 1982). An anterior invagination is also observed in several species. The subacrosomal substance may extend into this anterior fossa, for example, in *Haliotis* (Lewis et al., 1980) and in some bivalves (see Popham, 1979).

The processes most commonly suggested as being responsible for determining the shape of the nucleus are (1) the condensation of the chromatin (e.g., Fawcett et al., 1971; West, 1978b) and (2) the formation of a manchette of microtubules around the nucleus (e.g., Kubo and Ishikawa, 1981). The manchette can extend anteriorly over the acrosome, and posteriorly into the midpiece (see Shileiko and Danilova, 1979; Kohnert, 1980). In *Biomphalaria,* the manchette disappears when the nucleus has attained its final spiral shape (de Jong-Brink et al., 1977). These authors believe, therefore, that the manchette plays an active part in the spiralization of the nucleus and of the mitochondrial sheath. In a *Drosophila* mutant lacking the microtubules that normally surround the nucleus during elongation, the nuclei do not elongate and the chromatin is not packed normally (Wilkinson et al., 1974). This supports the view that microtubules play a role in shaping the nucleus. However, as pointed out by Fawcett et al. (1971) and West (1978b), there are species where the manchette does not appear until the nucleus has almost assumed its final shape, so its role in determining nuclear shape remains controversial (see also Risley et al., 1982). Dan and Takaichi (1979) argue that the function of the manchette is to enclose and hold in place the essential organelles while the superfluous material is discarded during spermiogenesis. Myles and Hepler (1982) point out that a distinction should be made between factors responsible for shape generation and those affecting shape determination. Microtubules and chromatin condensation do not provide the motive force for shape generation, but they may be important in determining the final shape, for example, by imposing restrictions on the motive force.

3. Mitochondria

In mature sperm of the primitive type, the middle piece contains one or a few large mitochondria which are probably formed by the fusion of several smaller ones. In modified type sperm, the mitochondria develop into a more elaborate structure called the chondriome (see André, 1962). In some species, an intermediate stage in the development of the chondriome is the so-called Nebenkern. This is a large aggregation of mitochondria adjacent to the nucleus, hence the name Nebenkern (secondary nucleus) (see Baccetti and Afzelius, 1976).

Chondriome formation starts with the fusion of the mitochondria, during the early spermatid stages, into a number of large mitochondria: four in *Viviparus* (Griffond, 1980) and nine in *Ocenebra (Purpura)* (Féral, 1977) and *Bithynia* (Kohnert, 1980). These mitochondria elongate posteriorly. They may form one or more spirals around the flagellar axoneme or a cylindrical sheath (Fig. 1). The cristae are reorganized into a periodic array. In several prosobranchs and pulmonates, the chondriome contains a large amount of paracrystalline material which, according to Dan and Takaichi (1979), is formed by the modification of mitochondrial cristae. This variety of chondriome is commonly referred to as the mitochondrial derivative. In *Paludina,* this material constitutes 95% of the total sperm volume (Favard and André, 1970). The Krebs cycle has been shown to be localized in the paracrystalline protein (Ritter and André, 1975; Anderson and Personne, 1976).

4. Centrioles

At the start of spermiogenesis, two centrioles are present. In primitive sperm, both centrioles are conserved. They are positioned at right angles to each other, the proximal one being oriented perpendicular to the axoneme and the distal one in line with the axoneme. The distal centriole forms the basal body of the flagellum. Deviations from this general picture are sometimes encountered. For example, in mature sperm of the bivalve *Laternula limicola,* the proximal centriole moves to the lateral side of the distal one in such a way that the two centrioles are parallel (Kubo and Ishikawa, 1978).

In modified sperm, several modifications have evolved. The proximal centriole is thought to disintegrate in many species (Gall, 1961; Walker and Mac-

Gregor, 1968; Garreau de Loubresse, 1971; Kohnert, 1980). In *Littorina,* the proximal centriole rotates until it is coaxial with the distal centriole and then the two fuse (Buckland-Nicks and Chia, 1976). In the sperm of the pulmonate *Discus* (Maxwell, 1976), no centriole could be detected at all. Koike and Nishiwaki (1980) observed that in *Conomurex* and *Lambis,* the nine triplets of the distal centriole disintegrate in mature sperm. In *Euhadra,* the centriole triplets first become doublets, which then degenerate (Dan and Takaichi, 1979). In some atypical sperm, the number of centrioles may increase enormously, each one acting as the basal body of a cilium. In the giant sperm of *Epitonium* and *Opalia,* for instance, the number of centrioles may approach 3000 (Bulnheim, 1968). A structure called the ring centriole in the literature is not a true centriole at all. In *Littorina,* the ring centriole is a dense band, produced by the distal centriole which then migrates to the base of the chondriome (Buckland-Nicks and Chia, 1976). A similar annulus has been observed in the prosobranchs *Bithynia* (Kohnert, 1980), *Conomurex,* and *Lambis* (Koike and Nishiwaki, 1980), but its origin in these species remains to be established.

Several types of structures may be associated with the centrioles. In *Euhadra* (Dan and Takaichi, 1979) and *Spurilla* (Eckelbarger and Eyster, 1981), a layer of dense material, the centriole adjunct, temporarily surrounds the centriole. It forms nine strands, which are continuous with the peripheral coarse fibers (Dan and Takaichi, 1979). In the oyster, fibers extending from the distal centriole apparently act as an apparatus that anchors the centriole to the plasma membrane (Galtsoff and Philpott, 1960). In *Colus,* the centriole, which is located at the anterior end of the intranuclear fossa, is attached to the nucleus by means of an electron-dense connecting piece (West, 1978a). In *Trichia* and *Succinea,* the centriolar region is filled with a homogeneous dense material called a centriolar derivative (Shileiko and Danilova, 1979). No centriolar microtubules can be distinguished and it is not known whether or not a normal centriole is present.

5. Flagellum

The typical sperm of molluscs possess one flagellum. Atypical sperm may develop very large numbers of flagella (see Bulnheim, 1968). In the atypical sperm of *Conomurex* and *Lambis* (Koike and Nishiwaki, 1980), a pair of undulating membranes is formed instead of flagella. These membranes run along the lateral sides of the spindle-shaped sperm and they contain 70–100 parallel axonemes. The axoneme within the flagellum normally consists of a central pair of microtubules surrounded by nine doublets. A few deviations from this general pattern have been reported in molluscs. In *Littorina,* a supernumerary tubule links the outer arm of doublet five to the basal body (Buckland-Nicks, 1973). In *Colus,* the end piece of the flagellum lacks an axonemal tubule (West, 1978a). In some gastropods, such as *Helix* (Anderson and Personne, 1969), an additional set of nine accessory fibers surrounds the axoneme. A peculiar configuration is found in *Theodoxus* (Giusti and Selmi, 1982). In this species, the flagellum folds backward at the point where the cytoplasmic sheath ends. As a consequence, movement is abnormal: the tail precedes the head. The flagella of many gastropods may contain large amounts of glycogen, which is generally located around the axoneme (see Anderson and Personne, 1976).

III. Oogenesis

The initial stages of oogenesis are very similar to the first phases of spermatogenesis. Primordial germ cells give rise to primary oogonia. In *Sypharochiton* (Selwood, 1968), these cells divide four times, giving rise to 16 secondary oogonia. These may enter into the first meiotic division and become primary oocytes, or they may differentiate into auxiliary cells.

A. Premeiotic Phase

Oogenesis is often divided into three stages: the premeiotic, previtellogenic, and vitellogenic phase. Different or more detailed subdivisions have been made by several authors (e.g., Ubbels, 1968; Ford, 1972). The premeiotic phase is defined two different ways. Some authors designate the period of oogonial proliferation as the premeiotic phase (Ford, 1972; Grant, 1978). In Raven's (1961) terminology, however, the premeiotic phase is the earliest phase of oocyte differentiation, when the chromosomes undergo a series of changes that represent the prophase of the first meiotic division. The successive stages of this prophase are the leptotene, zygotene, pachytene, and diplotene (see Raven, 1961; Grant, 1978). These early nuclear phenomena are called the premeiotic phenomena by Raven, and the early meiotic phenomena by others. At the end of this phase, meiosis is temporarily suspended and during diplotene the oocyte starts to grow and differentiate.

B. Previtellogenic Phase

During previtellogenesis, the oocyte starts growing. Both the nucleus and the cytoplasm increase in volume. The nucleus swells to form a germinal vesicle whose final volume is many times that of the original nucleus. In *Lymnaea*, an increase of 162 times the original volume is reported by Bretschneider and Raven (1951). The high density of pore complexes in the nuclear envelope, 60 pores/μm^2 in germinal vesicles of *Spisula* (Maul, 1980), suggests an intensive transport activity. The germinal vesicle appears to function as a selective storage device for a variety of substances, which are synthesized in the cytoplasm and subsequently migrate selectively into the germinal vesicle (Merriam and Hill, 1976; Wassarman and Mrozak, 1981). This storage function may be the reason for its giant proportions. The only molluscan species in which the nuclear proteins of the germinal vesicle have been analyzed is *Spisula* (Maul, 1980; Pederson and Jeffery, 1981, cited in Chapter 7, this volume). It is not known whether in molluscs the germinal vesicle contains any morphogenetic determinants, as has been shown to be the case in amphibians (Briggs and Cassens, 1966; Malacinski, 1974).

A large amount of RNA is synthesized in the germinal vesicle. Part of this RNA is stored, in either the germinal vesicle or the cytoplasm, to be used during early development. Another part becomes engaged in protein synthesis during oogenesis. There is little information on the temporal program of RNA synthesis, and still less on the sequential patterns of protein synthesis during oogenesis. In the following subsections, the synthesis of RNA and the production of cytoplasmic organelles will be discussed in more detail.

The mechanisms that provide for the uptake of materials needed by the growing oocyte are poorly known. There is some evidence that yolk proteins are ingested by pinocytosis (see later discussion). Nurse cells, which are connected to the oocyte by means of cytoplasmic bridges, seem to play a negligible role in molluscan oogenesis. In the older literature nurse cells are reported in several species (see Raven, 1966), but additional electron-microscopic studies are needed to verify the existence of cytoplasmic bridges.

Recent reports on nurse cells in *Agriolimax* (*Deroceras*) by Hill and Bowen (1976) and Hill (1977) mention the existence of cytoplasmic bridges but do not show any evidence.

1. Ribosomal RNA Synthesis

In molluscs, very few data are available on ribosomal RNA synthesis during oogenesis. The major features seem to correspond with the data obtained from lower vertebrates such as *Xenopus*.

The structural RNAs present in ribosomes in the ratio 1:1:1 have sedimentation coefficients of 5 S, 18 S, and 28 S in vertebrates. In the bivalve *Mulinia* (Kidder, 1976) and in the prosobranch *Ilyanassa* (Koser and Collier, 1971), the heaviest rRNA is reported to be a 26-S species.

The genes that code for the ribosomal RNAs are present in a large number of copies. In *Mulinia,* the haploid genome contains about 120 rRNA cistrons (Kidder, 1976), in *Spisula* there are 195 copies (Vincent, cited in Collier, 1971), in *Mytilus* 220 copies (Vincent, cited in Collier, 1971), and in *Ilyanassa* 395 copies (Collier, 1971; the value of 790 reported in this paper has been corrected by Kidder, 1976).

The whole sequence of 18-S and 26-S genes is amplified by selective replication in oocytes. Amplification of rRNA genes is twofold in *Mulinia* (Kidder, 1976), and fivefold in *Spisula* (Brown and Dawid, 1968). In *Mulinia,* the 18-S and 26-S genes are localized in a high-density satellite DNA (Kidder, 1976).

In many animal species, amplification of rRNA genes takes place during the pachytene stage (see Davidson, 1976). The additional copies of rDNA function during oogenesis and are subsequently rendered nonfunctional. When the germinal vesicle ruptures, the amplified rDNA is dispersed into the cytoplasm (Steinert et al., 1976). The extra rDNA is still present in unfertilized eggs, but is not replicated during cleavage and no longer play a role during embryonic development.

In vertebrate oocytes, the different ribosomal RNAs are not produced synchronously. A large amount of 5-S rRNA is produced during the previtellogenic stage and stored in the cytoplasm in 8-S and 42-S particles in which the RNA is associated with proteins. The 42-S particle contains 5-S rRNA and transfer RNA. The intense basophilia of previtellogenic oocytes, which is also observed in molluscs (see Ubbels, 1968), is probably due to these storage particles (see Denis, 1977). The 18-S and 28-S rRNA is synthesized later, during vitellogenesis. At that time the 5-S rRNA is released from the storage particles, and it probably returns to the nucleus to be assembled into ribosomes in the nucleolus.

None of these details have been studied in any molluscan species. In molluscs, nucleoli are generally already present in very young oocytes. Their size and morphology change during oogenesis (see Raven, 1961). Considerable differences exist between different species (see Franc, 1951). The nucleolus may grow to 200 times its original volume, and the cyclic formation of buds is often observed (see Raven, 1961). In many molluscan oocytes, such as *Bithynia* (Bottke, 1973b), *Patella* (Bolognari et al., 1976), and *Ilyanassa* (McCann-Collier, 1979), the nucleolus develops a separate zone during the previtellogenic phase. This is an azure B-negative zone and it is called the proteid portion. This type of nucleolus is known as a bipartite or amphinucleolus. An autoradiographic study of *Ilyanassa* oocytes by McCann-Collier (1979) shows that incorporation of RNA precursors, presumably into rRNA, occurs only in the azure B-positive zone. During the previtellogenic phase, there is a predominance of nucleolar RNA synthesis; at the beginning of vitellogenesis nonribosomal RNA synthesis increases. Comparable results were obtained in oocytes of *Lymnaea* (Kielbowna and Koscielski, 1974) and *Littorina* (Romanova and Gazarian, 1966). In many molluscs, the extrusion of nucleolar material through the nuclear membrane into the cytoplasm has been observed (see Raven, 1961; Ubbels, 1968; Silberzahn, 1979; Kessel, 1982). Even intact nucleoli have been reported to be extruded, for example, in *Physa* (Terakado, 1975; see also Raven, 1961). At the end of vitellogenesis the nucleoli generally disappear. They reappear during cleavage (see Chapter 7, this volume).

2. Nonribosomal RNA Synthesis

The mature oocyte contains a large, heterogeneous stockpile of RNA molecules. Aside from ribosomal RNA, which constitutes the bulk of RNA, there is heterogeneous nuclear RNA, messenger RNA, transfer RNA, a diverse set of repetitive sequence transcripts, and possibly other types as well.

The composition of the different classes of RNA and timing of their synthesis are poorly known in molluscs. There are only a few data on maternal mRNA. We may expect that in molluscs, as in other animals, mRNA is synthesized both for immediate use during oogenesis and for use during early development. The mechanisms that dictate whether messengers are translated or stored are largely unknown. The length of the associated poly (A) chains may have some influence (Cabada et al., 1977; Duncan and Humphreys, 1981). In *Spisula*, there is evidence that the selective repression of translation of maternal mRNAs is achieved by some phenol-soluble component, probably a protein (Rosenthal et al., 1980).

A few of these maternal messengers have been identified in molluscs. Raff et al. (1976) have shown that microtubule proteins in *Ilyanassa* embryos are translated from maternal mRNA. A special category of maternal mRNAs are the histone mRNAs. These are generally present in large quantities. They are transcribed from DNA sequences that may be reiterated several hundred times (see Kedes, 1979). In *Spisula* eggs, the presence of maternal mRNA coding for H1 histone has been demonstrated by Gabrielli and Baglioni (1975). The mRNAs coding for the other histones could not be demonstrated in this species. Apparently, the other histone mRNAs are newly synthesized during the cleavage period.

Evidence for the functioning of maternal messengers in early development has been obtained mainly by inhibiting transcription or translation during early development. Translation appears to be absolutely required for normal development, but the almost complete inhibition of transcription of the embryonic genome does not prevent the eggs from developing normally until gastrulation and even further. It is concluded from these experiments that the messengers needed to sustain the earliest part of the developmental program are maternal messengers. These experiments have been carried out with eggs of *Spisula* (Firtel and Monroy, 1970), *Acmaea* (Karp and Whiteley, 1973), *Ilyanassa* (Raff et al., 1976), and *Lymnaea* (Morrill et al., 1976). Experiments by Verdonk (1973), who studied the time of effect of recessive lethal genes induced by X irradiation in *Lymnaea*, lead to the same conclusion. The incorporation of labeled amino acids into proteins in isolated polar lobes of *Ilyanassa* also supports this view (Clement and Tyler, 1967; Geuskens, 1968).

In amphibian and echinoderm eggs, in addition to mature mRNAs, a class of maternal RNA has been found which consists of message sequences covalently linked to other kinds of RNA sequences (Thomas et al., 1981). These RNAs resemble unprocessed or partially processed mRNA precursors and constitute as much as 70% of the stored polyadenylated transcripts. Their fate and function are unknown and their presence in molluscan eggs has not been investigated.

Probably in all animals the genome contains diverse sets of repetitive sequences interspersed between single-copy sequences. The DNA-sequence organization can be investigated by studying the reassociation kinetics of denatured DNA. This kind of study has been carried out with the following molluscs: *Ilyanassa* (Davidson et al., 1971; Collier and Tucci, 1980), *Crassostrea* (McLean and Whiteley, 1973; Goldberg et al., 1975), *Spisula* (Goldberg et al., 1975), *Aplysia* (Angerer et al., 1975), and *Acmaea* (Karp and Whiteley, 1973). In all of these species, repetitive sequences could be demonstrated (see Chapter 7, this volume). The transcripts from these sequences have not been studied in molluscs. Costantini et al. (1978) have studied these transcripts in sea urchin oocytes by hybridization of labeled repetitive DNA with RNA from mature oocytes. They found that the number of transcripts from different repeat families varied from a few thousand to over 100,000 per oocyte. Both strands of the repeat sequences are generally transcribed.

An intriguing problem in the study of transcription during oogenesis is the role of the lampbrush chromosomes. In molluscs, these have been observed in *Ilyanassa* (Davidson, 1976), *Bithynia, Planorbarius, Lymnaea* (Bottke, 1973b), and in the cephalopod *Sepia* (Ribbert and Kunz, 1969). During the previtellogenic phase, the chromosomes become greatly extended and take up a characteristic form reminiscent of a lampbrush. A large number of lateral loops, extending from the chromosome axis, arise by the uncoiling of chromomeres. These loops are separated by condensed chromomeres. The total length of the loops represents about 5% of the genome in amphibians. The lampbrush chromosomes are very actively transcribed in amphibians. The overall rate of nuclear RNA synthesis is about 1000 times that observed in a typical *Xenopus* embryo cell (Anderson and Smith, 1978). The transcripts from the lampbrush chromosomes consist of a complex set of poly(A) RNAs similar in structure and sequence content to the maternal poly(A) RNA stored in the oocyte at the end of oogenesis (Anderson et al., 1982). Although the turnover of these transcripts is low, a very high rate of synthesis is still required for the accumulation of the large pool of maternal poly(A) RNA present in the full-grown oocyte. The special configuration of the chromosomes may optimize transcriptional efficiency.

A peculiar aspect of lampbrush-chromosome transcription is the production of mRNAs that normally serve in terminally differentiated cells of the adult organism. For example, Perlman et al. (1977) found that both tadpole and adult globin mRNA are transcribed from lampbrush chromosomes and are each present at about 200,000 copies per mature oocyte.

3. Cytoplasmic Organelles

The growth of the cytoplasm is largely due to the accumulation of organelles and inclusions. The organelles, particularly mitochondria, seem to be produced mainly during previtellogenesis. It is generally accepted that mitochondria are self-replicating organelles that arise by binary fission of preexistent ones. The maternally inherited mitochondria may constitute the only source of mitochondria of the organism. In the bivalve *Laternula* (Kubo et al., 1979), the paternal mitochondria are eliminated from the sperm before it enters the egg. Szollosi (1965) observed that in rat eggs the paternal mitochondria start swelling during cleavage of the egg and he presumes that they finally disintegrate. The results from experiments in which two species of *Xenopus* were hybridized also support the view that the embryonic mitochondria derive from the egg and not from the sperm (see Dawid, 1972). Primordial germ cells may inherit a substantial store of mitochondria from the egg, for example, in *Sphaerium* (Woods, 1932). During oogenesis the number of mitochondria increases considerably.

Mitochondria often occur in one or more clusters. These clusters might be sites where replication takes place. Sometimes a so-called intermitochondrial cement or nuage interconnects the mitochondria of a cluster. According to Clérot (1979), who investigated the intermitochondrial cement in germ cells of teleosts, this material derives at least partly from the nucleus and produces proteins needed for the elaboration of new mitochondria. This does not seem to be a universal mechanism, as in *Xenopus,* for instance, there is certainly no association of nuage with mitochondria in the previtellogenic phase when mitochondrial replication is at a peak (Billett, 1979). The distribution of mitochondria seems to follow a definite pattern during the early phases of oogenesis (see Raven, 1961). At first they are concentrated on one side of the nucleus, associated with Balbiani's body. Then they form a ring around the nucleus and finally they disperse.

A characteristic structure in many oocytes is the *yolk nucleus* or *Balbiani's body*. It shows considerable diversity in organization and composition between different species (see Raven, 1961; Nørrevang, 1968; Guraya, 1979). Three main types are distinguished: the ergastoplasmic, mitochondrial, and Golgi type (Weakly, 1967). This body develops adjacent to the nucleus in young oocytes and usually disintegrates before vitellogenesis. Guraya (1979) suggests that it may act as an initial center for the formation, multiplication, and accumulation of organelles and inclusions (lipid droplets). Yolk granules do not originate in the yolk nucleus. For this reason, the term yolk nucleus should be abandoned. According to Longo and Anderson (1970), the term Balbiani's body should also be abandoned because it refers to a variety of components and therefore conveys no information.

Yolk nuclei have been observed in various molluscs. For example, in *Ilyanassa* (Gérin, 1976), *Spisula* (Rebhun, 1956), *Mytilus* (Reverberi, 1966),

Bembicium (Bedford, 1966), and *Viviparus* (Bottke, 1973a) an ergastoplasmic type is present. In *Lepidochitona* (Richter, 1976), it consists of ribosomes surrounded by mitochondria. Generally, the yolk nucleus occurs as a single structure in the oocyte, but in *Ilyanassa* Gérin (1976) describes the presence of many of these structures, all of them presumably engaged in the production of lipid droplets. In a few cases, yolk nuclei are observed during cleavage of the egg. In *Nassarius,* Schmekel and Fioroni (1974) describe two types of yolk nuclei that occur in all blastomeres. These structures seem to be newly formed during early cleavage stages.

A very important but poorly known component of oocytes is the endoplasmic reticulum (ER). In young oocytes it often consists of small vesicles usually without ribosomes. The amount of ER almost certainly increases during growth of the oocytes, but this is hard to measure because of the irregular and changing configurations of the ER. The continuity of the rough ER (RER) with the outer nuclear membrane, which is observed in many cells, suggests that the RER originates from the nuclear membrane. Structural evidence for such a process in oocytes is rare. Kessel (1968) describes the blebbing of the outer nuclear membrane in a crustacean oocyte and suggests that this may be the source of new vesicles of ER. The smooth ER (SER) probably originates from the RER by the loss of ribosomes, but in cells where SER predominates, its origin is not clear (see Threadgold, 1976). Structural differentiation of the ER during previtellogenesis is practically restricted to the formation of the membranous type of Balbiani's body and to the formation of Golgi complexes.

C. Vitellogenic Phase

During this period the oocyte grows very rapidly, mainly because of the accumulation of yolk and other nutritive substances such as glycogen and lipid. Aside from these products, a number of organelles are being produced, such as cortical granules, pigment granules, and annulate lamellae. The basophilia of the cytoplasm disappears. RNA synthesis continues during vitellogenesis, but the pattern of synthesis changes. In *Ilyanassa,* rRNA synthesis is predominant in previtellogenesis, and nonribosomal RNA synthesis increases during vitellogenesis (McCann-Collier, 1979).

During vitellogenesis, the polarity of the oocyte becomes manifest in many species by the migration of the nucleus to an apical position (Raven, 1961; Selwood, 1968). Indications of polarity can often be seen at earlier stages. Griffond (1978) observed that oogonia of *Viviparus* are already polarized, as expressed by the localization of Golgi complexes in an indentation of the nucleus. In *Ilyanassa,* the early localization of mitochondria and Golgi complexes between the nucleus and the area of attachment to the ovarian wall also indicates the polar structure of the oocyte (Taylor and Anderson, 1969). A very clear example of polarity is seen in bivalve oocytes. In *Anodonta,* for example, the oocytes are attached by a narrow stalk which contains a large number of parallel microtubules. At an early stage, the nucleus migrates to the distal region of the oocyte, the future animal pole, and the opposite pole develops a micropyle after the oocyte detaches from the stalk (Beams and Sekhon, 1966).

The determining factor of oocyte polarity may be provided by its position relative to the follicle cells and the basement membrane of the acinar wall (see Section II). Several types of relationships can be distinguished (see Huebner and Anderson, 1976). In cephalopods, the oocytes are completely surrounded by follicle cells. In *Lymnaea* (Rigby, 1979), *Biomphalaria* (de Jong-Brink et al., 1976), and *Planorbarius* (Starke, 1971) the basal part of the oocyte rests on the basement membrane of the gonad, whereas the remaining part of the oocyte is surrounded by follicle cells. In the latter species, the previtellogenic oocyte is still completely surrounded by follicle cells. In *Viviparus* (Griffond, 1977) and *Ilyanassa* (Taylor and Anderson, 1969), the young oocyte is completely surrounded by follicle cells, but later the apex of the oocyte becomes exposed to the lumen of the acinus. In bivalves, the oocytes form a long narrow stalk that connects them to the acinar wall (Beams and Sekhon, 1966) (Fig. 2).

The microenvironment created by the basement membrane and the follicle cells may result in local modifications of the plasma membrane which then act to

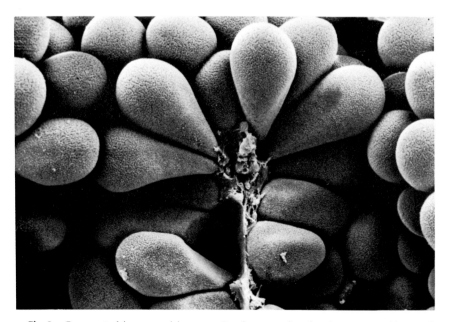

Fig. 2.   Fragment of the ovary of the oyster *Gryphaea angulata* showing oocytes with typical stalks.

polarize the egg. The apicobasal polarity of the oocyte corresponds to the animal–vegetative polarity of the egg. This correspondence is clearly demonstrated in several species, such as *Anodonta* (Beams and Sekhon, 1966) and *Nassarius* (Dohmen and van der Mey, 1977). The importance of the environment of the oocyte for the generation of spatial coordinates in the egg is beautifully illustrated by the *dicephalic* mutant of *Drosophila* (Lohs-Schardin, 1982). This mutation affects embryonic patterning and egg-shell polarity. The anomalous development can be traced back to abnormal follicle development, which causes nurse cells to appear at both poles of the oocyte instead of only at the anterior pole. Another type of spatial organization of the oocyte which might be induced by the follicle cells is the formation of subcortical accumulations in *Lymnaea*. The number and position of the accumulations correspond with the number and position of the nuclei of the follicle cells (Raven, 1970; Ubbels et al., 1969). This issue is discussed further in Section IV,C and in Chapter 5, this volume.

In the following subsections, the formation and distribution of various constituents that accumulate during the vitellogenic phase will be discussed in more detail.

1. Formation of Glycogen, Lipid, and Yolk

Glycogen synthesis seems to be partly associated with SER (see Threadgold, 1976). In molluscan oocytes, the deposition of glycogen is highly diverse. In *Crepidula*, all available space between the organelles and inclusions is filled with α-glycogen rosettes. In *Ilyanassa*, the glycogen in the animal hemisphere is localized in discrete clusters, whereas in the vegetal hemisphere it fills all the space between the other cytoplasmic components. In *Lymnaea*, no granular glycogen can be observed, although cytochemical reactions are positive (Bretschneider and Raven, 1951). There are no data available on the temporal pattern of glycogen synthesis.

Lipid droplets are generally present in very large numbers in molluscan eggs. They may be formed in different ways. In some species, no specific association between lipid droplets and cytoplasmic organelles can be observed, for example, in *Lymnaea* (Bretschneider and Raven, 1951), *Barnea* (Pasteels and de Harven, 1963), and *Sypharochiton* (Selwood, 1968). In these species, the lipid may be produced outside the oocyte and subsequently incorporated by pinocytosis or phagocytosis. The presence of large lipid droplets in the follicle cells, for example, in *Lymnaea* (Ubbels, 1968) and *Bithynia* (M. R. Dohmen, unpublished), suggests that exocytosis of lipid by the follicle cells and endocytosis by the oocyte might be the mechanism for lipid accumulation in a number of species. In other species, the lipid droplets seem to arise in Balbiani's bodies by the transformation of mitochondria and other organelles, for example, in *Ilyanassa* (Gérin, 1976). Lipid droplets surrounded by concentric lamellae of ER were observed in oocytes of *Ilyanassa* (Gérin, 1976), *Acmaea* (Kessel, 1982), and in the polyplacophoran *Trachydermon* (Durfort, 1976). A close association between lipid droplets and mitochondria can be seen in *Bembicium* (Bedford, 1966).

The *proteid yolk* granules show considerable structural differences between different species. In some species, two types of yolk granules are distinguished, for example, in *Lymnaea* (Ubbels, 1968; van der Wal, 1976a), *Planorbis* (Favard and Carasso, 1958), and *Biomphalaria* (de Jong-Brink et al., 1976). Histochemical investigations show that yolk granules may contain a variety of substances, such as glyco- or mucoproteins, phospholipids, and ferritin (Ubbels, 1968). In *Aplysia,* nucleolar buds seem to aggregate into yolk granules after their extrusion from the nucleus (Bolognari and Licata, 1976). In several freshwater pulmonate snails, the yolk granules contain paracrystalline inclusions consisting of dense spherical particles with a diameter ranging from 5 to 7 nm. According to Bottke and Sinha (1979), these particles represent the iron-containing protein ferritin. Terakado (1974) suggests that in *Physa* the crystal particles are formed by the transformation of parallel lamellae of dense material. The illustrations shown by Terakado suggest that the lamellae are formed by the fusion of membranes. The reverse of this process is observed during mobilization of the yolk in *Lymnaea,* when crystal particles are transformed back into membranes (van der Wal, 1976a).

It is not known whether the enzymes that act in breaking down or transforming the yolk during embryogenesis are synthesized during oogenesis and included in the yolk granule in a latent form or penetrate later from production sites in the cytoplasm. The latter possibility seems most likely, because van der Wal (1976b) has demonstrated by radioautography that incorporation of labelled leucine in yolk granules already occurs in uncleaved eggs. A plausible interpretation of these results is that yolk-transforming enzymes are synthesized in the cytoplasm and transferred into the yolk granules. However, other interpretations such as protein-synthetic activity within the granules cannot be ruled out.

Some of the nutritive substances that form yolk granules may be synthesized outside the oocyte. This has been clearly demonstrated for ferritin by Bottke and Sinha (1979) and Bottke et al. (1982) in the snails *Planorbarius* and *Lymnaea.* This protein is taken up by pinocytosis and the resulting vesicles fuse with immature yolk granules. Pinocytosis has also been observed along the basal part of the plasma membrane in oocytes of *Sypharochiton* (Selwood, 1968). The basal plasma membrane of the oocyte is close to the bloodstream and therefore it is likely that macromolecular substances are incorporated directly from the blood. In the vitellogenic oocytes of *Agriolimax (Deroceras),* a large number of pinocytotic tubules are present in the peripheral cytoplasm. They are thought to be involved in yolk production (Hill and Bowen, 1976). A very rigorous way of incorporating exogenous material is seen in *Sphaerium* (Woods, 1931), *Lamellaria* (Renault, 1965), and *Crepidula* (Silberzahn, 1979) where follicle cells are completely resorbed by the oocyte.

The assembly of endogenous and exogenous products into yolk granules can take place in different ways. In some species, there is evidence for the transformation of mitochondria into yolk granules, for example, in *Planorbis* (Favard and Carasso, 1958; Albanese and Bolognari, 1964) and *Mytilus* (Humphreys, 1962). In *Anodonta*, yolk granules are formed by the incorporation of small vesicles, probably derived from Golgi complexes, into multivesicular bodies (Beams and Sekhon, 1966). Golgi complexes are also involved in yolk formation in many other species, such as *Bembicium* (Bedford, 1966), *Mopalia*, and *Chaetopleura* (Anderson, 1969). The role of ER in yolk formation is not clear. Concentric rings of ER cisternae around yolk granules are often observed (Anderson, 1969), but their function is unknown. In some species, yolk granules are closely associated with yolk nuclei, such as in *Spisula* (Rebhun, 1960), but often the yolk nuclei disappear before vitellogenesis starts. Guraya (1979) concludes, therefore, that yolk bodies do not originate directly from yolk nuclei.

2. Special Plasms and Organelles

In oocytes of several molluscan species, special organelles or plasms are found that do not normally occur in other cell types. Some of these are transient structures that may function only in oogenesis itself, others develop into structures that are retained in the egg but their role in embryonic development remains to be established in most cases.

In the polyplacophoran *Katharina*, nuage material is present in the oocyte (Eddy, 1975). This material is thought to originate from the nucleus. A similar substance is thought to function in the synthesis of new mitochondria in teleost oocytes (Clérot, 1979). A close association between mitochondria and perinuclear fibrogranular bodies is observed in *Acmaea* oocytes (Kessel, 1982). In *Ilyanassa* oocytes, several uncommon components have been found. In young oocytes, a number of large, randomly situated *basophilic regions* are present. They are absent in mature eggs (Anderson, 1969). During previtellogenesis, a large number of granular bodies accumulate around the nucleus. According to Gérin (1972), these *perinuclear corpuscles* are enveloped in ER and develop into *double-membrane vesicles* that are present in large numbers in the egg. Gérin's account of the formation of these vesicles does not include the *polymerosome* stage that is likely to be a precursor of the double-membrane vesicle (McCann-Collier, 1977). Taylor and Anderson (1969) described another type of perinuclear bodies in *Ilyanassa* oocytes. These are spherical, dense bodies surrounded by a membrane. They appear at the onset of vitellogenesis and maintain their integrity and perinuclear position throughout this period. Pucci-Minafra et al. (1969) described large *ribosome clusters* in the vegetal hemisphere of *Ilyanassa* eggs. These structures probably represent glycogen (Geuskens and de Jonghe d'Ardoye, 1971; Dohmen and Verdonk, 1979a). In *Bithynia*, a peculiar structure, consisting of a large number of small vesicles (Fig. 4, Chapter 6, this volume), is present at the vegetal pole of the egg and has thus been called the *vegetal body* (Dohmen and Verdonk, 1974). This body becomes incorporated into the polar lobe and most probably contains the lobe-specific morphogenetic determinants (Chapter 6, this volume).

At the beginning of vitellogenesis, the first aggregates of vesicles appear and gradually more vesicles are added until, at the end of vitellogenesis, a large disk-shaped aggregate is formed at the basal pole of the oocyte (Fig. 3). This disc assumes a cup shape during the first meiotic division and this shape is maintained until the second cleavage. The vesicles that make up the vegetal body are formed by budding from cisternae of SER. These cisternae are almost electronlucent, but the budding vesicles are rather dense. The vegetal body is very rich in RNA (Dohmen and Verdonk, 1979a), probably associated with the vesicles. The small number of ribosomes that are present in between the vesicles cannot account for the very intense staining in cytochemical tests. The type of RNA and its pathway into the vesicles are unknown.

In many molluscs, *annulate lamellae* appear during vitellogenesis, for example, in *Spisula* (Rebhun, 1960), *Ilyanassa* (Taylor and Anderson, 1969), *Lepidochitona* (Richter, 1976), *Mopalia* and *Chaetopleura* (Anderson, 1969), and *Agriolimax (Deroceras)* (Hill and Bowen, 1976). Annulate lamellae do not occur exclusively in oocytes, but also in spermatocytes, neoplastic cells and other cells (see Wischnitzer, 1970; Kessel, 1973). They probably originate from the nuclear envelope, because they seem to be formed parallel to it. The pores present in the lamellae are very similar to the pores in the nuclear envelope, and it has been proposed that they have a similar function, the assembly of ribosomal sub-units into functional polyribosomes (Kessel, 1981a,b). Another function of the annulate lamellae might be the production of ER, which is also a function of the nuclear envelope. This is suggested by the frequently observed continuity between stacks of annulate lamellae and cisternae of the ER. Both the morphological aspect of the annulate lamellae and their postulated functions suggest that they serve as an amplification of the nuclear envelope, temporarily needed to meet the high demands of assembling during vitellogenesis. The reverse of this view has been put forward by Dhainaut (1973), who suggests that annulate lamellae may also be a form of degeneration of the ergastoplasm under certain conditions.

A characteristic kind of vesicle that is found only in oocytes is the *cortical granule*. These granules are present in several species of molluscs, such as *Dentalium* (Reverberi, 1970), *Mytilus* (Humphreys, 1967), *Spisula* (Longo and Anderson, 1970; Sachs, 1971), *Barnea* (Pasteels, 1966), *Bankia* (Popham, 1975), *Mopalia, Chaetopleura* (Anderson, 1969), *Lepidochitona* (Richter, 1976) and *Laternula* (Kubo et al., 1979). Their function is not understood, as they do not seem to extrude their contents to form a fertilization membrane. Several authors suggest that the cortical granules slowly release mucus to the perivitelline

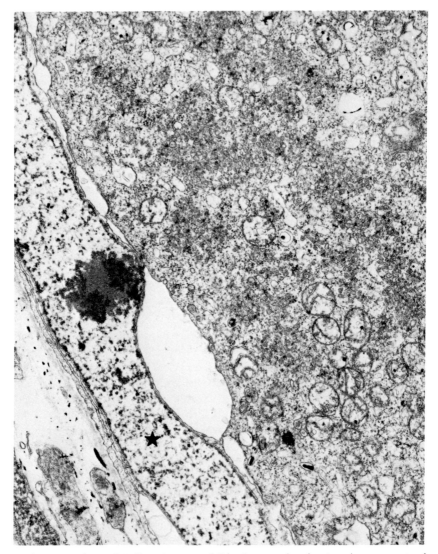

Fig. 3. Basal part of vitellogenic oocyte of *Bithynia tentaculata* showing a large aggregate of small vesicles that constitute the developing vegetal body. Star indicates the nucleus of a follicle cell.

coat (Humphreys, 1967). Popham (1975) observed that in *Bankia*, the cortical granules disappear in regions where supernumerary sperm are present in a high concentration. The acrosome reactions probably result in damage to the vitelline membrane, as Humphreys (1962) demonstrated that sperm extract causes the vitelline membrane to disappear in *Mytilus* eggs. At the same time, the cortical granules become invisible. They may extrude their contents in order to repair the damaged envelope. In *Laternula*, opening of the cortical granules is frequently observed in the region of sperm attachment. Kubo et al. (1979) suggest that they may play a role in the formation of the indentation of the egg surface where the sperm head is engulfed. The origin of cortical granules is poorly known (see Nørrevang, 1968). Anderson (1969) describes vacuolar bodies, containing acid mucopolysaccharide, that migrate from an adnuclear position to the peripheral ooplasm.

In *Bithynia*, a large number of vesicles having extremely dense contents are formed during vitellogenesis. Upon centrifugation, these vesicles constitute the heaviest stratum in the egg. They may be storage vesicles for mineral deposits such as calcium or phosphate. The formation and ultrastructure of calcareous granules in the oocytes of *Helix aspersa* have been documented by McGee-Russell (1968); other land snail genera also possess similar oocyte spherules (A. S. Tompa, personal communication). Similar vesicles are found in adult molluscan hepatopancreas, where they are formed by the Golgi complex (Abolins-Krogis, 1970). The amorphous mineral deposits in these granules consist mainly of calcium, magnesium, carbonate, and phosphate (Howard et al., 1981).

3. Egg Envelopes

The egg envelopes are generally classified after Ludwig (cited in Wilson, 1928) in (1) primary envelopes that are secreted by the oocyte, (2) secondary envelopes produced by follicle cells, and (3) tertiary envelopes formed by the oviduct or other maternal structures.

The primary envelope or vitelline membrane (Fig. 4) is comparable to the glycocalyx of other cell types. There may be a strong adherence between the microvilli and the vitelline membrane. In *Mytilus* (Humphreys, 1962) and *Mactra* (Sawada, 1964, cited in Nørrevang, 1968), microvilli may be torn away from the egg surface by the vitelline membrane. The composition of the primary envelope has been investigated in the gastropods *Megathura* (Heller and Raftery, 1976) and *Tegula* (Haino and Kigawa, 1966). In *Megathura*, it consists of polypeptide chains cross-linked by disulfide bonds and composed to a large extent of closely spaced threonine residues. Almost every threonine residue is linked to a saccharide moiety. The carbohydrate includes galactosamine, galactose, and fucose. The envelope of *Megathura* does not contain sulfate and phosphate, which are found in the envelope of *Tegula*. The material constituting

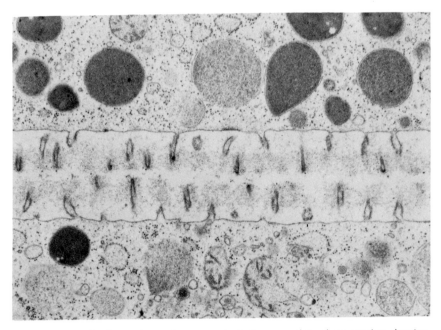

Fig. 4. Superficial part of two adjacent oocytes in the ovary of *Gryphaea angulata,* showing the fibrous vitelline membrane in which the microvilli are embedded.

the vitelline membrane seems to be produced by Golgi complexes and deposited on the oocyte surface by exocytosis (Selwood, 1970).

A secondary envelope or chorion is present on oocytes of Cephalopoda and some Polyplacophora (see Pearse, 1979). According to Anderson (1969), the material of the secondary envelope in chitons is produced in follicle cells by cisternae of RER. These cisternae, which are filled with a dense substance, become associated with the plasma membrane, fuse, and release their contents onto the basal surface of the follicle cells. In *Sypharochiton* (Selwood, 1970), the secondary envelope consists of two layers, an inner layer of protein and an outer layer of lipid. Both layers are secreted by follicle cells. Richter (1976) assumes that in *Lepidochitona,* the secondary envelope is secreted by the microvilli of the oocyte. Beams and Sekhon (1966) describe a thick *second envelope* on oocytes of the bivalve *Anodonta,* but there is no evidence that this layer is secreted by cells other than the oocyte itself.

A discussion of the tertiary envelope is more suitable in chapters on reproduction strategies and the anatomy of the reproductive organs (see later volume on reproduction, this treatise). Detailed accounts are given in Giese and Pearse (1977, 1979) and Fretter and Graham (1962).

IV. Egg Structure

The older literature on the structure of molluscan eggs has been reviewed extensively by Raven (1966). Since then a considerable number of studies have been published on the structure of molluscan eggs. Many of these studies are aimed primarily at the search for structural features that might give a clue as to which factors are involved in determining the course of development. The basic questions are as follows: what is the nature of morphogenetic determinants, how are they stored in the egg, and which mechanism brings about their localization in certain cytoplasmic compartments? Although these questions still remain to be answered for the most part, some progress has been made, as discussed next.

A. Compartmentation of the Egg

Cellular compartmentation relies in many cases on the presence of cytoplasmic membranes (Sitte, 1980). The nucleus, the mitochondria, the ER and its derivatives are examples of membrane-bounded compartments. For our purpose, a more important aspect of compartmentation is the positioning of cellular constituents, which might be called second-order compartmentation. For example, mitochondria may be concentrated in a definite region of the cell. This region can be considered to be a specific supercompartment. Similar considerations apply to the positioning of the nucleus and of every other cellular constituent.

The available data suggest strongly that in molluscan eggs at least three supercompartments exist, namely, the animal pole region, the vegetal pole region, and the cortical region (see also Chapters 5 and 6, this volume).

A cortical compartment is clearly demonstrated in many species by the accumulation of cortical granules in a layer close to the plasma membrane. These granules cannot be displaced by centrifugation, for example, in the eggs of *Mytilus* (Humphreys, 1962) and *Dentalium* (Dohmen and Verdonk, 1979a). As the cortical granules are not directly attached to the plasma membrane, the cortical cytoplasm must have structural properties that differ from the rest of the cytoplasm enabling it to resist disruption by centrifugation. The cortical compartment need not be homogeneous over the whole surface of the egg. In *Dentalium* eggs, for instance, cortical granules are absent in the vegetal pole region (Reverberi, 1970).

The cortical regions at the animal and vegetal pole seem to have special properties. At the vegetal pole, the cortex is characterized in many species by the presence of morphogenetic determinants. This is most clearly demonstrated in polar-lobe-forming eggs by experimental methods, as discussed in Chapter 6, this volume. Also, the cortex at the animal pole may have distinct properties. The first maturation spindle, which first lies near the center of the egg in most molluscan eggs, moves towards the animal pole. It is apparently attracted by this part of the egg surface (Raven, 1966). In *Lymnaea* eggs, a distinct animal pole plasm becomes visible about 1 h before the first cleavage. This plasm can be disrupted by centrifugation, but it is subsequently reformed at the same place (Raven and van der Wal, 1964). This phenomenon points to the existence of specific attractive actions exerted upon certain cytoplasmic components by factors in a cortical compartment at the animal pole.

There are several other reports mentioning the reestablishment of the original cytoarchitecture after centrifugation (see Raven, 1966). This suggests that factors in the cortical compartments act as primers for the organization of cytoplasmic compartments. How this is done remains unknown.

The existence of cytoplasmic compartments is evident in many species. In the ripe unfertilized molluscan egg, the distribution of the various constituents is often more or less homogeneous throughout the egg. From ovulation until first cleavage, ooplasmic segregation takes place along the animal–vegetative axis of the egg. In most cases mitochondria, lipid droplets, and yolk granules are the inclusions that are conspicuously involved, but other structures may be segregated as well. Typical examples are the eggs of *Ilyanassa* and *Aplysia,* where separate layers of mitochondria, lipid, and yolk can be distinguished (Clement and Lehmann, 1956; Attardo, 1957). The occurrence of segregation in intermediate positions between the poles of the egg, such as in *Ilyanassa* and *Aplysia,* suggests that the polar compartments may be the extremes of a gradient. Cortical factors at the poles might somehow maintain an animal–vegetative gradient in the cytoplasm. This gradient might establish several cytoplasmic compartments in which the inclusions may accumulate differentially. Species-specific variations in the properties of the gradient and of the cytoplasmic inclusions may determine whether sharply delineated zones or a more diffuse distribution pattern will be formed.

B. Morphogenetic Plasms

A special category of cytoplasmic constituents is formed by the morphogenetic determinants. The existence of these substances has been inferred from experiments such as deletion of polar lobes, manipulation of the cleavage process, and so on (see Chapter 6, this volume). The experimental results have demonstrated that specific morphogenetic determinants are localized in definite compartments. The compartment that has been most extensively studied in molluscan eggs is the vegetal pole plasm, particularly in polar-lobe-forming eggs. Many efforts have been devoted to identifying the nature of the determinants in polar lobes, but definite conclusions cannot yet be drawn. The experimental data suggest that the polar lobes exert their control of development by regulating the translation of preformed mRNA coding for cell lineage-specific proteins (see Dohmen and Verdonk, 1979a; Chapter 7, this volume).

In several species, putative storage devices for morphogenetic determinants have been observed. The most conspicuous is found in the eggs of *Bithynia*, where the polar lobe contains a large structure consisting of an aggregate of vesicles (Fig. 3). This aggregate is called the vegetal body (Dohmen and Verdonk, 1974). Experimental results provide strong evidence that this body represents a morphogenetic plasm that contains the lobe-specific determinants (see Chapter 6, this volume). Cytochemical investigations demonstrate that the vegetal body contains a large amount of RNA. This suggests that the lobe-specific determinants either consist of RNA or are associated with RNA.

Putative morphogenetic plasms reminiscent of the vegetal body have been found in several other species with small polar lobes. In *Crepidula*, the polar lobe contains a few small aggregates of vesicles (Dohmen and Lok, 1975). In *Buccinum*, a large number of electron-dense vesicles fill the polar lobe but it is not known whether or not they form a coherent plasm (Dohmen and Verdonk, 1979a).

Another specific component that has been observed in some polar lobes is a germ plasmlike structure. In several groups of animals, distinct germ plasms are segregated into the germ cell line and there is evidence that they are involved in the determination of germ cells (see Mahowald et al., 1979). In the polar lobes of *Crepidula* and *Buccinum* eggs, the putative germ plasms consist of a complex of small vesicles and dense granules (Dohmen and Verdonk, 1979b). Preliminary evidence indicates that, in *Crepidula*, this plasm is segregated into the 4d cell from which the gonads most probably originate (see Raven, 1966). It should be worthwhile to investigate the ultrastructure of *Sphaerium* eggs, in which Woods (1931, 1932) has described the segregation of a cloud of mitochondria, originally localized in the vegetal hemisphere of the uncleaved egg, into the 4d cell and then into the primordial germ cells. According to Woods, the mitochondria should not be considered a causal factor in the determination of the germ cells, but they serve to mark the segregation of a particular plasm that most probably contains a determinant for germ cells.

In large polar lobes, no structures resembling the special plasms found in small polar lobes have been detected. In *Ilyanassa*, two kinds of special structures have been suggested to contain the lobe-specific determinants, namely, double-membrane vesicles and *ribosome clusters*. Double-membrane vesicles (DMV) were first described by Crowell (1964) in *Ilyanassa obsoleta* and by Schmekel and Fioroni (1975) in *Nassarius reticulatus*.[1] The organelle consists of a central

---

[1]*Nassarius obsoletus* was removed from the genus *Nassarius* (Dumeril, 1806) by Stimpson and placed into the new genus *Ilyanassa* (Stimpson, 1865) on the basis that *Ilyanassa obsoleta* has a different operculum and lacks the caudal pedal cirri or metapodial tentacles characteristic of the genus *Nassarius*. Although embryologists tend to use the correct term *Ilyanassa obsoleta* (Say), scientists in other fields often still use the now outmoded designation, *Nassarius obsoletus*.

vacuole filled with an electron-dense substance, surrounded by a compartment in which a dense body is situated on one side. This organelle occurs in large numbers in the vegetal hemisphere of the egg, but it is not restricted exclusively to the polar lobe. When *Nassarius* or *Ilyanassa* eggs are centrifuged "upside-down" according to the technique devised by Clement (1968), the DMV are displaced towards the animal pole together with the yolk granules. This argues strongly against a possible morphogenetic role for the DMV as discussed by Dohmen and Verdonk (1979a).

Biochemical analysis of polar lobes and lobeless eggs has revealed some quantitative differences in chemical composition (see Chapter 7, this volume). An interesting finding is that the nucleotide content per unit volume is considerably higher in the polar lobe, both in *Ilyanassa* (Collier, 1976) and in *Nassarius* (van Dongen et al., 1981). This indicates that the developmental defects that occur after deletion of the polar lobe might result partly from a diminished capacity for nucleic acid synthesis, due to a lack of precursors.

C. The Mechanism of Localization

A number of studies indicate that cytoskeletal elements are involved in the positioning of cellular components. Microtubules probably serve as guiding structures for the transport of organelles, and microfilaments probably provide the driving force. However, the results from treatment with inhibitors of cytoskeletal elements do not all agree with this view. The movements of cytoplasmic organelles can generally be inhibited by treatment with cytochalasin B, for example, in eggs of the ascidian *Ciona* (Zalokar, 1974; Sawada and Osanai, 1981) and in zygotes of the brown alga *Fucus* (Brawley and Quatrano, 1979). In the eggs of the annelid *Chaetopterus,* however, cytochalasin B does not block ooplasmic segregation and localizing movements (Eckberg and Kang, 1981). Colchicine inhibits localizing movements in *Chaetopterus* (Eckberg and Kang, 1981), but not in ascidian eggs (Zalokar, 1974).

Intermediate filaments are thought to be involved in the positioning of the nucleus, as they form whorls around it in several cell types (Lazarides, 1980; Lehto et al., 1978; Virtanen et al., 1979; Wang et al., 1979). The first results of intermediate filament research in eggs are not encouraging. In mouse eggs, neither the fibers themselves nor any known intermediate filament proteins could be detected between the two-cell and early morula stage (Jackson et al., 1980). Quite another type of driving force for the movement of cytoplasmic organelles has been proposed by Jaffe et al. (1974). They suggested that electrical currents may serve to electrophorese cytoplasmic components, such as negatively charged vesicles, and thus bring about cytoplasmic segregation. Such currents exist in a number of developing or regenerating systems, and also in eggs, for example, in *Fucus* and *Xenopus* eggs (see Borgens et al., 1979).

After the organelles have been transported into the appropriate compartments, they are fixed more or less firmly in this position. Often, this is not the final step in the segregation process. During subsequent development of the egg, a shift in the position of segregated plasms may occur.

It is not clear whether a distinction should be made between the mechanisms that bring about segregation of cytoplasmic constituents and those that fix these constituents in the appropriate position. There are very few data on the mechanisms that maintain the spatial order in the cell. Hylander and Summers (1981) found that in sea urchin eggs the attachment of the cortical granules to the plasma membrane can be disrupted by treatment with ethyl carbamate, procaine, tetracaine, and ammonia. This indicates that cytoskeletal elements might be involved in attaching the cortical granules. Among molluscan eggs, those of *Bithynia* seem to be a suitable model system to study the mechanism of cortical binding. The vegetal body present at the vegetal pole in this species displays a very strong coherence and a strong attachment to the cortex of the egg. During centrifugation, the vegetal body is first stretched in the direction of the centrifugal force (Fig. 5a) and then torn away as a whole from the cortex (Fig. 5b). Only relatively strong centrifugal forces (1400 g) are capable of disrupting the bond between the vegetal body and the cortex. It is evident from ultrastructural studies that the

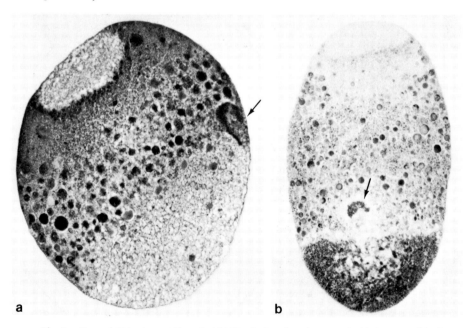

a   b

Fig. 5. Eggs of *Bithynia* centrifuged at 1400 g before first cleavage. In (**a**) the vegetal body (arrow) is still attached to the cortex, although it is stretched out in the direction of the centrifugal force. In (**b**) the vegetal body (arrow) has been displaced from its original site.

coherence of the vegetal body is not due to a surrounding membrane and that its attachment to the cortex is not caused by bundles of fibers. If fibers are involved at all, they must be organized into a kind of network. A likely mechanism that might serve to hold the vesicles of the vegetal body together and bind the whole structure to the egg surface might be a lattice of microtrabeculae. These are fine filaments, 3–6 nm in diameter, first demonstrated in fish erythrophores (Byers and Porter, 1977). This lattice is clearly seen in cells that are critical-point dried and viewed as whole mounts with a high-voltage electron microscope. Conventional thin sections of plastic-embedded cells are less suitable for visualizing microtrabeculae (Wolosewick and Porter, 1979).

In *Bithynia* eggs, only thin sections have been studied. The cortex at the vegetal pole is a dense layer, about 2–3 μm thick, containing a large number of ribosome-like particles and a poorly defined "flocculent ground substance" which might be made of microtrabeculae. No microtubules or any ordered array of fibers have been observed.

If a microtrabecular lattice were responsible for binding the vegetal body to the cortex, it is still not clear why this body adheres specifically to the vegetal pole cortex and not elsewhere. In cells thus far examined (all of them somatic cells), the lattice is found throughout the cytoplasm; this is presumably the case in egg cells as well. Unless the lattice itself possesses regional specificity, we will have to assume the existence of additional factors that confer the characteristic properties to the various cortical and cytoplasmic compartments.

A distinct cortical layer is found at the vegetal pole of *Buccinum* eggs (Dohmen and van der Mey, 1977). It consists of a complex network of fibers that seem to run parallel to the egg surface. Occasionally, microtubules are observed within or just beneath this cortical layer.

The executive agents that bring about spatial order in the egg cell are not to be identified with the determinants of this process, as has been beautifully demonstrated by Solomon (1980) in neuroblastoma cells. In these cells, the neurites can be made to retract by treatment with Nocodazole, an agent that depolymerizes microtubules but also affects the arrangement of intermediate filaments and microfilaments. Upon removal of the drug, most of the cells reextend the same number of neurites, at the same relative positions, as before. Apparently, the determinants of morphology are not affected by the disassembly of microtubules.

Many phenomena related to cytoplasmic localization can be explained by assuming that local areas of the egg surface act as determinants for the organization of a cytoplasmic compartment. This can be inferred from experiments on systems in which the position of cytoplasmic or cortical compartments can be varied by applying external stimuli, for example, in zygotes of brown algae and in ascidian eggs. In these systems, the external stimulus probably acts by inducing a local change in the plasma membrane. This area may thus become a primer for the organization of a polar structure in the cortex or the cytoplasm.

In eggs of brown algae, the polar axis can be shifted by altering the direction of illumination. As a result, ion-transporting channels in the plasma membrane accumulate at the dark side of the zygote, giving rise to a current of calcium ions entering the presumptive apical pole and leaving the basal pole. This membrane-generated electrical current is accompanied by the directional transport of cell wall material to the presumptive basal pole (Robinson and Jaffe, 1975; Nuccitelli, 1978: Quatrano et al., 1979: Brawley and Quatrano, 1979).

In ascidian eggs, ooplasmic segregation is directed by a local modification of the plasma membrane, resulting in a local increase in intracellular calcium, as has been demonstrated by treating the eggs with the calcium ionophore A 23187 (Jeffery, 1982).

The view that modifications of the plasma membrane can influence the structure of the cytoplasm is supported by the observation that integral membrane proteins can interact with the underlying cytoskeleton (e.g., Baumgold et al., 1981; Bennett, 1982). In molluscan eggs, cytoplasmic localization is generally related to the animal–vegetal axis. It is obvious to assume an organizing principle at one or both poles of the egg, capable of generating the compartmentation of the egg. This organizing principle might be, for example, an organizer of cytoskeletal structure or the source of a gradient that provides positional information for the self-organization of cytoplasmic compartments.

Apparently, the polar axis is initially determined by the position of the oocyte in the ovary, as the apicobasal axis of the oocyte corresponds with the later animal–vegetal axis of the egg (see Raven, 1966; Chapter 5, this volume). If the contact of the basal pole of the oocyte with the ovarian wall is the determining factor for polarity, one might suppose that this contact results in an "imprinting" of special properties upon the future vegetal pole surface, resulting, for example, in a special structure of the plasma membrane or the cortical cytoskeleton. Regional differences in membrane properties related to the polar axis have indeed been shown to exist in *Dentalium* eggs (Jaffe and Guerrier, 1981). Measurements of electrical excitability showed that polar lobes isolated at the trefoil stage were more excitable than the blastomeres. Similar experiments with *Ilyanassa* did not show any regional differences in membrane properties (Moreau and Guerrier, 1981). A clear demonstration of the regional differentiation at the vegetal pole is provided by the striking differences in surface architecture between the vegetal pole area and the rest of the egg surface in several polar lobe-forming gastropod eggs (Dohmen and van der Mey, 1977). In *Nassarius,* the basal area of fully grown oocytes, artificially released by opening the ovary, is devoid of microvilli. This area is the part of the oocyte that remains in contact with the follicle cells, whereas the remaining surface of the oocyte is either associated with other oocytes or exposed to the lumen of the ovary (Taylor and Anderson, 1969). In this initially smooth area, very long microvilli develop. The time when this occurs is not known exactly. They are already present during the maturation divisions and are retained during the early cleavages. In later cleavage stages, they disappear and are replaced by other excrescences that still give the vegetal area a characteristic appearance until gastrulation.

The observation that in *Nassarius* eggs, the former contact area of the oocyte with the follicle cells develops a characteristic surface architecture supports the view that the imprinting of special properties on the future vegetal pole surface may result from such a contact. This phenomenon is very pronounced in *Nassarius reticulatus*, but much less so in *Ilyanassa obsoleta*. In the other species in which the vegetal pole surface shows a typical architecture, such a strict correspondence with the former contact area in the ovary has not been established. The surface features observed at the vegetal pole may differ considerably between different species. Very long microvilli, like those in *Nassarius* eggs, also occur in *Bithynia*. In the eggs of *Buccinum, Crepidula* and *Nucella*, the vegetal pole is characterized by ridges consisting of cytoplasmic outgrowths that bear microvilli (Dohmen and van der Mey, 1977). These outgrowths are arranged in such a way as to form a typical pattern. In *Buccinum*, they form bifurcating ridges and in *Nucella*, they form an irregular pattern on the vegetal pole and a spiral configuration of long streaks that may extend from the vegetal pole well past the equator of the egg. In *Crepidula*, they form a bilaterally symmetrical pattern which may have very interesting implications, as discussed later.

In *Bithynia* and *Crepidula*, observations *in vivo*, either with phase optics or after labeling with fluorescent lectins, show that the surface architecture at the vegetal pole is static, although very slow alterations cannot be ruled out. The waves of elongation of microvilli that have been reported by Pasteels (1966) to occur in *Barnea* eggs probably represent a different phenomenon because these waves are not localized in a specific area of the egg. The significance of the surface features at the vegetal pole of polar lobe-forming eggs is still unknown. They may be related either to the animal–vegetal polarity of the egg or to the polar lobe-specific determinants or both. The fact that they are segregated into the D quadrant together with the lobe factors suggests a relationship with these factors. The same segregation makes it seem unlikely that they have any function related to the polarity of the egg. This polarity exists independently in each of the four quadrants of the cleaved egg, as demonstrated in isolated blastomeres of *Dentalium* and *Patella* (Wilson, 1904). When blastomeres of *Dentalium* are isolated at the two- or four-cell stage, the original polarity remains visible because of the presence of pole plasms. The isolated blastomeres will form micromeres at the animal pole. This demonstrates that the original polarity is retained and is still functional after isolation, independent of the special surface region that is segregated onto the D cell.

The possible relationship between the vegetal egg surface and the lobe-specific determinants has been investigated in *Bithynia*, where the determinants, presumably localized in the vegetal body, can be displaced from the vegetal pole by centrifugation. Normal development may ensue, even if the polar lobe, which is now devoid of the vegetal body, is deleted (see Chapter 6, this volume). This deletion removes the vegetal egg surface, whose specific properties are obviously no longer needed for normal development. Apparently, vegetal surface differentiations are important at an earlier stage, if they are important at all for development.

Quite a different aspect of surface differentiation at the vegetal pole can be studied in the eggs of *Crepidula*. Until now this is the only species in which the surface architecture displays a bilaterally symmetrical pattern (Fig. 6). It turns out that the axis of symmetry of this pattern has a definite relationship with the direction of the first cleavage plane, the angle being variable between 45° and 60° (Fig. 6). The pattern of ridges is established very early, probably during the first maturation division. This means that the first cleavage plane is predetermined at an early stage. Experiments intended to uncouple this relationship between cleavage plane and surface pattern—by centrifugation or otherwise—have not yet been carried out. These experiments could give information on the structural basis of this determination. It should be interesting to investigate this system in more detail, as the direction of the first cleavage plane may be very important, particularly in unequally cleaving eggs.

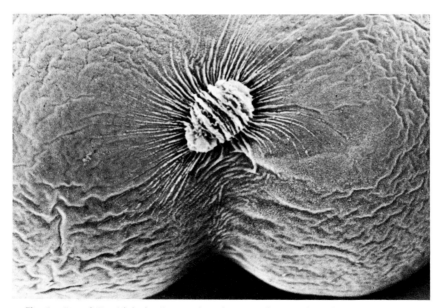

Fig. 6. Egg of *Crepidula fornicata* at first cleavage, showing the polar lobe which is very small in this species. The lobe is covered with an array of ridges displaying a bilateral symmetrical pattern. The axis of symmetry makes an angle of about 45° with the plane of first cleavage.

V. Concluding Remarks

The study of gametogenesis touches on almost every aspect of molecular, cell, and developmental biology. Molluscs provide material very suitable for the investigation of a number of problems. One of these problems is the origin of polarity of the germ cells. Polarity is probably induced by intercellular contacts in the gonads. The induction may operate by imprinting upon the germ cells a local differentiation of the plasma membrane, which then acts as an organizer of the polar structure of the cell. The initial polarity of the oocyte is still labile; it can be overruled experimentally at early stages until it is definitively fixed during the maturation of the egg (see Chapter 5, this volume).

In a number of prosobranch gastropods, one of the expressions of polarity of the egg is a local differentiation of the egg surface at the vegetal pole (see Section III,C). This structure might be related to the hypothetical imprinting of the oocyte surface during oogenesis. Currently, investigations with freeze-fracture methods and FRAP (fluorescence recovery after photobleaching) are undertaken to see whether the plasma membrane of the egg shows a polar organization. Definitive results are not yet available. If polarity at the level of the plasma membrane can be established, these investigations should be extended to earlier stages, during oogenesis, in order to determine the time at which this polarity originates. These data might give a clue as to when and how polarity is induced and thus provide a basis for further research. Because there is no reason to suppose that the polarity of sperm and eggs is induced in different ways, the proposed investigations could be performed with both male and female germ cells. Male germ cells may even be preferred, because the polar structure of the plasma membrane of sperm has already been clearly demonstrated (see Friend, 1982). Unfortunately, the technique of culturing eggs and sperm from primordial germ cells has not yet been developed; otherwise one could experimentally manipulate the intercellular relationships during gametogenesis, which would undoubtedly provide valuable data on the induction of polarity.

The cellular mechanisms resulting in a polar structure of the germ cells and in a corresponding distribution of cellular constituents probably rely on an interaction between the plasma membrane and the various classes of cytoskeletal elements. Molluscan oocytes may be used profitably to investigate these mechanisms because of the conspicuous cytoplasmic localizations present in a number of species. Immunocytochemical visualization of cytoskeletal elements may reveal structural relationships between segregated plasms, cytoskeletal elements, and plasma membrane domains. The fate of these plasms following experimental manipulations of the cytoskeleton may provide a convenient parameter in assessing the functional relationship between the two.

In molluscs, as compared with amphibians and sea urchins, there are very few data on the synthetic processes taking place during oogenesis, particularly regarding the accumulation of maternal mRNA, which is the most interesting product from a developmental biologist's point of view. This interest derives from the fact that a number of maternal mRNAs might be regulatory molecules involved in directing the course of early development. In molluscs, there is evidence that maternal mRNAs, localized in a specific compartment of the egg, namely, the polar lobe, may function as morphogenetic factors (Raff et al., 1976). In one animal, *Bithynia tentaculata*, the lobe-specific morphogenetic determinants are concentrated in a conspicuous RNA-rich body (see Chapter 6, this volume). At present, research is aimed at isolating and identifying the RNA from this body and assaying its morphogenetic potential by injecting it into an AB blastomere, which does not receive this material in normal development. If these experiments should prove successful, they may contribute greatly to our understanding of how development can be regulated by maternal factors.

A rather neglected aspect of molluscan reproduction and development is the topic of food eggs and atypical sperm. It will be hard to approach the problem of the determination of atypical sperm as long as *in vitro* culture methods are not available. The factors that cause aberrant development of food eggs, however, can be investigated more easily, for example, by transplanting cytoplasm from food eggs into normal eggs, by introducing nuclei from normal eggs into food eggs, or by inducing parthenogenetic development. It should be worthwhile to start the experimental study of this system, which is comparable to a mutant.

References

Abolins-Krogis, A. (1970). Electron microscope studies of the intra-cellular origin and formation of calcifying granules and calcium spherites in the hepatopancreas of the snail, *Helix pomatia* L. *Z. Zellforsch. Mikrosk. Anat.* **108**, 501–515.

Albanese, M. P., and Bolognari, A. (1964). Mitocondri, zone del Golgi e globuli vitellini negli ovociti in accrescimento di *Planorbis corneus* L. (Moll. Gast. Polm.) *Experientia* **20**, 29.

Anderson, D. M., and Smith, L. D. (1978). Patterns of synthesis and accumulation of heterogeneous RNA in lampbrush stage oocytes of *Xenopus laevis* (Daudin). *Dev. Biol.* **67**, 274–285.

Anderson, D. M., Richter, J. D., Chamberlin, M. E., Price, D. H., Britten, R. J., Smith, L. D., and Davidson, E. H. (1982). Sequence organization of the poly (A) RNA synthesized and accumulated in lampbrush chromosome stage *Xenopus laevis* oocytes. *J. Mol. Biol.* **155**, 281–309.

Anderson, E. (1969). Oocyte-follicle cell differentiation in two species of amphineurans (Mollusca), *Mopalia muscosa* and *Chaetopleura apiculata*. *J. Morphol.* **129**, 89–126.

Anderson, W. A., and Personne, P. (1969). Structure and histochemistry of the basal body derivative, neck and axoneme of spermatozoa of *Helix aspersa*. *J. Microsc. Biol. Cell.* **8**, 87–96.

Anderson, W. A., and Personne, P. (1976). The molluscan spermatozoon: Dynamic aspects of its structure and function. *Am. Zool.* **16**, 293–313.

André, J. (1962). Contribution à la connaissance du chondriom: Etude de ses modifications ultra-structurales pendant la spermatogénèse. *J. Ultrastruct. Res., Suppl.* **3**, 1–85.

Angerer, R. C., Davidson, E. H., and Britten, R. J. (1975). DNA sequence organization in the mollusc *Aplysia californica*. *Cell* **6**, 29–40.

Ankel, W. E. (1930). Über das Vorkommen und die Bedeutung zwittriger Geschlechtszellen bei Prosobranchiern. *Biol. Zentralbl.* **50,** 513–532.

Attardo, C. (1957). I mitochondri e la citochromo-ossidasi nello sviluppo di *Aplysia. Acta Embryol. Morphol. Exp.* **1,** 65–70.

Baccetti, B., and Afzelius, B. A. (1976). "The Biology of the Sperm Cell," Monogr. Dev. Biol., Vol. 10. Karger, Basel.

Baumgold, J., Gallant, P., Terakawa, S., and Pant, H. (1981). Tetrodotoxin affects submembranous cytoskeletal proteins in perfused squid giant axons. *Biochem. Biophys. Res. Commun.* **103,** 653–658.

Beams, H. W., and Kessel, R. G. (1974). The problem of germ cell determinants. *Int. Rev. Cytol.* **39,** 413–479.

Beams, H. W., and Sekhon, S. S. (1966). Electron microscope studies on the oocyte of the freshwater mussel (*Anodonta*), with special reference to the stalk and mechanism of yolk deposition. *J. Morphol.* **119,** 477–502.

Bedford, L. (1966). The electron microscopy and cytochemistry of oogenesis and the cytochemistry of embryonic development of the prosobranch gastropod *Bembicium nanum* L. *J. Embryol. Exp. Morphol.* **15,** 15–37.

Bennett, V. (1982). The molecular basis for membrane-cytoskeleton association in human erythrocytes. *J. Cell. Biochem.* **18,** 49–65.

Billett, F. S. (1979). Oocyte mitochondria. *In* "Maternal Effects in Development" (D. R. Newth and M. Balls, eds.), pp. 147–166. Cambridge Univ. Press, London and New York.

Bloch, D., and Hew, H. (1960). Schedule of spermatogenesis in the pulmonate snail *Helix aspersa,* with special reference to histone transition. *J. Biophys. Biochem. Cytol.* **7,** 515–531.

Bolognari, A., and Licata, A. (1976). On the origin of the protein yolk in the oocytes of *Aplysia depilans* (Gastropoda, Opisthobranchia). *Experientia* **32,** 870–871.

Bolognari, A., Licata, A., and Ricca, M. B. (1976). Primary nucleolus and amphinucleoli in the oocytes of *Patella coerulea* L. (Moll. Gast.). *Experientia* **32,** 1008–1009.

Borgens, R. B., Vanable, J. W., Jr., and Jaffe, L. F. (1979). Bioelectricity and regeneration. *BioScience* **29,** 468–474.

Bottke, W. (1973a). Zur Ultrastruktur des Ovars von *Viviparus contectus* (Millet, 1813), (Gastropoda, Prosobranchia). II. Die Oocyten. *Z. Zellforsch. Mikrosk. Anat.* **138,** 239–260.

Bottke, W. (1973b). Lampenbürstenchromosomen und Amphinukleolen in Oocytenkernen der Schnecke *Bithynia tentaculata* L. *Chromosoma* **42,** 175–190.

Bottke, W., and Sinha, J. (1979). Ferritin as exogenous yolk protein in snails. *Wilhelm Roux's Arch. Dev. Biol.* **186,** 71–75.

Bottke, W., Sinha, I., and Keil, I. (1982). Coated vesicle-mediated transport and deposition of vitellogenic ferritin in the rapid growth phase of snail oocytes. *J. Cell Sci.* **53,** 173–191.

Bounoure, L. (1939). "L'origine des cellules reproductrices et le problème de la lignée germinale." Gauthier-Villars, Paris.

Brawley, S. H., and Quatrano, R. S. (1979). Sulfation of fucoidin in *Fucus* embryos. IV. Autoradiographic investigations of fucoidin sulfation and secretion during differentiation and the effect of cytochalasin treatment. *Dev. Biol.* **73,** 193–205.

Bretschneider, L. H., and Raven, C. P. (1951). Structural and topochemical changes in the egg cells of *Limnaea stagnalis* L. during oogenesis. *Arch. Neerl. Zool.* **10,** 1–31.

Briggs, R., and Cassens, G. (1966). Accumulation in the oocyte nucleus of a gene product essential for embryonic development beyond gastrulation. *Proc. Natl. Acad. Sci. U.S.A.* **55,** 1103–1109.

Brisson, P. (1973). Observation ultrastructurale des cellules germinales chez l'embryon d'*Acroloxus lacustris* (L.) (Gastéropode pulmoné basommatophore). *C. R. Hebd. Seances Acad. Sci., Ser. D.* **277,** 2205–2208.

Brown, D. D., and Dawid, I. B. (1968). Specific gene amplification in oocytes. *Science* **160**, 272–280.

Buckland-Nicks, J. A. (1973). The fine structure of the spermatozoon of *Littorina* (Gastropoda, Prosobranchia) with special reference to sperm motility. *Z. Zellforsch. Mikrosk. Anat.* **144**, 11–29.

Buckland-Nicks, J. A., and Chia, F. S. (1976). Spermatogenesis of a marine snail, *Littorina sitkana*. *Cell Tissue Res.* **170**, 455–475.

Buckland-Nicks, J. A., and Chia, F. S. (1977). On the nurse cell and the spermatozeugma in *Littorina sitkana*. *Cell Tissue Res.* **179**, 347–356.

Bulnheim, H.-P. (1962). Elektronenmikroskopische Untersuchungen zur Feinstruktur der Atypischen und Typischen Spermatozoen von *Opalia crenimarginata* (Gastropoda, Prosobranchia). *Z. Zellforsch. Mikrosk. Anat.* **56**, 371–386.

Bulnheim, H.-P. (1968). Atypische Spermatozoenbildung bei *Epitonium tinctum*. Ein Beitrag zum Problem des Spermatozoendimorphismus der Prosobranchia. *Helgol. Wiss. Meeresunters.* **18**, 232–253.

Byers, H. R., and Porter, K. R. (1977). Transformation in the structure of the cytoplasmic ground substance in erythrophores during pigment aggregation and dispersion. I. A study using whole-cell preparations in stereo high voltage electron microscopy. *J. Cell Biol.* **75**, 541–558.

Cabada, M. O., Darnbrough, C., Ford, P. J., and Turner, P. C. (1977). Differential accumulation of two classes of poly (A) associated with messenger RNA during oogenesis in *Xenopus laevis*. *Dev. Biol.* **57**, 427–439.

Clement, A. C. (1968). Development of the vegetal half of the *Ilyanassa* egg after removal of most of the yolk by centrifugal force, compared with the development of animal halves of similar visible composition. *Dev. Biol.* **17**, 165–186.

Clement, A. C., and Lehmann, F. E. (1956). Über das Verteilungsmuster von Mitochondrien und Lipoidtropfen während der Furchung des Eies von *Ilyanassa*. *Naturwissenschaften* **43**, 478–479.

Clement, A. C., and Tyler, A. (1967). Protein-synthesizing activity of the anucleate polar lobe of the mud snail *Ilyanassa obsoleta*. *Science* **158**, 1457–1458.

Clérot, J. C. (1979). Les groupements mitochondriaux des cellules germinales des poissons téléostéens cyprinidés. II. Etude autoradiographique à haute résolution de l'incorporation de phénylalanine ^{3}H et d'uridine ^{3}H. *Exp. Cell Res.* **120**, 237–244.

Collier, J. R. (1971). Number of ribosomal RNA cistrons in the marine mud snail *Ilyanassa obsoleta*. *Exp. Cell Res.* **69**, 181–184.

Collier, J. R. (1976). Nucleic acid chemistry of the *Ilyanassa* embryo. *Am. Zool.* **16**, 483–500.

Collier, J. R., and Tucci, J. (1980). The reassociation kinetics of the *Ilyanassa* genome. *Dev., Growth Differ.* **22**, 741–748.

Colom, J., and Subirana, J. A. (1981). Presence of H2b histone in spermatozoa from marine gastropoda. *Exp. Cell Res.* **131**, 462–465.

Costantini, F. D., Scheller, R. H., Britten, R. J., and Davidson, E. H. (1978). Repetitive sequence transcripts in the mature sea urchin oocyte. *Cell* **15**, 173–187.

Crowell, J. (1964). The fine structure of the polar lobe of *Ilyanassa obsoleta*. *Acta Embryol. Morphol. Exp.* **7**, 225–234.

Dan, J. C., and Takaichi, S. (1979). Spermiogenesis in the pulmonate snail, *Euhadra hickonis*. III. Flagellum formation. *Dev., Growth Differ.* **21**, 71–86.

Davidson, E. H. (1976). "Gene Activity in Early Development" 2nd ed. Academic Press, New York.

Davidson, E. H., Hough, B. R., Chamberlin, M. E., and Britten, R. J. (1971). Sequence repetition in the DNA of *Nassaria (Ilyanassa)obsoleta*. *Dev. Biol.* **25**, 445–463.

Dawid, I. B. (1972). Cytoplasmic DNA. *In* ''Oogenesis'' (J. D. Biggers and A. W. Schuetz, eds.), pp. 215–226. Univ. Park Press, Baltimore.

de Jong-Brink, M., de Wit, A., Kraal, G., and Boer, H. H. (1976). A light and electron microscope study on oogenesis in the freshwater pulmonate snail *Biomphalaria glabrata*. *Cell Tissue Res.* **171**, 195–219.

de Jong-Brink, M., Boer, H. H., Hommes, T. G., and Kodde, A. (1977). Spermatogenesis and the role of Sertoli cells in the freshwater snail *Biomphalaria glabrata*. *Cell Tissue Res.* **181**, 37–58.

Denis, H. (1977). Accumulation du RNA dans les oocytes des vertébrés inférieurs. *Biol. Cell.* **28**, 87–92.

Dhainaut, A. (1973). Mode de formation des lamelles annelées résultant de l'évolution, en condition anhormonale, des ovocytes de *Nereis diversicolor* (Annélide Polychète). *Z. Zellforsch. Mikrosk. Anat.* **137**, 481–492.

Dohmen, M. R., and Lok, D. (1975). The ultrastructure of the polar lobe of *Crepidula fornicata* (Gastropoda, Prosobranchia). *J. Embryol. Exp. Morphol.* **34**, 419–428.

Dohmen, M. R., and van der Mey, J. C. A. (1977). Local surface differentiations at the vegetal pole of the eggs of *Nassarius reticulatus, Buccinum undatum*, and *Crepidula fornicata. Dev. Biol.* **61**, 104–113.

Dohmen, M. R., and Verdonk, N. H. (1974). The structure of a morphogenetic cytoplasm, present in the polar lobe of *Bithynia tentaculata* (Gastropoda, Prosobranchia). *J. Embryol. Exp. Morphol.* **31**, 423–433.

Dohmen, M. R., and Verdonk, N. H. (1979a). The ultrastructure and role of the polar lobe in development of molluscs. *In* ''Determinants of Spatial Organization'' (S. Subtelny and I. R. Konigsberg, eds.), pp. 3–27. Academic Press, New York.

Dohmen, M. R., and Verdonk, N. H. (1979b). Cytoplasmic localizations in mosaic eggs. *In* ''Maternal Effects in Development'' (D. H. Newth and M. Balls, eds.), pp. 127–145. Cambridge Univ. Press, London and New York.

Duncan, R., and Humphreys, T. (1981). Multiple oligo (A) tracts associated with inactive sea urchin mRNA maternal sequences. *Dev. Biol.* **88**, 211–219.

Durfort, M. (1976). Relation entre les lamelles annelées et le réticulum endoplasmique granulaire dans les ovocytes de *Trachydermon cinereus*, Thiele (Mollusque, Polyplacophore). *Ann. Sci. Nat., Zool. Biol. Anim.* [12] **18**, 449–457.

Eckberg, W. R., and Kang, Y.-H. (1981). A cytological analysis of differentiation without cleavage in cytochalasin *B-* and colchicine-treated embryos' of *Chaetopterus pergamentaceus. Differentiation (Berlin)* **19**, 154–160.

Eckelbarger, K. J., and Eyster, L. S. (1981). An ultrastructural study of spermatogenesis in the nudibranch mollusc *Spurilla neapolitana. J. Morphol.* **170**, 283–299.

Eddy, E. M. (1975). Germ plasm and the differentiation of the germ cell line. *Int. Rev. Cytol.* **43**, 229–280.

Fain-Maurel, M. A. (1966). Acquisitions récentes sur les spermatogénèses atypiques. *Ann. Biol. Anim., Biochim., Biophys.* **5**, 513–564.

Favard, P., and André, J. (1970). The mitochondria of spermatozoa. *In* ''Comparative Spermatology'' (B. Baccetti, ed.), pp. 415–429. Academic Press, New York.

Favard, P., and Carasso, N. (1958). Origine et ultrastructure des plaquettes vitellines de la planorbe. *Arch. Anat. Microsc. Morphol. Exp.* **47**, 211–234.

Fawcett, D. W. (1961). Intercellular bridges. *Exp. Cell Res., Suppl.* **8**, 174–187.

Fawcett, D. W., and Bedford, J. M., eds. (1979). ''The Spermatozoon, Maturation, Motility, Surface Properties and Comparative Aspects.'' Urban & Schwarzenberg, Baltimore.

Fawcett, D. W., Anderson, W. A., and Phillips, D. M. (1971). Morphogenetic factors influencing the shape of the sperm head. *Dev. Biol.* **26**, 220–251.

Féral, C. (1977). Etude de la spermatogénèse typique chez *Ocenebra erinacea*, Mollusque Gastéropode, Prosobranche. *Bull. Soc. Zool. Fr.* **102**, 25–30.

Firtel, R. A., and Monroy, A. (1970). Polysomes and RNA synthesis during early development of the surf clam *Spisula solidissima*. *Dev. Biol.* **21**, 87–104.

Ford, P. J. (1972). Ribonucleic acid synthesis during oogenesis in *Xenopus laevis*. *In* "Oogenesis" (J. D. Biggers and A. W. Schuetz, eds.), pp. 167–191. Univ. Park Press, Baltimore.

Franc, A. (1951). Ovogénèse et évolution nucléolaire chez les Gastéropodes Prosobranches. *Ann. Sci. Nat., Zool. Biol. Anim.* [11] **13**, 135–143.

Franzén, A. (1955). Comparative morphological investigations into the spermiogenesis among Mollusca. *Zool. Bidr. Uppsala* **30**, 399–456.

Fretter, V., and Graham, A. (1962). "British Prosobranch Molluscs. Their Functional Anatomy and Ecology." Ray Society, London.

Friend, D. S. (1982). Plasma-membrane diversity in a highly polarized cell. *J. Cell Biol.* **93**, 243–249.

Gabrielli, F., and Baglioni, C. (1975). Maternal messenger RNA and histone synthesis in embryos of the surf clam *Spisula solidissima*. *Dev. Biol.* **43**, 254–263.

Gall, J. G. (1961). Centriole replication. A study of spermatogenesis in the snail *Viviparus*. *J. Biophys. Biochem. Cytol.* **10**, 163–193.

Galtsoff, P. S., and Philpott, D. E. (1960). Ultrastructure of the spermatozoon of the oyster, *Crassostrea virginica*. *J. Ultrastruct. Res.* **3**, 241–253.

Garreau de Loubresse, N. (1971). Spermiogénèse d'un Gastéropode prosobranche: *Nerita senegalensis;* evolution du canal intranucléaire. *J. Microsc. Biol. Cell.* **12**, 425–440.

Gérin, Y. (1972). Morphogénèse des vésicules à double membrane du lobe polaire d'*Ilyanassa obsoleta* Say. Etude ultrastructurale. *J. Microsc. Biol. Cell.* **13**, 57–66.

Gérin, Y. (1976). Origin and evolution of some organelles during oogenesis in the mud snail *Ilyanassa obsoleta*. II. The yolk nucleus and the lipochondria. *Acta Embryol. Exp.* **1**, 27–35.

Geuskens, M. (1968). Mise en évidence au microscope électronique de polysomes actifs dans les lobes polaires isolés d'*Ilyanassa*. *Exp. Cell Res.* **54**, 263–266.

Geuskens, M., and de Jonghe d'Ardoye, V. (1971). Metabolic patterns in *Ilyanassa* polar lobes. *Exp. Cell Res.* **67**, 61–72.

Giese, A. C., and Pearse, J. S., eds. (1977). "Reproduction of Marine Invertebrates," Vol. 4. Academic Press, New York.

Giese, A. C., and Pearse, J. S., eds. (1979). "Reproduction of Marine Invertebrates," Vol. 5. Academic Press, New York.

Giusti, F., and Selmi, M. G. (1982). The morphological peculiarities of the typical spermatozoa of *Theodoxus fluviatilis* (L.) (Neritoidea) and their implication for motility. *J. Ultrastruct. Res.* **78**, 166–177.

Goldberg, R. B., Crain, W. R., Ruderman, J. V., Moore, G. P., Barnett, T. R., Higgins, R. C., Gelfand, R. A., Galan, G. G., Britten, R. J., and Davidson, E. H. (1975). DNA sequence organization in the genomes of five marine invertebrates. *Chromosoma* **51**, 225–251.

Gomot, L. (1973). Etude du fonctionnement de l'appareil génital de l'escargot *Helix aspersa* par la méthode des cultures d'organes. *Arch. Anat. Histol. Embryol.* **56**, 131–160.

Gould, H. N. (1917). Studies on sex in the hermaphrodite mollusc *Crepidula plana*. I. History of the sexual cycle. *J. Exp. Zool.* **23**, 1–69.

Grant, P. (1978). "Biology of Developing Systems." Holt, Rinehart & Winston, New York.

Griffond, B. (1977). Individualisation et organogénèse de la gonade embryonnaire de *Viviparus viviparus* L. (Mollusque gastéropode à sexes séparés). *Wilhelm Roux's Arch. Dev. Biol.* **183**, 131–147.

Griffond, B. (1978). Sexualisation de la gonade de *Viviparus viviparus* L. (Mollusque gastéropode prosobranche à sexes séparés). *Wilhelm Roux's Arch. Dev. Biol.* **184**, 213–231.

Griffond, B. (1980). Etude ultrastructurale de la spermatogenèse typique de *Viviparus viviparus* L., Mollusque Gastéropode. *Arch. Biol.* **91**, 445–462.

Griffond, B. (1981). Étude ultrastructurale de la spermatogénèse atypique de *Viviparus viviparus* L., mollusque gastéropode. *Arch. Biol.* **92**, 275–286.

Griffond, B., and Bride, J. (1981). Etude histologique et ultrastructurale de la gonade d'*Helix aspersa* Muller a l'eclosion. *Reprod. Nutr. Dev.* **21**, 149–161.

Guraya, S. S. (1979). Recent advances in the morphology, cytochemistry, and function of Balbiani's vitelline body in animal oocytes. *Int. Rev. Cytol.* **59**, 249–321.

Haino, K., and Kigawa, M. (1966). Studies on the egg-membrane lysin of *Tegula pfeifferi:* Isolation and chemical analysis of the egg membrane. *Exp. Cell Res.* **42**, 625–633.

Healy, J. M., and Jamieson, B. G. M. (1981). An ultrastructural examination of developing and mature paraspermatozoa in *Pyrazus ebeninus* (Mollusca, Gastropoda, Potamididae). *Zoomorphol.* **98**, 101–119.

Heller, E., and Raftery, M. A. (1976). The vitelline envelope of eggs from the giant keyhole limpet *Megathura crenulata.* I. Chemical composition and structural studies. *Biochemistry* **15**, 1194–1198.

Hill, R. S. (1977). Studies on the ovotestis of the slug *Agriolimax reticulatus* (Müller). 2. The epithelia. *Cell Tissue Res.* **183**, 131–141.

Hill, R. S., and Bowen, J. P. (1976). Studies on the ovotestis of the slug *Agriolimax reticulatus* (Müller). I. The oocyte. *Cell Tissue Res.* **173**, 465–482.

Hochpöchler, F. (1979). Vergleichende Untersuchungen über die Entwicklung des Geschlechtsapparates der Stylommatophora (Gastropoda). *Zool. Anz.* **202**, 289–306.

Hogg, N. A. S., and Wijdenes, J. (1979). A study of gonadal organogenesis, and the factors influencing regeneration following surgical castration in *Deroceras reticulatum* (Pulmonata: Limacidae). *Cell Tissue Res.* **198**, 295–307.

Howard, B., Mitchell, P. C. H., Ritchie, A., Simkiss, K., and Taylor, M. (1981). The composition of intracellular granules from the metal-accumulating cells of the common garden snail (*Helix aspersa*). *Biochem. J.* **194**, 507–511.

Huebner, E., and Anderson, E. (1976). Comparative spiralian oogenesis - structural aspects: An overview. *Am. Zool.* **16**, 315–343.

Humphreys, W. J. (1962). Electron microscope studies on eggs of *Mytilus edulis. J. Ultrastruct. Res.* **7**, 467–487.

Humphreys, W. J. (1967). The fine structure of cortical granules in eggs and gastrulae of *Mytilus edulis. J. Ultrastruct. Res.* **17**, 314–326.

Hylander, B. L., and Summers, R. G. (1981). The effect of local anesthetics and ammonia on cortical granule–plasma membrane attachment in the sea urchin egg. *Dev. Biol.* **86**, 1–11.

Jackson, B. W., Grund, C., Schmid, E., Bürki, K., Franke, W. W., and Illmensee, K. (1980). Formation of cytoskeletal elements during mouse embryogenesis. *Differentiation (Berlin)* **17**, 161–179.

Jaffe, L. A., and Guerrier, P. (1981). Localization of electrical excitability in the early embryo of *Dentalium. Dev. Biol.* **83**, 370–373.

Jaffe, L. F., Robinson, K. R., and Nuccitelli, R. (1974). Local cation entry and self-electrophoresis as an intracellular localization mechanism. *Ann. N.Y. Acad. Sci.* **238**, 372–389.

Jeffery, W. R. (1982). Calcium ionophore polarizes ooplasmic segregation in ascidian eggs. *Science,* **216**, 545–547.

Joosse, J., and Reitz, D. (1969). Functional anatomical aspects of the ovotestis of *Lymnaea stagnalis. Malacologia* **9**, 101–109.

Karp, G. C., and Whiteley, A. H. (1973). DNA-RNA hybridization studies of gene activity during the development of the gastropod, *Acmaea scutum. Exp. Cell Res.* **78**, 236–241.

Kedes, L. H. (1979). Histone genes and histone messengers. *Annu. Rev. Biochem.* **48**, 837–870.

Kessel, R. G. (1968). Mechanisms of protein yolk synthesis and deposition in crustacean oocytes. *Z. Zellforsch. Mikrosk. Anat.* **89**, 17–38.

Kessel, R. G. (1973). Structure and function of the nuclear envelope and related cytomembranes. *Prog. Surf. Membr. Sci.* **6**, 243–329.

Kessel, R. G. (1981a). Origin, differentiation, distribution and possible functional role of annulate lamellae during spermatogenesis in *Drosophila melanogaster. J. Ultrastruct. Res.* **75**, 72–96.

Kessel, R. G. (1981b). Annulate lamellae and polyribosomes in young oocytes of the rainbow trout, *Salmo gairdneri. J. Submicrosc. Cytol.* **13**, 231–252.

Kessel, R. G. (1982). Differentiation of *Acmaea digitalis* oocytes with special reference to lipid–endoplasmic reticulum–annulate lamellae–polyribosome relationships. *J. Morphol.* **171**, 225–243.

Kidder, G. M. (1976). The ribosomal RNA cistrons in clam gametes. *Dev. Biol.* **49**, 132–142.

Kielbowna, L., and Koscielski, B. (1974). A cytochemical and autoradiographic study on oocyte nucleoli in *Lymnaea stagnalis* L. *Cell Tissue Res.* **152**, 103–111.

Kohnert, R. (1980). Zum Spermiendimorphismus der Prosobranchier: Spermiogenese und ultrastruktureller Aufbau der Spermien von *Bithynia tentaculata* (L.). *Zool. Anz.* **205**, 145–161.

Koike, K., and Nishiwaki, S. (1980). The ultrastructure of dimorphic spermatozoa in two species of the Strombidae (Gastropoda: Prosobranchia). *Venus (Jpn. J. Malacol.)* **38**, 259–274.

Koser, R. B., and Collier, J. R. (1971). The molecular weight and thermolability of *Ilyanassa* ribosomal RNA. *Biochim. Biophys. Acta* **254**, 272–277.

Kubo, M. (1977). The formation of a temporary-acrosome in the spermatozoon of Laternula limicola (Bivalvia, Mollusca). *J. Ultrastruct. Res.* **61**, 140–148.

Kubo, M., and Ishikawa, M. (1978). Organizing process of the temporary-acrosome in spermatogenesis of the bivalve *Lyonsia ventricosa. J. Submicrosc. Cytol.* **10**, 411–421.

Kubo, M., and Ishikawa, M. (1981). Organization of the acrosome and helical structures in sperm of the aplysiid, *Aplysia kurodai* (Gastropoda, Opisthobranchia). *Differentiation (Berlin)* **20**, 131–140.

Kubo, M., Ishikawa, M., and Numakunai, T. (1979). Ultrastructural studies on early events in fertilization of the bivalve *Laternula limicola. Protoplasma* **100**, 73–83.

Lazarides, E. (1980). Intermediate filaments as mechanical integrators of cellular space. *Nature (London)* **283**, 249–256.

Le Gall, S., and Streiff, W. (1975). Protandric hermaphroditism in prosobranch gastropods. *In* "Intersexuality in the Animal Kingdom" (R. Reinboth, ed.), pp. 170–178. Springer-Verlag, Berlin and New York.

Lehto, V.-P., Virtanen, I., and Kurki, P. (1978). Intermediate filaments anchor the nuclei in nuclear monolayers of cultured human fibroblasts. *Nature (London)* **272**, 175–177.

Lewis, C. A., Leighton,D. L., and Vacquier, V. D. (1980). Morphology of Abalone spermatozoa before and after the acrosome reaction. *J. Ultrastruct. Res.* **72**, 39–46.

Lohs-Schardin, M. (1982). Dicephalic—a *Drosophila* mutant affecting polarity in follicle organization and embryonic patterning. *Wilhelm Roux's Arch. Dev. Biol.* **191**, 28–36.

Longo, F. J., and Anderson, E.(1970). An ultrastructural analysis of fertilization in the surf clam *Spisula solidissima.* I. Polar body formation and development of the female pronucleus. *J. Ultrastruct. Res.* **33**, 495–514.

Luchtel, D. (1972a). Gonadal development and sex determination in pulmonate molluscs. I. *Arion circumscriptus. Z. Zellforsch. Mikrosk. Anat.* **130**, 279–301.

Luchtel, D. (1972b). Gonadal development and sex determination in pulmonate molluscs. II. *Arion ater rufus* and *Deroceras reticulatum. Z. Zellforsch. Mikrosk. Anat.* **130**, 302–311.

McCann-Collier, M. (1977). An unusual cytoplasmic organelle in oocytes of *Ilyanassa obsoleta. J. Morphol.* **153**, 119–127.

McCann-Collier, M. (1979). RNA synthesis during *Ilyanassa* oogenesis: An autoradiographic study. *Dev., Growth Differ.* **21**, 391–399.

McGee-Russell, S. (1968). The method of combined observations with light and electron micro-scopes applied to the study of histochemical colourations in nerve cells and oocytes. *In* "Cell Structure and Its Interpretation" (S. McGee-Russell, and K. Ross, eds.), p. 433. Arnold, London.

McLean, K. W., and Whiteley, A. H. (1973). Characteristics of DNA from the oyster, *Crassostrea gigas. Biochim. Biophys. Acta* **335**, 35–41.

Mahowald, A. P., Allis, C. D., Karrer, K. M., Underwood, E. M., and Waring, G. L. (1979). Germ plasm and pole cells of *Drosophila. In* "Determinants of Spatial Organization" (S. Subtelny and I. R. Konigsberg, eds.), pp. 127–146. Academic Press, New York.

Malacinski, G. M. (1974). Biological properties of a presumptive morphogenetic determinant from the amphibian oocyte germinal vesicle nucleus. *Cell Differ.* **3**, 31–44.

Maul, G. G. (1980). Determination of newly synthesized and phosphorylated nuclear proteins in mass-isolated germinal vesicles of *Spisula solidissima. Exp. Cell Res.* **129**, 431–438.

Maxwell, W. L. (1976). The neck region of the spermatozoon of *Discus rotundatus* (Müller) (Pulmonata, Stylommatophora). *J. Morphol.* **150**, 299–305.

Melone, G., Lora Lamia Donin, C., and Cotelli, F. (1980). The paraspermatic cell (atypical sper-matozoon) of Prosobranchia: A comparative ultrastructural study. *Acta Zool. (Stockholm)* **61**, 191–201.

Merriam, R. W., and Hill, R. J. (1976). The germinal vesicle nucleus of *Xenopus laevis* oocytes as a selective storage receptacle for proteins. *J. Cell Biol.* **69**, 659–668.

Moor, B. (1977). Zur Embryologie von *Bradybaena (Eulota) fruticum* Müller (Gastropoda, Pul-monata Stylommatophora). *Zool. Jahrb., Abt. Anat. Ontog. Tiere* **97**, 323–399.

Moreau, M., and Guerrier, P. (1981). Absence of regional differences in the membrane properties from the embryo of the mud snail *Ilyanassa obsoleta. Biol. Bull. (Woods Hole, Mass.)* **161**, 320–321.

Morrill, J. B., Rubin, R. W., and Grandi, M. (1976). Protein synthesis and differentiation during pulmonate development. *Am. Zool.* **16**, 547–561.

Myles, D. G., and Hepler, P. K. (1982). Shaping of the sperm nucleus in *Marsilea:* A distinction between factors responsible for shape generation and shape determination. *Dev. Biol.* **90**, 238–252.

Nieuwkoop, P. D., and Sutasurya, L. A. (1979). "Primordial Germ Cells in the Chordates." Cambridge Univ. Press, London and New York.

Nieuwkoop, P. E., and Sutasurya, L. A. (1981). "Primordial Germ Cells in the Invertebrates." Cambridge Univ. Press, London and New York.

Nishiwaki, S. (1964). Phylogenetical study on the type of the dimorphic spermatozoa in Pros-obranchia. *Sci. Rep. Tokyo Kyoiku Daigaku, Sect. B* **11**, 237–275.

Nørrevang, A. (1968). Electron microscopic morphology of oogenesis. *Int. Rev. Cytol.* **23**, 113–186.

Nuccitelli, R. (1978). Ooplasmic segregation and secretion in the *Pelvetia* egg is accompanied by a membrane-generated electrical current. *Dev. Biol.* **62**, 13–33.

Odintsova, N. A., Svinarchuk, F. P., Zalenskaya, I. A., and Zalenskii, A. O. (1981). Partial fractionation and certain characteristics of the basic chromatin proteins of bivalve mollusk sperm. *Biokhimiya* **46**, 404–409.

Pasteels, J. J. (1966). Les mouvements corticaux de l'oeuf de *Barnea candida* (Mollusque Bivalve) étudiés au microscope électronique. *J. Embryol. Exp. Morphol.* **16**, 311–319.

Pasteels, J. J., and de Harven, E. (1963). Etude au microscope électronique du cytoplasme de l'oeuf vierge et fecondé de *Barnea candida* (Mollusque Bivalve). *Arch. Biol.* **74**, 415–437.

Pearse, J. S. (1979). Polyplacophora. *In* "Reproduction of Marine Invertebrates" (A. C. Giese and J. S. Pearse, eds.), Vol. 5, pp. 27–85. Academic Press, New York.

Pearse, J. S., and Woollacott, R. M. (1979). Chiton sperm: No acrosome? *Am. Zool.* **19**, 956.

Perlman, S. M., Ford, P. J., and Rosbash, M. (1977). Presence of tadpole and adult globin RNA sequences in oocytes of *Xenopus laevis*. *Proc. Natl. Acad. Sci. U.S.A.* **74**, 3835–3839.

Popham, J. D. (1975). The fine structure of the oocyte of *Bankia australis* (Teredinidae, Bivalvia) before and after fertilization. *Cell Tissue Res.* **157**, 521–534.

Popham, J. D. (1979). Comparative spermatozoon morphology and bivalve phylogeny. *Malacol. Rev.* **12**, 1–20.

Portmann, A. (1927). Die Nähreier-Bildung durch atypische Spermien bei *Buccinum undatum* L. *Z. Zellforsch. Mikrosk. Anat.* **5**, 230–243.

Pucci-Minafra, I., Minafra, S., and Collier, J. R. (1969). Distribution of ribosomes in the egg of *Ilyanassa obsoleta*. *Exp. Cell Res.* **57**, 167–178.

Puigdoménech, P., Martinez, P., Cabré, O., Palau, J., Bradbury, E. M., and Crane-Robinson, C. (1976). Studies on the role and mode of operation of the very lysine-rich histones in eukaryote chromatin. Nuclear-magnetic-resonance studies on nucleoproteins and histone φ 1. DNA complexes from marine invertebrate sperm. *Eur. J. Biochem.* **65**, 357–363.

Quatrano, R. S., Brawley, S. H., and Hogsett, W. E. (1979). The control of the polar deposition of a sulfated polysaccharide in *Fucus* zygotes. In "Determinants of Spatial Organization" (S. Subtelny and I. R. Konigsberg, eds.), pp. 77–96. Academic Press, New York.

Raff, R. A., Newrock, K. M., Secrist, R. D., and Turner, F. R. (1976). Regulation of protein synthesis in embryos of *Ilyanassa obsoleta*. *Am. Zool.* **16**, 529–545.

Raven, C. P. (1961). "Oogenesis. The Storage of Developmental Information." Pergamon, Oxford.

Raven, C. P. (1966). "Morphogenesis. The Analysis of Molluscan Development," 2nd ed. Pergamon, New York.

Raven, C. P. (1970). The cortical and subcortical cytoplasm of the *Lymnaea* egg. *Int. Rev. Cytol.* **28**, 1–44.

Raven, C. P., and van der Wal, U. P. (1964). Analysis of the formation of the animal pole in the eggs of *Limnaea stagnalis*. *J. Embryo. Exp. Morphol.* **12**, 123–139.

Rebhun, L. I. (1956). Electron microscopy of basophilic structures of some invertebrate oocytes. II. Fine structure of the yolk nuclei. *J. Biophys. Biochem. Cytol.* **2**, 159–170.

Rebhun, L. I. (1960). Some electron microscope observations on membranous basophilic elements of invertebrate eggs. *J. Ultrastruct. Res.* **5**, 208–225.

Reinke, E. E. (1914). The development of the apyrene spermatozoa of *Strombus bituberculatus*. *Carnegie Inst. Washington Publ.* **183**, 195–239.

Renault, L. (1965). Observations sur l'ovogénèse et sur les cellules nourricières chez *Lamellaria perspicua* (L.) (Mollusque Prosobranche). *Bull. Mus. Natl. Hist. Nat., Zool.* **37**, 282–284.

Reverberi, G. (1966). Electron microscopy of some cytoplasmic structures of the oocytes of *Mytilus*. *Exp. Cell Res.* **42**, 392–394.

Reverberi, G. (1970). The ultrastructure of the ripe oocyte of *Dentalium*. *Acta Embryol. Exp.* pp. 255–279.

Ribbert, D., and Kunz, W. (1969). Lampenbürsten-Chromosomen in den Oocytenkernen von *Sepia officinalis*. *Chromosoma* **28**, 93–106.

Richter, H. P. (1976). Feinstrukturelle Untersuchungen zur Oogenese der Käferschnecke *Lepidochitona cinereus* (Mollusca, Polyplacophora). *Helgol. Wiss. Meeresunters.* **28**, 250–303.

Rigby, J. E. (1979). The fine structure of the oocyte and follicle cells of *Lymnaea stagnalis*, with special reference to the nutrition of the oocyte. *Malacologia* **18**, 377–380.

Risley, M. S., Eckhardt, R. A., Mann, M., and Kasinsky, H. E. (1982). Determinants of sperm nuclear shaping in the genus *Xenopus*. *Chromosoma* **84**, 557–569.

Ritter, C., and André, J. (1975). Presence of a complete set of cytochromes despite the absence of cristae in the mitochondrial derivative of snail sperm. *Exp. Cell Res.* **92**, 95–101.

Robinson, K. R., and Jaffe, L. F. (1975). Polarizing fucoid eggs drive a calcium current through themselves. *Science* **187**, 70–72.

Romanova, L. G., and Gazarian, K. G. (1966). Studies of RNA and protein metabolism in the nucleoli of the mollusc's oocytes. *Citologiya* **8**, 648–652.

Roosen-Runge, E. C. (1977). "The Process of Spermatogenesis in Animals." Cambridge Univ. Press, London and New York.

Rosenthal, E. T., Hunt, T., and Ruderman, J. V. (1980). Selective translation of mRNA controls the pattern of protein synthesis during early development of the surf clam, *Spisula solidissima*. *Cell* **20**, 487–494.

Sabelli, B., and Scanabissi, F. S. (1980). Sexualization of germ cells in *Papillifera papillaris* (Pulmonata, Stylommatophora). *Monit. Zool. Ital.* **14**, 19–26.

Sachs, M. I. (1971). A cytological analysis of artificial parthenogenesis in the surf clam *Spisula solidissima*. *J. Ultrastruct. Res.* **36**, 806–823.

Sawada, T., and Osanai, K. (1981). The cortical contraction related to the ooplasmic segregation in *Ciona intestinalis* eggs. *Wilhelm Roux's Arch. Dev. Biol.* **190**, 280–214.

Schmekel, L., and Fioroni, P. (1974). The ultrastructure of the yolk nucleus during early cleavage of *Nassarius reticulatus* L. (Gastropoda, Prosobranchia). *Cell Tissue Res.* **153**, 79–88.

Schmekel, L., and Fioroni, P. (1975). Cell differentiation during early development of *Nassarius reticulatus* L. (Gastropoda, Prosobranchia). I. Zygote to 16-cell stage. *Cell Tissue Res.* **159**, 503–522.

Selwood, L. (1968). Interrelationships between developing oocytes and ovarian tissues in the chiton *Sypharochiton septentriones* (Ashby) (Mollusca, Polyplacophora). *J. Morphol.* **125**, 71–104.

Selwood, L. (1970). The role of the follicle cells during oogenesis in the chiton *Sypharochiton septentriones* (Ashby) (Polyplacophora, Mollusca). *Z. Zellforsch. Mikrosk. Anat.* **104**, 179–192.

Shileiko, L. V., and Danilova, L. V. (1979). Ultrastructure of spermatozoa in pulmonate molluscs *Trichia hispida* and *Succinea putris*. *Ontogenez* **10**, 437–447.

Silberzahn, N. (1979). Les cellules de la lignée femelle chez un hermaphrodite protandre *Crepidula fornicata*, mollusque prosobranche. *Ann. Soc. Fr. Biol. Dev., Paris* pp. 17–18.

Sitte, P. (1980). General principles of cellular compartmentation. *In* "Cell Compartmentation and Metabolic Channeling" (L. Nover, F. Lynen, and K. Mothes, eds.), pp. 17–32. Elsevier, Amsterdam.

Solomon, F. (1980). Neuroblastoma cells recapitulate their detailed neurite morphologies after reversible microtubule disassembly. *Cell* **21**, 333–338.

Staiger, R. (1951). Cytologische und morphologische Untersuchungen zur Determination der Nähreier bei Prosobranchiern. *Z. Zellforsch. Mikrosk. Anat.* **35**, 495–549.

Starke, F. J. (1971). Elektronenmikroskopische Untersuchung der Zwittergonadenacini von *Planorbarius corneus* L. (Basommatophora). *Z. Zellforsch. Mikrosk. Anat.* **119**, 483–514.

Steinert, G., Thomas, C., and Brachet, J. (1976). Localization by *in situ* hybridization of amplified ribosomal DNA during *Xenopus laevis* oocyte maturation (a light and electron microscopy study). *Proc. Natl. Acad. Sci. U.S.A.* **73**, 883–836.

Subirana, J. A., Cozcolluela, C., Palau, J., and Unzeta, M. (1973). Protamines and other basic proteins from spermatozoa of molluscs. *Biochim. Biophys. Acta* **317**, 364–379.

Szollosi, D. G. (1965). The fate of sperm middle-piece mitochondria in the rat egg. *J. Exp. Zool.* **159**, 367–377.

Takaichi, S. (1978). Spermiogenesis in the pulmonate snail, *Euhadra hickonis*. II. Structural changes of the nucleus. *Dev., Growth Differ.* **20**, 301–315.

Takaichi, S., and Dan, J. C. (1977). Spermiogenesis in the pulmonate snail, *Euhadra hickonis*. I. Acrosome formation. *Dev., Growth Differ.* **19**, 1–14.

Taylor, G. T., and Anderson, E. (1969). Cytochemical and fine structural analysis of oogenesis in the gastropod, *Ilyanassa obsoleta*. *J. Morphol.* **129**, 211–248.

Terakado, K. (1974). Origin of yolk granules and their development in the snail, *Physa acuta*. *J. Electron Microsc.* **23**, 99–106.

Terakado, K. (1975). Extrusion of intact nucleolus in oocytes of *Physa acuta*. *J. Electron Microsc.* **24**, 295–297.

Thomas, T. L., Posakony, J. W., Anderson, D. M., Britten, R. J., and Davidson, E. H. (1981). Molecular structure of maternal RNA. *Chromosoma* **84**, 319–335.

Thompson, T. E., and Bebbington, A. (1969). Structure and function of the reproductive organs of three species of *Aplysia* (Gastropoda: Opisthobranchia). *Malacologia* **7**, 347–380.

Threadgold, L. T. (1976). "The Ultrastructure of the Animal Cell," 2nd ed. Pergamon, Oxford.

Tochimoto, T. (1967). Comparative histochemical study on the dimorphic spermatozoa of the Prosobranchia with special reference to polysaccharides. *Sci. Rep. Tokyo Kyoiku Daigaku, Sect. B* **13**, 75–109.

Ubbels, G. A. (1968). A cytochemical study of oogenesis in the pond snail *Limnaea stagnalis*. Ph.D. Thesis, University of Utrecht, Utrecht, The Netherlands.

Ubbels, G. A., Bezem, J. J., and Raven, C. P. (1969). Analysis of follicle cell patterns in dextral and sinistral *Limnaea peregra*. *J. Embryol. Exp. Morphol.* **21**, 445–466.

van der Wal, U. P. (1976a). The mobilization of the yolk of *Lymnaea stagnalis* (Mollusca). I. A structural analysis of the differentiation of the yolk granules. *Proc. K. Ned. Akad. Wet., Ser. C* **79**, 393–404.

van der Wal, U. P. (1976b). The mobilization of the yolk of *Lymnaea stagnalis*. II. The localization and function of the newly synthesized proteins in the yolk granules during early embryogenesis. *Proc. K. Ned. Akad. Wet., Ser. C* **79**, 405–420.

van Dongen, C. A. M., Mikkers, F. E. P., de Bruyn, C., and Verheggen, T. P. E. M. (1981). Molecular composition of the polar lobe of first cleavage stage embryos in comparison with the lobeless embryo in *Nassarius reticulatus* (Mollusca) as analyzed by isotachophoresis. *In* "Analytical Isotachophoresis" (F. M. Everaerts, ed.), pp. 207–216. Elsevier, Amsterdam.

Verdonk, N. H. (1973). Gene expression in early development of *Lymnaea stagnalis*. *Dev. Biol.* **35**, 29–35.

Virtanen, I., Kurkinen, M., and Lehto, V.-P. (1979). Nucleus-anchoring cytoskeleton in chicken red blood cells. *Cell Biol. Int. Rep.* **3**, 157–162.

Walker, M., and MacGregor, H. C. (1968). Spermatogenesis and the structure of the mature sperm in *Nucella lapillus* (L.). *J. Cell Sci.* **3**, 95–104.

Wang, E., Cross, R. K., and Choppin, P. W. (1979). Involvement of microtubules and 10-nm filaments in the movement and positioning of nuclei in syncytia. *J. Cell Biol.* **83**, 320–337.

Wassarman, R. M., and Mrozak, S. C. (1981). Program of early development in the mammal: Synthesis and intracellular migration of histone H4 during oogenesis in the mouse. *Dev. Biol.* **84**, 364–371.

Weakly, B. S. (1967). "Balbiani's body" in the oocyte of the golden hamster. *Z. Zellforsch. Mikrosk. Anat.* **83**, 582–588.

West, D. L. (1978a). Reproductive biology of *Colus stimpsoni* (Prosobranchia: Buccinidae). II. Spermatogenesis. *Veliger* **21**, 1–9.

West, D. L. (1978b). Ultrastructural and cytochemical aspects of spermiogenesis in *Hydra hymanae*, with reference to factors involved in sperm head shaping. *Dev. Biol.* **65**, 139–154.

Wilkinson, R. F., Stanley, H. P., and Bowman, J. T. (1974). Genetic control of spermiogenesis in *Drosophila melanogaster:* The effects of abnormal cytoplasmic microtubule populations in mutant ms (3) 10 R and its colcemid-induced phenocopy. *J. Ultrastruct. Res.* **48**, 242–258.

Wilson, E. B. (1904). Experimental studies in germinal localization. I. The germ-regions in the eggs of *Dentalium*. *J. Exp. Zool.* **1**, 1–74.

Wilson, E. B. (1928). "The Cell in Development and Heredity," 3rd ed. Macmillan, New York.

Wischnitzer, S. (1970). The annulate lamellae. *Int. Rev. Cytol.* **27**, 65–100.

Wolosewick, J. J., and Porter, K. R. (1979). Microtrabecular lattice of the cytoplasmic ground substance. Artifact or reality. *J. Cell Biol.* **82**, 114–139.

Woods, F. H. (1931). History of the germ cells in *Sphaerium striatinum* (Lam.). *J. Morphol.* **51**, 545–595.

Woods, F. H. (1932). Keimbahn determinants and continuity of the germ cells in *Sphaerium striatinum* (Lam.). *J. Morphol.* **53**, 345–365.

Yamasaki, M. (1966). On the mitochondria and Golgi apparatus in spermatogenesis of the pond snail (*Cipangopaludina japonica iwakawa* Pilsbry). *Sci. Rep. Tohoku Univ., Ser. 4* **32**, 237–249.

Yasuzumi, G. (1974). Electron microscope studies on spermiogenesis in various animal species. *Int. Rev. Cytol.* **37**, 53–119.

Zalensky, A. O., and Zalenskaya, I. A. (1980). Basic chromosomal proteins of marine invertebrates. III. The proteins from sperm of bivalvia molluscs. *Comp. Biochem. Physiol. B* **66B**, 415–419.

Zalokar, M. (1974). Effect of colchicine and cytochalasin B on ooplasmic segregation of ascidian eggs. *Wilhelm Roux's Arch. Entwicklungsmech. Org.* **175**, 243–248.

2

Meiotic Maturation and Fertilization[1]

FRANK J. LONGO

Department of Anatomy
The University of Iowa
Iowa City, Iowa

[1]During the preparation of this review, the author's laboratory was supported by grants from the NIH and the NSF.

THE MOLLUSCA, VOL. 3
Development

I. Introduction

Investigations of meiotic maturation and fertilization in molluscs go back to well before the turn of the century (Wilson, 1925; Raven, 1966). As a result of this long history, a considerable body of information concerning the interaction of molluscan eggs and sperm at all levels of observation has been assembled. Light microscopic investigations, providing general and specific details of meiotic maturation and fertilization in molluscs, are described by Conklin (1902), Wilson (1925), Raven (1964a, 1966), Kume and Dan (1968), Brachet (1960), Austin (1965), Giese and Pearse (1979), and Reverberi (1971a). Ultrastructural and biochemical aspects of meiotic maturation and fertilization are reviewed in publications by Austin (1968), Monroy (1965), Monroy and Tyler (1967), Pasteels and de Harven (1962, 1963), Pasteels (1965a,b), Longo (1973, 1976a), Collier (1976), Kidder (1976), McLean (1976), Raff et al. (1976), and Sastry (1979). In this chapter, events of fertilization and meiotic maturation in molluscs, commencing with the activation of the spermatozoon and culminating in the association of the paternally and maternally derived chromosomes within the zygote are examined.

The principal aim of this chapter is to discuss aspects of molluscan gametes and their interaction that are both common to and different from those of other organisms. Thus, mechanisms for the regulation of meiotic maturation and fertilization can be compared and contrasted. In this way, trends and patterns can be discerned which, in turn, may have a bearing on taxonomic and evolutionary relationships. This chapter is not meant to be all inclusive; aspects concerning spawning, sexual behavior, reproductive periods, etc, have been omitted. Moreover, not all molluscs are discussed; for example, meiotic maturation and fertilization in cephalopods is left to previous studies.

II. Meiotic Stage of the Molluscan Egg at Insemination

In molluscs, as in most other organisms, meiosis of the ovum is delayed until after its ovulation and insemination. The stages of meiotic maturation at which molluscan ova inseminate are (1) the primary oocyte, that is, the egg containing a large prophase nucleus or germinal vesicle, and (2) the oocyte at the first metaphase of meiosis (Fig. 1). In either case, molluscan eggs are fertilized prior to the extrusion of the first polar body. In the first instance, often referred to as the germinal vesicle stage, the egg nucleus (germinal vesicle) is restricted to meiotic prophase (usually diplotene) and, hence, is tetraploid. In the second case, there is no nucleus per se, since ova are inseminated after breakdown of the germinal vesicle. Insemination in this instance usually occurs when the maternally derived chromatin aligns on the metaphase plate of the first meiotic spindle (Fig. 1).

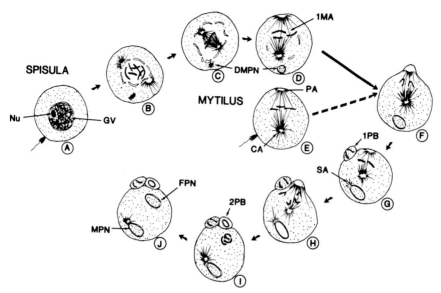

Fig. 1. Sequence of events involving meiotic maturation of fertilized *Spisula* and *Mytilus* eggs. Breakdown of the germinal vesicle (GV) and formation of the first meiotic apparatus (1MA) are depicted for inseminated *Spisula* eggs (**A–D**). Insemination of *Mytilus* eggs occurs at the first metaphase of meiosis (**E**). Continued meiotic maturation and development of the first and second polar bodies (1PB and 2PB) are shown in (**F**)–(**I**). The fertilized egg, having completed meiotic maturation, is shown in (**J**). DMPN, Developing male pronucleus; SA, sperm aster; PA and CA, peripheral and central asters; Nu, nucleolus; MPN and FPN, male and female pronuclei.

A listing of molluscs comprising each of the two groups is incomplete and there is some confusion as to which stage of meiosis the eggs of some organisms belong (Table 1). Furthermore, the relationship of the stage of meiotic maturation at insemination to the classification of molluscs is not apparent. The specific stage of meiosis at a particular period, such as spawning or insemination, has not always been clearly documented in many molluscs. This is particularly important because in a number of molluscs the ovum may undergo germinal vesicle breakdown and proceed to the first metaphase of meiosis simultaneously with spawning. The egg may stay arrested at this stage, only to resume meiosis following insemination. Hence, the assumption that the largest and most prevalent type of ovum present in the dissected ovary of a ripe mollusc represents the meiotic stage at which fertilization takes place is questionable.

Crassostrea gigas and *C. echinata* are molluscs in which spawning and insemination reportedly occur prior to germinal vesicle breakdown; however, fertilization can take place regardless of the condition of the germinal vesicle (Kume and Dan, 1968). Eggs dissected from the ovaries of *Crassostrea gigas, Spisula*

TABLE I

Meiotic Stage of Molluscan Eggs at Ovulation/Spawning and Fertilization[a]

Organism	Ovulation/Spawning	Fertilization	References
Chaetopleura apiculata	GVBD	—	Costello and Henley (1971)
Crepidula sp.	—	MI	Raven (1966)
Dentalium sp.	GV	GV	Reverberi (1971c)
Dentalium sp.	—	MI	Raven (1966)
Helix sp.	—	MI	Raven (1966)
Ilyanassa obsoleta	GV	GV	Cather (1963)
Limax sp.	—	MI	Raven (1966)
Lymnaea stagnalis	—	MI	Bretschneider (1948)
Crassostrea virginica	GV	—	Costello and Henley (1971)
Crassostrea virginica	GVBD[b]	MI	Longwell and Stiles (1968)
Crassostrea virginica	GV	GV	Galtsoff (1964), Raven (1966)
Cumingia tellinoides	MI	MI	Raven (1966)
Cumingia tellinoides	GV	MI	Costello and Henley (1971)
Argopecten irradians	MI	MI	Raven (1966)
Barnea candida	GV	GV	Raven (1966)
Ensis directus	GV	GV	Raven (1966), Costello and Henley (1971)
Crenomytilus grayanus	GV	MI	Drozdov (1979)
Mactra veneriformis	GV	GV	Iwata (1951)
Mytilus edulis	GV	MI	Field (1922)
Mytilus edulis	MI	MI	Raven (1966), Longo and Anderson (1969c)
Spisula solidissima	GV	GV	Costello and Henley (1971)
Venus mercenaria	MI	MI	Costello and Henley (1971)
Mya arenaria	MI	MI	Raven (1966)
Mya arenaria	GV	—	Costello and Henley (1971)

[a] GV, germinal vesicle; MI, ova having undergone germinal vesicle breakdown or at the first metaphase of meiosis; GVBD, germinal vesicle breakdown; —, data not provided.

[b] Occasionally eggs were at the germinal vesicle stage.

sachalinensis, and *Mactra sulcatoria* contain germinal vesicles and, when placed in seawater, undergo germinal vesicle breakdown. In some cases, they are able to mature to the first metaphase of meiosis and form the first polar body; fertilization reportedly can take place at any time during this period (Ginsburg, 1974).

Because insemination induces resumption of meiotic maturation and the formation of two polar bodies, meiosis is intimately associated with fertilization in molluscs. Later stages of meiotic maturation overlap events of fertilization (Austin, 1965; Longo, 1973). Such eggs were at one time referred to as having the *Ascaris* type of fertilization, in which the sperm normally enters the egg before it has completed its meiotic divisions (Wilson, 1925). Although the term *Ascaris*-type fertilization is seldom used today, it continues to relate events of maturation and fertilization in a meaningful manner and embodies concepts important to our understanding of the diversity and similarities of fertilization events among animals.

In the *Ascaris* type of fertilization, the incorporated sperm nucleus "pauses" within the egg cytoplasm until the polar bodies have formed, with the following results: (1) during the period from insemination through polar body formation the sperm nucleus differentiates into a male pronucleus, and (2) at the time of their association, both the male and female pronuclei give rise to a group of chromosomes for the ensuing cleavage division. The maternally and paternally derived chromosomes eventually intermix on the presumptive metaphase plate of the first cleavage spindle. These processes, which are classically looked upon as the concluding events of fertilization, represent in fact the early aspects of the first cleavage division of the embryo (prophase–metaphase). Hence, organisms having the *Ascaris* type of fertilization do not demonstrate pronuclear fusion and, consequently, do not possess a zygote or fusion nucleus (Wilson, 1925; Longo, 1973).

III. Sperm Chemotaxis

There is evidence that the sperm of some chitons are capable of a chemotactic response to ova and egg extracts (Miller, 1977). Spermatozoa show directed movements to maternally derived materials (e.g., seawater and alcohol extracts of eggs) in a non-species–specific manner. Earlier investigations considered sperm chemotaxis in other molluscs (von Medem, 1942, 1945).

Hylander and Summers (1977) have indicated that trapping of sperm within the jelly layers of pelecypod eggs does not occur. Initially, individual sperm appear to move in a random path into the egg jelly. As the sperm approaches the egg, it becomes oriented perpendicular to the ovum surface, then quickly moves to and binds with the vitelline layer. As a result of this movement and binding of sperm to the egg surface, the number of sperm in the immediate vicinity of the egg can be quite large. What role, if any, the accumulation of sperm in this manner has in the overall process of fertilization, such as increasing the probability of gamete fusion or reducing the possibility of polyspermy, has not been determined.

Agglutination of sperm by egg water (solutions in which ova are suspended) has been described for a number of molluscs (von Medem, 1942, 1945; Tyler and Fox, 1939, 1940). Although much has been written about egg water-induced sperm agglutination (Monroy, 1965; Metz, 1967), the actual role of this process in the fertilization of molluscan eggs has not been elucidated.

IV. Acrosomal Reaction

The structure of the acrosome and the acrosomal reaction in molluscs has been considered (see Franzen, 1955, 1956; Dan, 1967; Colwin and Colwin, 1967; Anderson and Personne, 1976, for reviews; see also Niijima and Dan, 1965a,b; Longo and Dornfield, 1967; Popham, 1974; Galtsoff, 1964; Galtsoff and Philpott, 1960; Daniels et al., 1971; Hylander and Summers, 1977; Longo and Anderson, 1969a; Lewis et al., 1980; Pasteels, 1965a,b). In molluscs, the acrosomal reaction consists of the following processes: (1) the opening of the acrosomal vesicle and the release of lytic substances, and (2) the exposure of the acrosomal process, a rodlike structure instrumental in gamete fusion (Fig. 2). Opening of the acrosomal vesicle involves fusion of the plasmalemma with the membrane delimiting the outer aspect of the acrosomal vesicle (outer acrosomal membrane). As a result of this fusion, the plasma membrane and the membrane delimiting the acrosomal vesicle become continuous with one another and the contents within the acrosomal vesicle are externalized (Fig. 2). Dan (1967) and Hylander and Summers (1977) indicated that in pelecypods, fusion occurs at the apex of the acrosome and the plasmalemma–acrosomal membrane complex that is produced is reflected to the apical aspect of the sperm nucleus (Fig. 2). Niijima and Dan (1965b) claimed that the "trigger" or point of initiation of the acrosomal reaction is located at the apex of the acrosomal vesicle in *Mytilus*. This suggestion is based primarily on observations demonstrating that this site is the first to be recognizably altered when the acrosomal reaction is induced.

A number of physical and chemical agents have been shown to stimulate or promote the acrosomal reaction in pelecypods and gastropods; in fact, over 20 species of molluscs have been observed (Dan, 1967). Conditions that induce the acrosomal reaction in molluscs include: increased external calcium (Wada et al., 1956; Lewis et al., 1980), increased alkalinity of the seawater suspension (Dan, 1956, 1967; Dan and Wada, 1955), and incubation of sperm at low temperatures (Wada et al., 1956). Both the calcium ionophore A23187 and aging of sperm have been shown to stimulate the acrosomal reaction in *Haliotis* (Lewis et al.,

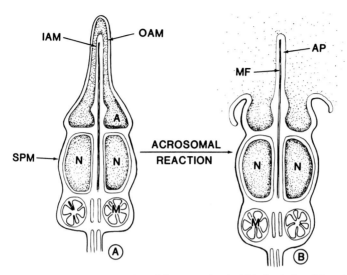

Fig. 2. Diagrammatic representation of the sperm head of *Mytilus* before (**A**) and after (**B**) the acrosomal reaction. SPM, Sperm plasma membrane; A, acrosome; IAM and OAM, inner and outer acrosomal membranes; N, nucleus; M, mitochondria; AP, acrosomal process; MF, actin microfilaments in the acrosomal process.

1980). The absence of calcium in suspensions of sperm inhibits the acrosome reaction in pelecypods and gastropods (Dan, 1967). Dan (1956) reported that bivalve sperm undergo an acrosomal reaction when exposed to egg water whereas sperm of gastropods do not.

The question of when the acrosome reaction occurs during gamete interaction has not been settled. Dan (1956) claimed that sperm undergo the acrosomal reaction as they move through the jelly layer of the egg. On contacting the ovum, flagellar activity continues with the penetration of the sperm into the egg. Similar movements of molluscan sperm have also been described by Wada et al. (1956), Dan and Wada (1955), and Hylander and Summers (1977). However, Hylander and Summers (1977) contend that the acrosome reaction occurs when the sperm contacts the surface of the egg.

Previous experiments have shown that germinal vesicle eggs, dissected from *Pinctada, Mercenaria,* and *Tapes,* do not fertilize when mixed with sperm (Wada and Wada, 1953; Loosanoff and Davies, 1963; Kume and Dan, 1968). When these oocytes were suspended in sea water with ammonia, germinal vesicle breakdown occurred and the eggs fertilized. Similar results were also obtained with oocytes of other molluscs (Kume and Dan, 1968; Hatanaka et al., 1943; Yamamoto and Nishioka, 1943; Wada, 1953). It was speculated that failure to fertilize germinal vesicle eggs was due to immaturity of the ovum and that incubation in alkaline sea water promoted maturation. In addition to inducing germinal vesicle breakdown and polar body formation (Ii and Rebhun, 1979; Finkel and Wolf, 1980), sea water containing ammonia may also induce sperm to undergo the acrosomal reaction, thereby increasing the likelihood of gamete fusion (Longo, 1977). Investigations demonstrating that immature ova fail to fertilize in these instances, owing to unique conditions of the egg or the sperm, do not seem to have been published.

One component of the sperm acrosomal vesicle in a number of molluscs has been shown to be lytic substances capable of lysing extracellular moieties surrounding the egg/embryo (Tyler, 1939; Berg, 1949, 1950; Dan, 1956, 1962; Wada et al., 1956; Lewis et al., 1980). Although the function of the lytic material at fertilization has not been demonstrated, it is believed to assist the sperm in its movements to the surface of the egg or to alter the vitelline layer such that gamete fusion is facilitated. Wada et al. (1956), Dan (1962), and Humphreys (1962) have shown that lytic agents isolated from the acrosome of *Mytilus* sperm are able to destroy the egg's vitelline layer. Hylander and Summers (1977) were unable to find dissolution or alteration of the vitelline layer in the vicinity of attached sperm of *Chama, Spisula,* and *Modiolus.* These results are difficult to reconcile in light of earlier studies showing dissolution of the vitelline layers of fertilized and unfertilized eggs (Berg, 1949, 1950; Humphreys, 1962; Dan, 1967).

In many of the molluscs examined, the acrosomal vesicle has been shown to surround a tubular subacrosomal structure (Dan, 1967; Lewis et al., 1980; Niijima and Dan, 1965a; Galtsoff, 1964; Popham, 1974). In *Mytilus,* the acrosomal vesicle consists of a hollow cone covering the apex of a bundle of microfilaments which insert on the membrane lining the inner surface of the acrosomal vesicle (Fig. 2; Longo and Dornfeld, 1967; Tilney, 1975). The filaments extend from this point of insertion through a canal in the nucleus toward the basal body of the flagellum. Tilney (1975) has demonstrated that this filament bundle contains actin and has a characteristic and uniform polarity. Whether or not the subacrosomal space of other molluscs contains actin has not been demonstrated.

As a result of the acrosomal reaction and the reflection of the plasmalemma–outer acrosomal membrane complex, the apex of the spermatozoon is seen to possess a projection, generally referred to as the acrosomal process (Fig. 2; Dan, 1967; Hylander and Summers, 1977). The acrosomal process is derived from subacrosomal structures of the unreacted spermatozoon. In pelecypods (Dan, 1967; Hylander and Summers, 1977; Longo and Anderson, 1969b, 1970b), it consists of a core of microfilaments which is delimited by membrane derived from the inner aspect of the acrosomal vesicle (Fig. 2). In these instances then, where sperm are mixed with eggs, the acrosomal process is uncovered by a dehiscence of the acrosomal vesicle.

In contrast to the situation in molluscs, formation of the acrosomal process in other marine invertebrates (e.g., *Limulus* and *Thyone*) may involve a reorganization of actin and an elongation of the acrosomal process (Tilney, 1975; Tilney et al., 1973, 1978). In molluscs, elongation of the acrosomal process is not apparent when sperm are mixed with eggs. For example, in *Mytilus*, when sperm undergo the acrosomal reaction in close proximity to the egg surface, the length of the acrosomal process (about 3 μm) is less than that of the intact acrosome and only slightly greater than that of the invagination at the base of the acrosomal vesicle, into which it fits in the unreacted state (Dan, 1967). When the acrosome reaction is induced with increased calcium or pH, the acrosomal process may attain a length of 10–14 μm (Dan, 1967; Niijima and Dan, 1965b; Longo, 1977). Similar results have also been reported for sperm of *Barnea* (Pasteels, 1965a). Noticeable elongation of the acrosomal process does not seem to occur during fertilization in *Chama, Spisula, Modiolus,* or *Bankia* (Hylander and Summers, 1977; Longo and Anderson, 1970b; Popham, 1974). Extension of the acrosomal process up to 7 μm has been reported in sperm suspension of the abalone, *Haliotis,* induced experimentally with calcium ionophore or increased calcium (Lewis et al., 1980). It is apparent from these observations that formation of the acrosomal process differs depending on whether the acrosomal reaction is induced by physiological conditions, such as mixing sperm with eggs, or by artificial means.

As previously indicated, the membrane delimiting the acrosomal process is derived, at least in part, from the inner aspect of the acrosomal vesicle (Fig. 2). It is this portion of the delimiting membrane of the spermatozoon that fuses with the egg plasma membrane (Colwin and Colwin, 1967). The observation that sperm failing to undergo the acrosome reaction are unable to fuse with eggs has led investigators to speculate that the inner acrosomal membrane of invertebrate sperm possesses a fusogenic site(s) that is (are) exposed with the acrosome reaction (Colwin and Colwin, 1967). Hence, failing the acrosome reaction, sperm are believed to be incapable of gamete fusion. Experiments inseminating *Arbacia* eggs with *Mytilus* sperm substantiate such a claim, because *Mytilus* sperm were only able to fuse with *Arbacia* eggs after they had been induced to undergo the acrosome reaction (Longo, 1977).

V. Sperm–Egg Binding

Mollusc sperm have been observed to approach and attach to the surface of the egg (Fig. 3A). This attachment, referred to as gamete binding (Hylander and Summers, 1977), has been observed primarily in pelecypods (Dan and Wada, 1955; Popham, 1975; Longo and Anderson, 1969b, 1970b; Longo, 1976a; Hylander and Summers, 1977). Attachment of the gametes, which appears to be species-specific, is manifested by a close association of the former basal aspect of the acrosome with portions of the vitelline layer. During this period, the

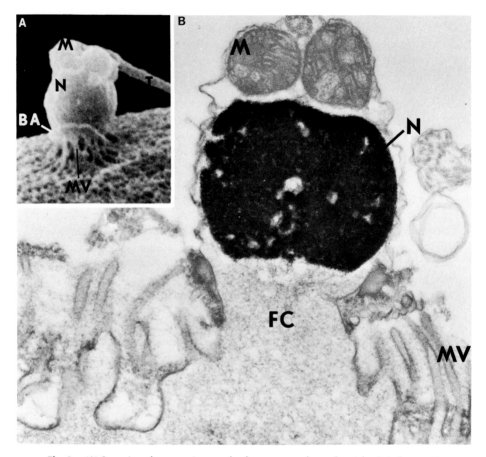

Fig. 3. (**A**) Scanning electron micrograph of a sperm on the surface of a *Spisula* egg. Note the microvilli (MV) attached to the base of the reacted acrosome (BA). (**B**) *Spisula* sperm, subsequent to gamete fusion, which is located within a fertilization cone (FC). Note the arched microvilli (MV) directed to the incorporated sperm nucleus (N). M, sperm mitochondria; T, sperm tail.

acrosomal process is observed interdigitating with microvilli of the egg. Hylander and Summers (1977) claimed that the strength of gamete binding is evidenced by the fact that agitation of sperm during processing for microscopic analysis pulls the microvilli partially out of the vitelline layer (Figs. 3A, B). However, it is also possible that the orientation of microvilli to the spermatozoon in this case is not an artifact of specimen preparation but represents an active process in the association of the egg and sperm (Longo and Anderson, 1970b; Popham, 1975; Longo, 1976a).

The association of sperm and eggs of molluscs is morphologically similar to that demonstrated in echinoids, where material from the acrosome has been shown to be involved in gamete binding (Moy and Vacquier, 1979). This acrosomal substance has been referred to as bindin. Bindin has been isolated from oyster sperm and there is evidence that it is a glycoprotein that attaches to glycoprotein of the ovum surface (Brandriff et al., 1978).

VI. Gamete Fusion, Sperm Incorporation

Subsequent to gamete binding, the membrane delimiting the acrosomal process fuses with the egg plasmalemma thereby forming a zygote (Longo and Anderson, 1969b, 1970b; Longo, 1982). Because many molluscan eggs possess a highly organized surface, consisting of numerous, regularly arranged microvilli, the question has been raised as to the actual site of gamete fusion. Does it occur at microvillar or intervillar regions of the egg surface (Longo, 1976a; Hylander and Summers, 1977)? Although gamete fusion and sperm incorporation have been documented in a number of pelecypods, ultrastructural observations have not firmly established the actual site of gamete fusion in molluscs (Pasteels, 1965a,b; Hylander and Summers, 1977; Longo and Anderson, 1969b, 1970b; Popham, 1974).

Raven (1966) claimed that in most molluscs sperm entered at any site along the surface of the egg; however, gamete fusion occurred at the vegetal pole in *Unio, Bulla,* and *Crepidula* (see also Guerrier, 1970). According to Jura (1960), the sperm of *Succinea* was often found near the vegetal pole following incorporation, seldom above the equator of the egg. Upon fusion of the gametes, the contents, as well as the plasma membranes of the egg and sperm, become continuous. Integration of gamete plasma membranes has been investigated in *Spisula,* and it has been shown that components of the egg plasma membrane (specifically concanavalin A receptors) are capable of moving into the sperm-derived membrane following fusion of the sperm and egg (Longo, 1982).

At the site of gamete fusion, a specialized protrusion of cytoplasm is generated that surrounds the contents of the incorporating spermatozoon. In molluscs, this modified portion of the egg cytoplasm, referred to as a fertilization cone, does not seem to enlarge as much as those formed in echinoderm zygotes (Raven, 1966; Krauss, 1950; Ikeda, 1930; Longo and Anderson, 1969b, 1970b; Longo, 1973, 1976a, 1982). Structurally, the fertilization cone contains a granulofibrillar material and ribosomes; few if any maternally derived organelles are present in this portion of the zygote cytoplasm (Fig. 3B). In addition, the microvilli characteristic of other portions of the zygote are absent at the site of sperm entry (Longo and Anderson, 1969b, 1970b; Longo, 1976a).

The fertilization cones of other organisms have been shown to possess microfilaments that contain actin (Longo and Anderson, 1968; Longo, 1973; Tilney and Jaffe, 1980). Microfilaments have not been consistently observed in the fertilization cones of molluscs. Whether or not this observation reflects an absence of such structures or a modification due to specimen preparation has not been determined. In molluscs, the fertilization cone increases in size to form a small protrusion on the zygote surface, which is eventually reabsorbed into the cortex of the egg. As a result, little if any sign of the site of gamete fusion/sperm incorporation remains along the periphery of the zygote.

On the basis of evidence from investigations with fertilized sea urchin eggs, it has been suggested that the fertilization cone participates in the normal progression of the sperm nucleus into the zygote cortex, specifically rotation (Longo, 1980). Evidence demonstrating that the fertilization cones of molluscan zygotes participate in sperm incorporation comes from studies of *Spisula* eggs incubated in seawater containing the plant lectin concanavalin A (Longo, 1978b). When *Spisula* eggs were inseminated in the presence of concanavalin A, gamete fusion occurred; however, processes involving the distention of the egg's surface and plasma membrane (e.g., polar body formation, cleavage, and morphogenesis of the fertilization cone) were inhibited (Longo, 1978b). As a result of the inhibition of the morphogenesis of the fertilization cone, movement of the sperm into the cortex of the zygote failed to occur. When eggs were treated with succinyl-concanavalin A, a bivalent lectin rather than the tetravalent, native concanavalin A, morphogenesis of fertilization cones occurred and sperm incorporation proceeded unimpeded. These observations, as well as those from investigations examining the effects of concanavalin A on somatic cells (Edelman, 1976; Zagyansky and Jard, 1979) suggest that concanavalin A "anchors" sites on the surface of the egg. Consequently, the plasmalemma and cortex become "rigid," unable to participate in the normal progression of sperm incorporation.

Accompanying the sperm nucleus into eggs of *Spisula* and *Mytilus* at fertilization are the sperm mitochondria, centrioles, and a portion of the axonemal complex of the sperm flagellum (Longo and Anderson, 1969b, 1970b; Longo, 1973). How much of the axonemal complex is incorporated has not been established. Raven (1966) claimed that the sperm tail is not incorporated into the zygotes of *Bulla, Crepidula,* and *Eulota.*

The pathway taken by the incorporating sperm nucleus into the cortex of molluscan zygotes is frequently one in which the apex of the sperm nucleus, formerly directed to the center of the ovum, becomes diverted to the surface of the fertilized egg. This process is usually referred to as sperm rotation (Longo, 1973), and has been observed in a number of molluscs (Raven, 1966; Meves, 1915; Longo and Anderson, 1969b, 1970b; Kostanecki and Wierzejski, 1896; Linville, 1900; de Larambergue, 1939). Following this movement, the sperm nucleus is often observed within the zygote cortex, lateral to the site of gamete fusion. At this site, the sperm nucleus becomes reorganized into a male pronucleus.

VII. Blocks of Polyspermy

Polyspermy-preventing mechanisms in molluscs have not been extensively studied and our understanding of the means by which physiological monospermy is assured in this group is rather poor. In some molluscs, there is apparently no block of incorporation of more than one sperm into the egg (Raven, 1966). For example, in *Lymnaea* (Crabb, 1927; Bretschneider, 1948; Horstmann, 1955), *Helix* (Garnault, 1888, 1889), and *Bulinus* (de Larambergue, 1939) several sperm reportedly enter the egg, but only one develops into a pronucleus and the supernumerary ones disappear.

The cortical granules in eggs of other organisms, namely echinoderms and vertebrates, have been shown to be involved in the development of a block of polyspermy (Schuel, 1978; Epel, 1978). In some molluscs, such as *Nassarius* (Schmekel and Fioroni, 1975) and *Ilyanassa* Taylor and Anderson, 1969) cortical granules are not observed.[2] The eggs of a number of molluscs, however, have been shown to contain cortical granules (Humphreys, 1964, 1967; Longo and Anderson, 1969b,c, 1970a,b; Reverberi, 1971b; Allen, 1953; Pasteels and de Harven, 1962; Drozdov, 1979; see also Anderson, 1974; Huebner and Anderson, 1976). Despite the presence of a prominent vitelline layer and cortical granules in molluscan eggs, there is little evidence demonstrating that either structure is involved in the prevention of polyspermy. A release of the entire population of cortical granules (*Crassostrea,* Osanai, 1969) or a selective release at the site of sperm incorporation (*Crenomytilus,* Drozdov, 1979) have been reported. In the latter instance, the number of cortical granules per length of cortical surface at the site of sperm incorporation was compared to adjacent areas. These observations did not eliminate the possibility that an apparent decrease in cortical granule number at the site of sperm incorporation may be due to their movement to adjacent areas and not to their dehiscence. In other molluscs, dehiscence of cortical granules has not been shown to be a part of the fertilization process (Reverberi and Mancuso, 1961; Humphreys, 1964, 1967; Longo and Anderson, 1969b,c, 1970a,b; Pasteels and de Harven, 1962).

Raven (1966) states, "Although occasionally a slight elevation of vitelline membrane after insemination has been described in various molluscan eggs, no distinct fertilization membrane is formed in this group" (see also, Raven, 1972; Popham, 1975). With respect to oyster eggs, this situation appears to be unresolved; although Galtsoff (1964) indicates that oyster eggs form a fertilization membrane at insemination, Osanai (1969) claims that this is not the case. En-

---

[2]*Nassarius obsoletus* was removed from the genus *Nassarius* (Dumeril, 1806) by Stimpson and placed into the new genus *Ilyanassa* (Stimpson, 1865) on the basis that *Ilyanassa obsoleta* has a different operculum and lacks the caudal pedal cirri or metapodial tentacles characteristic of the genus *Nassarius*. Although embryologists tend to use the correct term *Ilyanassa obsoleta* (Say), scientists in other fields often still use the now outmoded designation *Nassarius obsoletus*.

largement of the perivitelline space has been reported in *Spisula* (Rebhun, 1962a); however, this has not been substantiated in subsequent studies (Longo, 1976a). No morphological changes in the vitelline layer at the time of fertilization have been reported in *Mytilus* (Humphreys, 1962, 1964, 1967; Longo and Anderson, 1969b,c), *Barnea* (Pasteels and de Harven, 1962), *Bankia* (Popham, 1975), *Spisula* (Allen, 1953; Longo and Anderson, 1970a,b; Hylander and Summers, 1977), *Chama* (Hylander and Summers, 1977), *Mopalia,* or *Chaetopleura* (Anderson, 1969).

Changes in the properties of the vitelline layer of *Spisula* eggs, that is, its solubility in alkaline isotonic NaCl, have been demonstrated by Rebhun (1962b). What relation, if any, this alteration has to polyspermy prevention has not been determined. Changes in the distribution of intramembranous particles and in microvillar morphology have been shown in freeze-fracture replicas of *Spisula* zygotes in comparison to unfertilized eggs (Longo, 1976a,b). Whether or not these changes are involved with a block of polyspermy has not been established. Ziomek and Epel (1975) reported that cytochalasin B prevents the block of polyspermy in *Spisula,* presumably by altering the plasma membrane. More recent evidence, however, has indicated that although eggs incubated in cytochalasin B can be activated by sperm, sperm incorporation and polar body formation are inhibited (Longo, 1972, 1978a; Gould-Somero et al., 1977). The rigidity of the cortex of *Mactra* eggs reportedly increased with fertilization (Sawada, 1960a) but appeared to decrease in *Mytilus* (Humphreys, 1964). In *Mactra,* potassium and magnesium ions and calcium-free seawater diminished the cortical rigidity, whereas sodium and calcium produced no change in comparison to seawater (Sawada, 1960a). The relationship of these cytoplasmic alterations to the block of polyspermy has not been documented.

Despite the absence of gross structural changes in the cortex of molluscan eggs at activation, a block of polyspermy has been demonstrated in *Spisula* (Longo, 1976a; see also Longo, 1973). Reinsemination experiments, in which fertilized eggs were reexposed to sperm after their initial insemination, demonstrate that the block of polyspermy is established by 5 min postinsemination. Since similar results were obtained with artificially activated *Spisula* eggs, it is believed that the block of polyspermy in both cases involves similar transformations, rendering the activated ovum refractory to sperm. More recent investigations have demonstrated that *Spisula* eggs undergo an electrophysiologically detectable response within 5 sec of exposure to sperm (Finkel and Wolf, 1980). This alteration was characterized by a rapid and prolonged depolarization of the plasma membrane followed approximately 5 min later by the beginning of a steady hyperpolarization, which was completed by 10 min postinsemination. Depolarization in this case was believed to result from a transient increase in sodium conductance, which may be crucial in preventing polyspermy since the degree of polyspermy in *Spisula* was sensitive to external sodium ions.

Depolarization of the plasma membrane and its involvement in the block of polyspermy have been demonstrated in echinoderm eggs (Jaffe, 1976). Perhaps such a mechanism is a feature of molluscan eggs. Furthermore, because molluscs possess a block of polyspermy that is not manifested by overt changes in the morphology of the egg cortex, further study of this process may provide insight into mechanisms whereby gametes of other organisms may interact with one another prior to the elicitation of the cortical granule reaction.

VIII. Egg Activation

Changes in the egg that accompany gamete attachment and fusion initiate a sequence or program of events that results in the activation of embryonic development. For molluscs, evidence (albeit limited) suggests that morphogenetic changes of fertilized eggs are regulated in a manner similar to that described for echinoderms (Epel, 1978). In both groups of organisms, alterations in intracellular calcium and pH appear to play important roles in egg activation.

Earlier investigations of the activation of molluscan eggs demonstrated the presence of a substance in the body fluids of the adult that inhibited maturation divisions of *Mactra* eggs (Sawada, 1952; Iwata, 1951). Sawada (1954a) demonstrated that the material may be a polysaccharide. On the basis of the activation of *Mactra* eggs by periodate, Sawada (1954b) suggested that polysaccharides of the egg plasma membrane are oxidized, thereby increasing the permeability of the ovum to ions which then trigger activation.

Calcium ionophore A-23187 has been shown to induce the activation of *Acmaea* and *Spisula* eggs (Steinhardt et al., 1974; Schuetz, 1975). In *Spisula,* ionophore activation does not occur when eggs are incubated in calcium-free seawater, suggesting that calcium is transported across the plasma membrane and is involved in the resumption of meiotic maturation (Schuetz, 1975). Similar results have also been demonstrated by Ii and Rebhun (1979). With respect to investigations with other molluscs, Paul (1975) has indicated that acid release did not occur at fertilization in *Mytilus* and *Acmaea*. The basis for this difference is unknown.

Acid release from fertilized eggs, with a presumed increase in intracellular pH, has been shown to be essential for activation in *Spisula* (Ii and Rebhun, 1979; Finkel and Wolf, 1980; see also Shen and Steinhardt, 1978; Epel, 1978). Allen (1953) showed the release of acid from *Spisula* eggs at fertilization and artificial activation and speculated that maintenance of the germinal vesicle was due to a sodium–potassium antagonism. The nature of the acid that is released is unknown; it does not appear to be derived from cortical granules and is believed to be due to an efflux of protons (see also Gillies et al., 1980). Activated *Spisula* eggs incubated in sodium-free seawater do not undergo acid release and germinal vesicle breakdown (Ii and Rebhun, 1979, suggesting that a sodium–proton exchange occurs, leading to an increase in intracellular pH (Ii and Rebhun, 1979; Epel, 1978; Shen and Steinhardt, 1978). Finkel and Wolf (1980) have also demonstrated that agents giving rise to an increase in intracellular pH in *Spisula* eggs (e.g., ammonia) induce egg activation (germinal vesicle breakdown). These results suggest that a shift of internal pH might be a general mechanism of egg activation which in turn may be responsible for the activation of "dormant" enzymes and the change from a quiescent to a metabolically active stage (Epel et al., 1974; Epel, 1978). In regard to ammonia activation of molluscan eggs, it is important to point out that this agent brings about a transient increase in intracellular calcium and pH in sea urchin eggs (Zucker et al., 1978).

IX. Meiotic Maturation

The process of meiotic maturation in molluscs was studied by early cytologists (for reviews, see Wilson, 1925; Raven, 1966). There have been relatively few studies examining the regulation of this process in molluscs (see Longo, 1973, for a general ultrastructural review; Cather, 1963, for a schedule of meiotic events in *Ilyanassa*). Regulation of meiotic maturation has been studied in starfish, amphibian, and mammalian eggs (Masui and Clarke, 1979).

The sequence of events in molluscan eggs, fertilized at the germinal vesicle stage, differs from that of ova fertilized at metaphase I of meiosis. This variation primarily involves, in the former case, germinal vesicle breakdown, the formation of a meiotic spindle, and its organization at the periphery of the zygote. Other than these events, morphogenesis of fertilized eggs from each group, involving the later stages of meiotic maturation (e.g., polar body formation) are comparable.

A. Germinal Vesicle Breakdown

One of the most dramatic indications of egg activation in molluscs fertilized at meiotic prophase is the breakdown of the germinal vesicle. The germinal vesicle of the molluscan egg is a large, euchromatic, spheroid body containing a large nucleolus, often with several distinct regions (Longo and Anderson, 1970a). The nuclear envelope is distinguished by its smooth contour and numerous pores. Granular aggregations have been observed within the germinal vesicles of a number of molluscs (Popham, 1975; Taylor and Anderson, 1969; Reverberi, 1967). The composition of these structures has not been determined but they are thought to be RNA (Collier, 1966; Taylor and Anderson, 1969), possibly in transit to the cytoplasm.

Breakdown of the germinal vesicle is a rapid event; in *Spisula* and *Ensis* it occurs within 10 min of insemination at 20°C (Costello and Henley, 1971). The process of germinal vesicle breakdown, as examined by electron microscopy, has been described in *Spisula* (Longo and Anderson, 1970a) and *Bankia* (Popham, 1975), and is morphologically similar to that observed in mammalian ova (Calarco et al., 1972). Initiation of germinal vesicle breakdown is recognized when the surface of the nucleus becomes irregular and plicated. Internally, the chromosomes condense and the nucleolus disappears. Concomitantly, openings are seen along the nuclear envelope, presumably owing to multiple fusions of the inner and the outer membranes comprising the nuclear envelope. These breaks lead to the formation of cisternae composed of membrane derived from the inner and outer laminae of the nuclear envelope. Since the cisternae lack distinguishing morphological characteristics they are soon "lost" among other membranous elements within the zygote cytoplasm. As a result of this breakdown of the nuclear envelope, the condensing chromosomes are "mixed" with the surrounding cytoplasm.

Germinal vesicle breakdown is not inhibited by anerobic conditions or uncouplers of phosphorylation (Pasteels, 1935; Sawada, 1960b; Sawada and Hosokawa, 1959; Sawada and Rebhun, 1969). Conversely, such conditions and agents prevent the formation of the meiotic apparatus. Agents that depolymerize microtubules (e.g., colchicine) inhibit the formation of the meiotic apparatus and induce its disappearance. These observations suggest that physiological mechanisms involving the initial stages of nuclear activation, specifically germinal vesicle breakdown, differ from those involving the formation of the meiotic apparatus and its maintenance.

Early investigations by Delage (1899) demonstrated that breakdown of the germinal vesicle is a necessary condition for successful fertilization of *Dentalium* eggs. Anucleate fragments of eggs, formed prior to germinal vesicle breakdown, were incapable of fertilization. Fragments taken from eggs in which the germinal vesicle had broken down, however, were capable of insemination, suggesting that the germinal vesicle of *Dentalium* contains a substance affecting the fertilizability of the ovum. The possible effects of germinal vesicle materials on the fertilizability of other molluscan eggs has not been elaborated. Furthermore, germinal vesicle materials appear to be necessary for the egg cytoplasm to effectively initiate and complete processes characteristic of the mature ovum, e.g., transformation of the sperm nucleus into a male pronucleus (Longo and Kunkle, 1978; Longo, 1981). The concept of cytoplasmic maturation and its possible role in molluscan eggs has not been fully explored.

B. Formation of the Meiotic Apparatus

Concomitant with the breakdown of the germinal vesicle, asters make their appearance within the zygote cytoplasm (see Fig. 1). Early cytologists recognized that fertilized molluscan eggs may be initially associated with one (Jordan, 1910; Bretschneider, 1948) or two forming asters (Kostanecki, 1904; Garnault, 1888, 1889; Lams, 1910; see Raven, 1966, for a review). In *Spisula,* two asters become situated on either side of the disrupting germinal vesicle (Rebhun, 1959; Longo and Anderson, 1970a). In the center of each aster is located a pair of centrioles surrounded by some amorphorous material of unknown composition. Fascicles of microtubules project from the region surrounding the centrioles and are separated from each other by areas filled with yolk bodies and endoplasmic reticulum. This organization suggests that the centrally located centrioles and amorphorous material are in some way involved with the polymerization and organization of the microtubules that are a part of the asters. Microtubules appear within breaks of the nuclear envelope and become associated with condensing chromosomes and oriented to poles of the developing spindles. Concomitantly, the two asters move opposite one another and their centers define what become the poles of the meiotic spindle.

Burnside et al. (1973) analyzed the tubulin content of *Spisula* eggs and determined that it is slightly greater than 3% of the total protein of the ovum. Since unfertilized *Spisula* eggs lack morphologically distinguishable microtubules and are inseminated at the germinal vesicle stage, tubulin is apparently "stored" until activation, when it is presumably utilized in the formation of the meiotic spindles. Weisenberg and Rosenfeld (1975) have isolated tubulin from *Spisula* eggs and demonstrated that it is capable of assembling into microtubules *in vitro*. In contrast to unfertilized ova, activated *Spisula* eggs contain centrioles and granules that are capable of organizing microtubules about them to form asters *in vitro*. Because these elements (centrioles and granules) are able to organize microtubules into well-defined structures, such as asters, they have been referred to as microtubule-organizing centers (Pickett-Heaps, 1975). Weisenberg and Rosenfeld (1975) suggested that microtubule-organizing centers have a similar role in the development of meiotic asters *in vivo*.

The meiotic spindle is developed in the center of the zygote and then moves to what is identified as the animal pole and becomes positioned with its long axis normal to the surface of the ovum/zygote (see Fig. 1). Placement of the meiotic apparatus at the periphery of the molluscan zygote in this manner results in an asymmetry in its structure (Longo and Anderson, 1969c, 1970a; Longo, 1973). The aster located in the cortex of the zygote (the peripheral aster) is reduced in its dimensions when compared to the centrally located one (Wilson, 1925; Raven, 1966; Longo, 1973; Longo and Anderson, 1969c, 1970a). The surface of the zygote, where the peripheral aster is located, possesses microvilli; however, there is usually an absence of cortical granules (Longo and Anderson, 1969c, 1970a).

C. Polar Body Formation

Polar body formation seems to occur in much the same manner regardless of whether the egg is fertilized at the germinal vesicle stage or at metaphase I.

Earlier accounts of this process are found in Field (1922), Wilson (1925), and Raven (1966; see also Raven et al., 1958; Raven, 1959, 1964b). Ultrastructural investigations have been reviewed by Longo (1973; see also Pasteels and de Harven, 1962).

At anaphase I there is a separation of dyads. Those that are destined to occupy the first polar body move to the peripheral pole of the meiotic apparatus, which becomes situated within an elevated mass of cytoplasm. Eventually this projection, containing the peripheral aster and chromosomes, becomes separated from the zygote (see Pasteels and Harven, 1962; Longo, 1973). Separation is mediated by the formation of a contractile ring of microfilaments, presumably containing actin, that is located at the base of the projection in the area where the cleavage furrow develops. Structurally the cleavage furrow is similar to that described for mitotic cells (Longo, 1972; Schroeder, 1975). It is believed to function much in the manner of a purse string, separating the cytoplasmic projection (the presumptive first polar body) from the zygote.

In addition to the dyad chromosomes, the first polar body of *Mytilus* and *Spisula* contains all of the cytoplasmic components observed within the fertilized egg (Humphreys, 1964; Longo and Anderson, 1969c, 1970a). Usually the chromosomes do not disperse and become localized within a nucleus (Longo, 1973; Raven, 1966). Division of the first polar body reportedly occurs in *Loligo* (Hoadley, 1930). Such division has not been observed in *Lymnaea, Bulinus, Spisula,* and *Mytilus* (Raven, 1966; Longo and Anderson, 1969c, 1970a).

Immediately following the formation of the first polar body, the chromosomes remaining within the zygote became organized on the metaphase plate of the second meiotic apparatus (see also Raven et al., 1958; Raven, 1959, 1964b, 1966). This takes place without intervening telophase and prophase stages. At anaphase II, there is a separation of the chromosomes; those that move to the peripheral aster become confined within a protrusion of cytoplasm that is eventually constricted from the zygote and develops into the second polar body (Fig. 4). Organization of this protrusion of cytoplasm, and its constriction via a cleavage furrow occurs in much the same manner described for the formation of the first polar body (Raven et al., 1958; Raven, 1959; Longo and Anderson, 1969c, 1970a). The chromosomes remaining in the zygote aggregate within the animal pole and eventually become organized within a female pronucleus (see later).

In *Mytilus,* development of the second polar body takes place immediately subjacent to the first (Fig. 4); and it remains in this location, at least until cleavage (Longo and Anderson, 1969c). In *Spisula,* development of the second polar body also takes place subjacent to the first; however, it becomes localized lateral to the first polar body. Morphologically, the cytoplasm of the second polar body is similar to that of the first; however, in the second the chromosomes disperse and become organized within a nucleus (Humphreys, 1964; Longo and

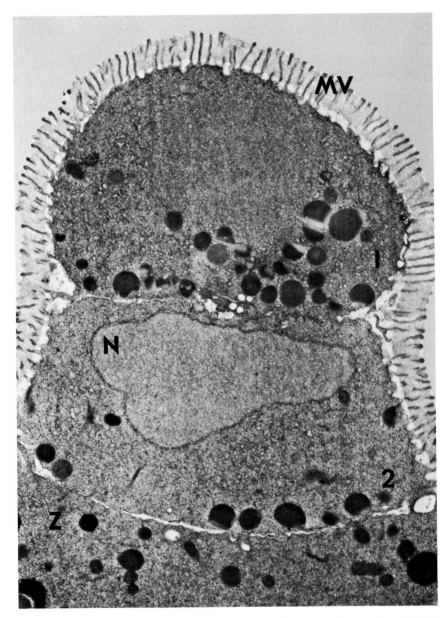

Fig. 4. First (1) and second (2) polar bodies of a *Mytilus* zygote. The second polar body contains a nucleus (N). Z, Zygote; MV, microvilli.

Anderson, 1969c, 1970a). What further role, if any, the polar bodies play in fertilization and embryogenesis has not been established.

Centrifugation studies of molluscan eggs prior to meiotic maturation (Raven, 1966) have indicated that the inequality of the meiotic division (i.e., the relatively small size of the polar bodies versus the relatively large size of the zygote) is not due to an inherent property of the maturation spindle but rather to the orientation of the meiotic apparatus within the egg. Morrill (1963) centrifuged *Lymnaea* eggs prior to the first and second meiotic divisions and obtained large polar bodies due to the movement of the spindle from the egg cortex (see also Conklin, 1917; Clement, 1935; Morgan, 1936; 1937). Normal embryonic development occurred when the polar bodies contained less than 25% of the egg volume.

Investigations of polar body formation have shown that cytochalasin B, a fungal metabolite that destroys microfilaments and alters the permeability of membranes (Pollard and Weihing, 1974), affects the formation of the cleavage furrow and, hence, constriction of the cytoplasmic mass that would normally develop into a polar body (Longo, 1972). Inhibition of polar body formation by cytochalasin B occurred at anaphase I or II with the same results, that is, chromosomes that would normally have been emitted are retained, thereby altering the ploidy of the zygote. When *Spisula* zygotes were incubated in cytochalasin B throughout the course of maturation, all of the chromosomes normally emitted with the polar bodies remained within the zygote and became organized into a variable number of nuclei (Longo, 1972).

As previously mentioned, concanavalin A also affects processes of the fertilized egg involving the deformation of its surface. Consequently, *Spisula* zygotes treated with this lectin fail to form polar bodies (Longo, 1978b). In this case, as with cytochalasin B, all of the chromosomes normally emitted with the first and second polar bodies remained within the zygote (Longo, 1978b).

When compared to cytokinesis of mitotic somatic cells, the process of polar body formation is seen to differ considerably (Mazia, 1961; Schroeder, 1975; Longo, 1973). First, polar body formation consists of a two-step process involving the extrusion of a portion of cytoplasm containing chromosomes, and its separation, via a cleavage furrow, from the zygote. Second, unlike somatic cells, where there is a bifurcation of the cytoplasm at the level of the metaphase plate, during polar body formation a cleavage furrow develops in a plane between the peripheral aster and the region formerly occupied by the metaphase plate (Longo and Anderson, 1969c, 1970b). Consequently, division of the zygote cytoplasm is unequal. The underlying basis for these differences has not been determined. Chambers (1917) has shown that cytoplasmic protrusions can be induced in meiotic zygotes by manipulating the cytoplasmic area subjacent to the plasma membrane. Conditions regulating the morphogenesis of polar bodies have not been determined and we have little insight into how this process is regulated.

Because the formation of polar bodies is morphologically similar to the development of polar lobes, it is not unreasonable to speculate that factors regulating the morphogenesis of polar lobes may be similar to those involving the extrusion of polar bodies (Conrad and Williams, 1974a,b; Conrad and Davis, 1977, 1980).

D. Cytoplasmic Alterations during Meiotic Maturation

During meiotic maturation and most often in concert with the formation of polar bodies, there is a modification of the vegetal pole of the zygote in the form of crenations (Longo and Anderson, 1970a,b), polar lobes (Cather, 1963; Conrad and Williams, 1974a,b; Conrad et al., 1973; Conrad and Davis, 1977, 1980; Raff, 1972; Clement, 1971, 1976; Dohmen and van der Mey, 1977), or the redistribution of cytoplasmic components throughout the zygote (Bedford, 1966; Hess, 1971; Timmermans et al., 1970; cf. Raven, 1966, 1972). These morphogenetic events, which are considered to be part of the general phenomenon of cytoplasmic localization/ooplasmic segregation, often result in the redistribution or localization of ooplasmic components (Davidson, 1976; Raven, 1972). Not all molluscs demonstrate these modifications, and the most dramatic examples are seen in eggs that develop polar lobes during meiotic maturation (e.g., *Ilyanassa*).

Many studies of polar lobe formation consider only those that form at the time of cleavage. Nevertheless, polar lobes also occur in many molluscs during meiotic maturation. For example, the first polar lobe of *Ilyanassa* forms when the first meiotic apparatus moves to the animal pole; it recedes upon formation of the first polar body. Just prior to the formation of the second meiotic apparatus, the second polar lobe forms and recedes following the formation of the second polar body (Cather, 1963).

Cytoplasmic streaming and the redistribution of cellular inclusions in zygotes of a number of molluscs undergoing meiotic maturation have been described (Raven, 1966). In general, these changes involved the movement of cytoplasmic inclusions, which are believed to be regulated by *cortical factors* (Raven, 1966; see also Luchtel, 1976). Bedford (1966) described the formation of radial and animal–vegetal gradients in fertilized *Bembicium* eggs which were chiefly associated with yolk bodies. How these changes are brought about and what their relationship is to maturation events of the egg and possible cytoskeletal alterations have not been established.

E. Development of the Female Pronucleus

Following the formation of the polar bodies, the maternally derived chromosomes remaining in the zygote become organized into a female pronucleus (see Fig. 1). This process occurs within the animal pole, close to the site of polar body formation, and often follows the development of the male pronucleus. The initial events in the formation of the female pronucleus are recognized when individual chromosomes become surrounded by cisternae of endoplasmic reticulum. The cisternae fuse to form a structure reminiscent of a nuclear envelope that surrounds individual dispersing chromosomes. The dispersing chromosomes surrounded by membrane are morphologically similar to structures formed at telophase of mitotically dividing cells and have been referred to as karyomeres (Longo and Anderson, 1969c, 1970a). Eventually, the karyomeres fuse together to form a female pronucleus that is initially irregular in conformation but soon becomes a smooth-surface spheroid or ellipsoid (Raven, 1966; Longo and Anderson, 1969c, 1970a). Nucleolus-like bodies are formed in the pronuclei of some molluscs, for example, *Ilyanassa* (Schmekel and Fioroni, 1976).

X. Development of the Male Pronucleus

While the maternally derived chromosomes are engaged in the completion of meiosis and the formation of a female pronucleus, the incorporated sperm nucleus transforms into a male pronucleus. Studies of molluscs have indicated that the sequence of events involving the metamorphosis of the sperm nucleus into a male pronucleus is similar to that described for other organisms (Longo, 1973, 1976a; Longo and Kunkle, 1978). In the organisms studied thus far, development of the male pronucleus has been shown to include: (1) breakdown of the sperm nuclear envelope, (2) chromatin dispersion, and (3) development of a pronuclear envelope. These processes occur while the sperm nucleus is located within the cortex of the zygote, in proximity to the sperm mitochondria, centrioles, and axonemal complex of the sperm tail (Longo and Anderson, 1969b, 1970b; Longo, 1973).

Development of the male pronucleus has been investigated in a number of molluscs (Longwell and Stiles, 1968; Cather, 1963; Pasteels, 1965a,b; Ginsburg, 1974; Longo and Anderson, 1969b, 1970b; Longo, 1973). These studies have indicated that soon after the sperm nucleus enters the egg cytoplasm it begins to undergo processes characteristic of male pronuclear development. Contrary to the claim of Raven (1966) that completion of male pronuclear development does not occur until after the second polar body is formed, a number of studies have shown that in some molluscs formation of the male pronucleus is completed either well before or immediately following anaphase II (Longwell and Stiles, 1968; Longo and Anderson, 1969b, 1970b; Cather, 1963).

A. Breakdown of the Sperm Nuclear Envelope

Breakdown of the sperm nuclear envelope has been studied in *Spisula, Mytilus,* and *Barnea* and occurs soon after the incorporation of the sperm nucleus (Pasteels, 1965a,b; Longo and Anderson, 1969b, 1970b). This process appears to involve multiple fusions of the inner and outer laminae of the nuclear envelope such that numerous vesicles are formed. Owing to the absence of distinguishing morphological characteristics the vesicles are soon "lost" in and among the cisternae of endoplasmic reticulum that are observed in the immediate area of the metamorphosing sperm nucleus. This process results in the direct exposure of the condensed chromatin to the egg cytoplasm. It has been speculated that this is necessary so that conditions or factors present in the egg cytoplasm gain access to and bring about morphological and chemical changes in the sperm chromatin (Longo and Kunkle, 1978).

B. Sperm Chromatin Dispersion

Dispersion of the condensed sperm chromatin follows the breakdown of the sperm nuclear envelope. During this process, the chromatin appears to transform from a condensed, electron-dense mass to a more dispersed form (Fig. 5). Investigations in a wide variety of organisms have demonstrated that during this stage of pronuclear development, the nucleoprotein content of the paternally derived chromatin is modified (Longo and Kunkle, 1978). Hence, structural alteration of the condensed chromatin is believed to be a manifestation of chemical changes of the paternally derived chromatin.

Chromatin dispersion has been examined ultrastructurally in *Barnea* (Pasteels, 1963, 1965a,b), *Mytilus* (Longo and Anderson, 1969b), and *Spisula* (Longo and Anderson, 1970b) and the series of events observed in each is morphologically similar. The chromatin along the periphery of the sperm nucleus changes from a dense mass to a diffuse granulofibrillar aggregation. Characteristically, the junction of the dense and diffuse chromatin is highly irregular. "Branches" of condensed chromatin project into the disperse chromatin, becoming thinner the deeper they extend into the diffuse chromatin (Fig. 5). Eventually, all of the chromatin is transformed into the diffuse form, resulting in a substantial increase in volume occupied by the paternally derived chromatin (Longo, 1977).

Studies providing insight into chemical modifications of the incorporated sperm nucleus during male pronuclear development in molluscs are limited. Histochemical investigations with gametes and zygotes of *Helix* have demonstrated that the protamine of the incorporated sperm nucleus is lost and replaced by basic proteins that differ from those found in nuclei of adult somatic cells and sperm nuclei (Bloch and Hew, 1960). Disulfide cross-links in the sperm nuclei of *Octopus* and the snail *Viviparus* have been investigated in order to understand chemical changes that may occur at the time of male pronuclear development (Bedford and Calvin, 1974). Octopus sperm nuclei are resistant to SDS owing to a high content of disulfide linkages; snail sperm nuclei, however, undergo immediate expansion and disintegration owing to fewer disulfide bonds. In those

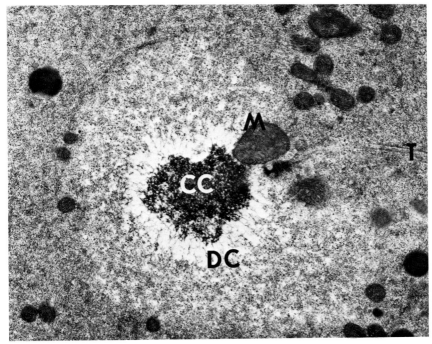

Fig. 5. Incorporated sperm nucleus of *Mytilus* lacking a nuclear envelope and undergoing chromatin dispersion. CC and DC, Condensed and dispersing chromatin; M, sperm mitochondria; T, portion of incorporated sperm tail.

forms with high concentrations of disulfide linkage, it has been speculated that these bonds are broken as a necessary prerequisite for the removal of sperm-specific nucleoproteins and the acquisition of cleavage histones (Longo and Kunkle, 1978; Longo, 1981).

C. Development of the Male Pronuclear Envelope

In some molluscs, such as *Spisula* (Longo and Anderson, 1970b), following its dispersion, the sperm chromatin remains exposed to the egg cytoplasm without an intervening nuclear envelope until meiotic maturation of the maternally derived chromatin is completed. A nuclear envelope is then developed along the periphery of the dispersed sperm chromatin; this process seems to take place in concert with the formation of the female pronuclear envelope. In other molluscs, such as *Mytilus* (Longo and Anderson, 1969b), dispersion of the sperm chromatin is followed immediately by development of a nuclear envelope. In either case, development of the membrane system that comes to delimit the dispersed, paternally derived chromatin is the same; the structure that is formed is referred to as the male pronuclear envelope (Longo, 1973).

Following or during chromatin dispersion, cisternae become aligned along the periphery of the dispersed chromatin and fuse together to form the male pronuclear envelope (Fig. 6). Although the origin of the cisternae that comprise the male pronuclear envelope has not been demonstrated unequivocally, Pasteels (1963, 1965a,b) claimed that they are derived from endoplasmic reticulum. Investigations with the sea urchin, *Arbacia,* have demonstrated that the endoplasmic reticulum is a major source of membrane for the male pronuclear envelope (Longo, 1976c). Upon formation of a nuclear envelope, the male pronucleus is completed. In molluscs, the male pronucleus is comparable in size and morphology to the female pronucleus, consequently, it is often difficult to distinguish the two pronuclei from each other (Fig. 6). Generally, the one situated

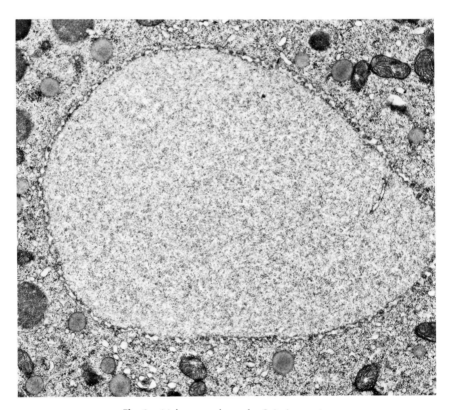

Fig. 6. Male pronucleus of a *Spisula* zygote.

closest to the animal pole represents the female pronucleus; the one closely associated with the incorporated axonemal complex is considered to be the male.

Aspects concerning the regulation of male pronuclear development in molluscs have been considered but only to a limited extent (Longo and Kunkle, 1978; Longo, 1982). Consequently, factors directing the transformation of the sperm nucleus into a male pronucleus in this group are relatively unexplored. Ultrastructural studies of cross-fertilization involving molluscs (sea urchin female × mussel male) indicate that *Mytilus* sperm that were incorporated into *Arbacia* eggs were capable of transforming into male pronuclei (Longo, 1977; see also Kupelwieser, 1909; Wada and Wada, 1953; Afzelius, 1972a,b). The male pronuclei that develop are morphologically similar to those that form in *Arbacia* eggs inseminated with *Arbacia* sperm. These results suggest that *Arbacia* eggs contain factors capable of transforming *Mytilus* sperm nuclei into pronuclei and, hence, lack species specificity. These substances may also regulate the actual form of the pronucleus.

XI. Development of the Sperm Aster

Generally, formation of the sperm aster occurs during the development of the male pronucleus and its morphogenesis has many similarities to that involving the formation of the meiotic asters (Raven et al., 1958; Raven, 1959). According to Raven (1959, 1966), the sperm aster may appear immediately after the incorporation of the sperm nucleus, that is, at the first meiotic division, at the second meiotic division, or at the completion of the second meiotic division (Cather, 1963).

The sperm aster forms around the centrioles brought into the egg with the spermatozoon (Longo and Anderson, 1969b, 1970b; Longo, 1973). Surrounding the centrioles is an amorphorous ground substance and some cisternae and microtubules; this area constitutes, at the ultrastructural level, what was referred to as the centrosphere by early light microscopists (Wilson, 1925). Fascicles of microtubules accumulate along the periphery of the centrosphere and project into the surrounding cytoplasm giving the aster its starlike appearance. The sperm aster increases in size with the development of the male pronucleus and appears to reach its maximum dimension with the migration of the pronuclei.

XII. Pronuclear Migration

According to Raven (1966), in gastropods, the female pronucleus remains at the animal pole where the two pronuclei become associated. In pelecypods, the male and female pronuclei migrate to the center of the zygote. In either case, the result of these movements is the close approximation of the two pronuclei. Just prior to and during pronuclear migration, the two pronuclei undergo DNA replication (Ito and Leuchtenberger, 1955; van den Biggelaar, 1971).

The function of the sperm aster in molluscan zygotes has not been demonstrated, however, there is evidence in sea urchins and mammals that this structure is involved in the migration of the pronuclei (Zimmerman and Zimmerman, 1967; Longo, 1976d). Presumably it has a comparable role in molluscs. That other factors may be involved in the migration of the pronuclei is suggested by observations of the movement of the female pronucleus in artificially activated *Spisula* eggs (Sachs, 1971; cf. Longo, 1973). In some cases, such as *Spisula,* microtubules are observed in association with the female pronucleus prior to migration; whether or not these elements are involved in the movements of the female pronucleus has not been established (Longo and Anderson, 1970a). Cytochalasin B does not appear to affect pronuclear migration. Colchicine has been shown to inhibit the development of the sperm aster and prevent pronuclear migration (Zimmerman and Zimmerman, 1967; Longo, 1973, 1976d).

XIII. Association of the Pronuclei

At the completion of their movements, the pronuclei are observed as two spheroids or ellipsoids closely associated with one another (Raven, 1966; Longo and Anderson, 1969b, 1970b; Longo, 1973). In some molluscs, such as *Mytilus,* the proximal surfaces of the opposed pronuclei interdigitate (Longo and Anderson, 1969b). In *Spisula,* the association of the two pronuclei is not as intimate (Longo and Anderson, 1970b).

Morphogenesis of the male and female pronuclei in pelecypods corresponds to the *Ascaris* type of fertilization (Longwell and Stiles, 1968; Longo and Anderson, 1969b, 1970b; cf. Raven, 1966; Longo, 1973). While closely associated, there is simultaneous condensation of chromosomes and breakdown of nuclear envelopes in both pronuclei (Fig. 7). Breakdown of the pronuclear envelopes seems to occur in the same manner described for the sperm nuclear envelope, resulting in the formation of cisternae that are scattered into the surrounding cytoplasm. Microtubules appear in regions where portions of the pronuclear envelopes disappear; they attach to condensing chromosomes and become oriented to the poles of the developing mitotic spindle, which are demarcated by developing asters. With the breakdown of the pronuclear envelopes the chromosomes are released into the cytoplasm and eventually, they become aligned on what becomes the metaphase plate of the first cleavage spindle (Fig. 7). Hence, in the molluscs that have been studied thus far, there is no fusion of male and female pronuclei to form a zygote nucleus as observed in sea urchins (Wilson, 1925; Longo and Anderson, 1968; Longo, 1973). Historically, the alignment of

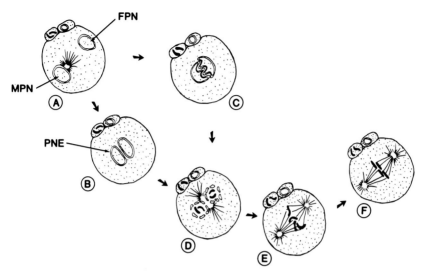

Fig. 7. Sequence of events involving the migration and association of the male (MPN) and female pronuclei (FPN) in *Mytilus* and *Spisula* zygotes. Association of the pronuclei in *Spisula* is depicted in (**B**), a more intimate association, involving an interdigitation of the proximal surfaces of the two pronuclei occurs in *Mytilus* (**C**). In (**D**), the pronuclear envelopes (PNE) break down forming cisternae that are scattered throughout the cytoplasm. Concomitantly, there is the development of a mitotic spindle and the condensation of maternally and paternally derived chromosomes shown in (**D**) and (**E**). Eventually the chromosomes, derived from the male and female pronuclei, intermix and become situated on the metaphase plate of the first cleavage spindle (**F**).

the maternally and paternally derived chromosomes on the metaphase plate of the first cleavage spindle has been considered to represent the conclusion of fertilization in zygotes demonstrating the *Ascaris* type of fertilization (Wilson, 1925; Longo, 1973).

XIV. Metabolic Changes at Meiotic Maturation/Fertilization

The synthesis of RNA upon fertilization has been investigated in a number of molluscs (Collier, 1966, 1976, Kidder, 1972a,b, 1976; Tapaswi, 1972, 1974; Brahmachary et al., 1968; Verdonk, 1973; McLean, 1976; McLean and Whitely, 1974; Firtel and Monroy, 1970). In *Crassostrea,* McLean and Whitely (1974) have demonstrated that the fertilized egg takes up uridine at a rate equivalent to the unfertilized egg. Furthermore, the rate of which uridine is incorporated into high molecular weight RNA is not altered at insemination. McLean and Whitely (1974) suggested that development during cleavage is under the control of information stored in the unfertilized egg. Although RNA synthesis is not markedly increased at fertilization in *Lymnaea,* all major forms of RNA are synthesized from oogenesis throughout fertilization (Tapaswi, 1974). In contrast to these results, eggs and early embryos of *Acmaea* reportedly do not incorporate uridine until late cleavage (Karp, 1973).

In *Spisula,* the unfertilized egg is active in protein synthesis and at fertilization there is an increase in the rate of amino acid incorporation by a factor of 3–4 (Bell and Reeder, 1967; Monroy and Tolis, 1964; Firtel and Monroy, 1970). Associated with this increase in rate of incorporation is an increase in amino acid uptake ten- to twelvefold at the completion of meiotic maturation (Bell and Reeder, 1967; Firtel and Monroy, 1970). Bell and Reeder (1967) indicated that the increase in protein synthesis in fertilized *Spisula* eggs occurs in the absence of RNA synthesis. In *Ilyanassa,* protein synthesis occurs in the unfertilized egg; approximately 15 min postfertilization, amino acid uptake and incorporation is increased. When the percentage of incorporation was calculated, protein synthetic activity of the zygote was found to be about 2.5 times greater than the unfertilized ovum (Mirkes, 1970).

Recent investigations of protein synthesis in *Spisula* eggs have demonstrated rapid and dramatic changes in the pattern after fertilization (Rosenthal et al., 1980). There is a reduction in the synthesis of prominent oocyte-specific proteins and an increase in the synthesis of at least three proteins whose labeling dominated the pattern of protein synthesis in early embryos. Rosenthal et al. (1980) suggested that these changes are modulated at the translational level. In connection with these observations, earlier investigations of *Ilyanassa* embryos have shown that synthesis of microtubular protein occurs in conjunction with mRNA synthesized during oogenesis and stored in the egg (Raff et al., 1975, 1976).

Firtel and Monroy (1970) have examined the regulation of protein synthesis in *Spisula* eggs and demonstrated that a small number of polysomes are active in protein synthesis. After fertilization, there was a progressive increase in the number of ribosomes that became associated with polysomes and by 30 min postinsemination the specific activity of polysomes was 2.5 times greater than that found in unfertilized ova. By the time the pronuclei were formed there were 4–5 times more polysomes than were present in the egg prior to maturation. Firtel and Monroy (1970) suggested that the increase in the number of polysomes at fertilization might be due to activation of stored, maternally derived mRNA and the formation of polysomes from newly synthesized mRNA. Eggs of *Mulinia* also undergo an increase in polysomes by 45 min postinsemination (Kidder, 1976). Furthermore, there did not seem to be a shift in the size distribution of the polysomes at the time of fertilization. Kidder (1976) pointed out that protein synthesis in *Spisula* and *Mulinia* was similar to that observed in sea urchins and that this relationship discourages speculation concerning differences in gene transcription patterns between early mosaic and regulative embryos.

Metabolic changes in mollusc eggs at fertilization have been reviewed (Monroy and Tyler, 1967; Raven, 1966, 1972). According to Raven (1966), upon fertilization, respiratory activity increased in *Mactra lateralis* and *Crassostrea virginica,* decreased in *Cumingia* (Whitaker, 1933), and did not change in *C. commercialis* and *C. gigas* (Cleland, 1950; Yasumasu et al., 1975; see also, Ballentine, 1940). Yasumasu et al. (1975) have studied glycolysis in fertilized *Crassostrea* eggs and demonstrated that it is regulated at steps of phosphorylase, phosphofructokinase, and pyruvate kinase. Phosphorylase is first activated, followed simultaneously by pyruvate kinase and phosphofructokinase. Because this pattern was similar to that found in sea urchin eggs, Yasumasu et al. (1975) speculated that these rate-limiting steps are regulated in the same manner as described for sea urchins. They suggested that carbohydrate utilization is enhanced at fertilization and glycolysis is activated. Results of Krane and Crane (1960) with fertilized *Spisula* eggs are consistent with the observation that the predominant pathway of glucose utilization is the hexose monophosphate shunt. Activation of NAD kinase, as inferred from elevation of total NADP levels in *Spisula* zygotes (Krane and Crane, 1960), also occurred immediately following insemination of sea urchin eggs and was regulated by calcium and calmodulin *in vitro* (Epel et al., 1981). ATP phosphohydrolase, NAD dehydrogenase, succinate dehydrogenase, and acid phosphatase have been studied in fertilized *Crassostrea* eggs (Kobayashi, 1968a,b). The activity of the latter two enzymes reportedly increased at fertilization. In fertilized *Barnea* eggs, a positive ATPase reaction surrounds the incorporated sperm nucleus and expands to the cytoplasm during meiotic maturation and pronuclear association (Dalcq and Pasteels, 1963).

XV. Concluding Remarks

As an experimental model, molluscan gametes provide an excellent system for investigating processes of meiotic maturation and fertilization. The eggs and sperm of a number of pelecypods are plentiful and easily manipulated for morphological, physiological, and chemical studies of reproductive phenomena. Investigators can take advantage of these traits in an effort to discern mechanisms and controls of meiosis and insemination that may be common to all animals.

For example, features of molluscan ova involving blocks of polyspermy have not been established and we have little insight into how the egg ensures monospermy. Further investigations hold promise of providing a better understanding of polyspermy blocks in other organisms as well. Although molluscan sperm have been shown to contain lytic substances within their acrosomes, the site of action of these agents has not been precisely established.

Because the eggs of molluscs are inseminated prior to the completion of meiosis, they represent a useful paradigm that may provide valuable insight into the relation of meiotic maturation and fertilization, and the reinitiation of meiosis. Furthermore, as in most organisms, the inseminated molluscan oocyte must complete meiosis and develop a female pronucleus following insemination. Although the maternally and paternally derived chromosomes are confined within the same cytoplasm, the two groups undergo processes that appear to be dissimilar or antagonistic to one another, that is, dispersion of the sperm-condensed chromatin appears to have no effect on the condensed or condensing meiotic chromosomes of the oocyte. Observations such as these help to demonstrate the specificity of the processes regulating morphogenesis of the maternally and paternally derived chromatin. How these processes are regulated awaits further investigation.

Acknowledgments

The author is grateful to Christopher Carron and Stephen Downs of the University of Iowa for their helpful suggestions, exchange of ideas, and reading the manuscript during its preparation. Special thanks are due to Fredrick So and Julie Anolik for their help in preparing this manuscript for publication.

References

Afzelius, B. A. (1972a). Reactions of the sea urchin oocyte to foreign spermatozoa. *Exp. Cell Res.* **72,** 25–33.

Afzelius, B. A. (1972b). Ultrastructure of species–foreign spermatozoa after penetrating the sea urchin oocyte. *Acta Embryol. Exp.,* Issue 1, pp. 123–133.

Allen, R. D. (1953). Fertilization and artificial activation in the egg of the surf-clam, *Spisula solidissima. Biol. Bull. (Woods Hole, Mass.)* **105,** 213–237.

Anderson, E. (1969). Oocyte–follicle cell differentiation in two species of Amphineurans (Mollusca), *Mopalia mucosa* and *Chaetopleura apiculata. J. Morphol.* **129,** 89–126.

Anderson, E. (1974). Comparative aspects of the ultrastructure of the female gamete. *Int. Rev. Cytol., Suppl.* **4,** 1–70.

Anderson, W. A., and Personne, P. (1976). The molluscan spermatozoon: Dynamic aspects of its structure and function. *Am. Zool.* **16,** 293–313.

Austin, C. R. (1965). "Fertilization." Prentice-Hall, Englewood Cliffs, New Jersey.

Austin, C. R. (1968). "Ultrastructure of Fertilization." Holt, Rinehart & Winston, New York.

Ballentine, R. (1940). Analysis of the changes in respiratory activity accompanying the fertilization of marine eggs. *J. Cell. Comp. Physiol.* **15,** 217–232.

Bedford, J. M., and Calvin, H. I. (1974). The occurrence and possible functional significance of —S—S— crosslinks in sperm heads, with particular reference to eutherian mammals. *J. Exp. Zool.* **188,** 137–156.

Bedford, L. (1966). The electron microscopy and cytochemistry of oogenesis and the cytochemistry of embryonic development of the prosobranch gastropod *Bembicium nanum* L. *J. Embryol. Exp. Morphol.* **15,** 15–37.

Bell, E., and Reeder, R. (1967). The effect of fertilization on protein synthesis in the egg of the surf-clam *Spisula solidissima*. *Biochim. Biophys. Acta* **142,** 500–511.

Berg, W. E. (1949). Some effects of sperm extracts on the eggs of *Mytilus*. *Am. Nat.* **83,** 221–226.

Berg, W. E. (1950). Lytic effects of sperm extracts on the eggs of *Mytilus edulis*. *Biol. Bull. (Woods Hole, Mass.)* **98,** 128–138.

Bloch, D. P., and Hew, H. Y. C. (1960). Changes in nuclear histones during fertilization, and early embryonic development in the pulmonate snail, *Helix aspersa*. *J. Cell Biol.* **8,** 69–81.

Brachet, J. (1960). "The Biochemistry of Development." Pergamon, Oxford.

Brahmachary, R. L., Banerjee, K. P., and Basu, T. K. (1968). Investigations on transcription in *Limnaea* embryos. *Exp. Cell Res.* **51,** 177–184.

Brandriff, B., Moy, G. W., and Vacquier, V. D. (1978). Isolation of sperm bindin from the oyster (*Crassostrea gigas*). *Gamete Res.* **1,** 89–99.

Bretschneider, L. H. (1948). Insemination in *Limnaea stagnalis* L. *Proc. K. Ned. Akad. Wet.* **51,** 358–362.

Burnside, B., Kozak, C., and Kafatos, F. C. (1973). Tubulin determination by an isotope dilution–vinblastine precipitation method. The tubulin content of *Spisula* eggs and embryos. *J. Cell Biol.* **59,** 755–762.

Calarco, P. G., Donahue, R. P., and Szollosi, D. (1972). Germinal vesicle breakdown in the mouse oocyte. *J. Cell Sci.* **10,** 369–385.

Cather, J. N. (1963). A time schedule of the meiotic and early mitotic stages of *Ilyanassa*. *Caryologia* **16,** 663–670.

Chambers, R. (1917). Microdissection studies. II. The cell aster: A reversible gelation phenomenon. *J. Exp. Zool.* **23,** 483–505.

Cleland, K. W. (1950). Respiration and cell division in developing oyster eggs. *Proc. Linn. Soc. N. S. W.* **75,** 282–295.

Clement, A. C. (1935). The formation of giant polar bodies in centrifuged eggs of *Ilyanassa*. *Biol. Bull. (Woods Hole, Mass.)* **69,** 403–414.

Clement, A. C. (1971). *Ilyanassa*. *In* "Experimental Embryology of Marine and Fresh-Water Invertebrates" (G. Reverberi, ed.), pp. 188–214. Am. Elsevier, New York.

Clement, A. C. (1976). Cell determination and organogenesis in molluscan development: A reappraisal based on deletion experiments in *Ilyanassa*. *Am. Zool.* **16,** 447–453.

Collier, J. R. (1966). The transcription of genetic information in the spiralion embryo. *Curr. Top. Dev. Biol.* **1,** 39–59.

Collier, J. R. (1976). Nucleic acid chemistry in the *Ilyanassa* embryo. *Am. Zool.* **16,** 483–500.

Colwin, L. H., and Colwin, A. L. (1967). Membrane fusion in relation to sperm–egg association. *In* "Fertilization" (C. B. Metz and A. Monroy, eds.), Vol. 1, pp. 295–368. Academic Press, New York.

Conklin, E. G. (1902). Karyokinesis and cytokinesis in the maturation, fertilization and cleavage of *Crepidula* and other Gastropoda. *Proc. Acad. Nat. Sci. Philadelphia* **21,** 1–121.

Conklin, E. G. (1917). Effects of centrifugal force on the structure and development of *Crepidula*. *J. Exp. Zool.* **22,** 311–419.

Conrad, G. W., and Davis, S. E. (1977). Microiontophoretic injection of calcium ions or of cyclic AMP causes rapid shape changes in fertilized eggs of *Ilyanassa obsoleta*. *Dev. Biol.* **61,** 184–201.

Conrad, G. W., and Davis, S. E. (1980). Polar lobe formation and cytokinesis in fertilized eggs of *Ilyanassa obsoleta*. III. Large bleb formation caused by Sr^{2+}, ionophores ×537A and A23187, and compound 48/80. *Dev. Biol.* **74,** 152–172.

Conrad, G. W., and Williams, D. C. (1974a). Polar lobe formation and cytokinesis in fertilized eggs of *Ilyanassa obsoleta*. I. Ultrastructure and effects of cytochalasin B and colchicine. *Dev. Biol.* **36,** 363–378.

Conrad, G. W., and Williams, D. C. (1974b). Polar lobe formation and cytokinesis in fertilized eggs of *Ilyanassa obsoleta*. II. Large bleb formation caused by high concentrations of exogenous calcium ions. *Dev. Biol.* **37**, 280–294.

Conrad, G. W., Williams, D. C., Turner, F. R., Newrock, K. M., and Raff, R. A. (1973). Microfilaments in the polar lobe constriction of fertilized eggs of *Ilyanassa obsoleta*. *J. Cell Biol.* **59**, 228–233.

Costello, D. P., and Henley, C. (1971), "Methods for Obtaining and Handling Marine Eggs and Embryos." Baker Mfg. Co., Printers, New Bedford, Massachusetts.

Crabb, E. D. (1927). The fertilization process in the snail, *Lymnaea stagnalis appressa* Say. *Biol. Bull. (Woods Hole, Mass.)* **53**, 67–108.

Dalcq, A. M., and Pasteels, J. J. (1963). La localisation d'enzymes de déphosphorylation dans les oeufs de quelques Invertébres. *Dev. Biol.* **7**, 457–487.

Dan, J. C. (1956). The acrosome reaction. *Int. Rev. Cytol.* **5**, 365–393.

Dan, J. C. (1962). The vitelline coat of the *Mytilus* egg. I. Normal structure and effect of acrosomal lysin. *Biol. Bull. (Woods Hole, Mass.)* **123**, 531–541.

Dan, J. C. (1967). Acrosome reaction and lysins. *In* "Fertilization" (C. B. Metz and A. Monroy, eds.), Vol. 1, pp. 237–294. Academic Press, New York.

Dan, J. C., and Wada, S. K. (1955). Studies on the acrosome. IV. The acrosome reaction in some bivalve spermatozoa. *Biol. Bull. (Woods Hole, Mass.)* **109**, 40–55.

Daniels, E. W., Longwell, A. C., McNiff, J. M., and Wolfgang, R. W. (1971). Ultrastructure of spermatozoa from the American oyster *Crassostrea virginica*. *Trans. Am. Micros. Soc.* **90**, 275–282.

Davidson, E. (1976). "Gene Activity in Early Development," 2nd ed. Academic Press, New York.

Delage, Y. (1899). Etudes sur la mérogonie. *Arch. Zool. Exp. Gen.* **7**, 383–417.

de Larambergue, M. (1939). Etude de l'autofécondation chez les Gastéropodes Pulmonés. Recherches sur l'aphallie et la fécondation chez *Bulinus* (*Isidora*) *contortus* Michaud. *Bull. Biol. Fr. Belg.* **73**, 21–231.

Dohmen, M. R., and van der Mey, J. C. (1977). Local surface differentiations at the vegetal pole of the eggs of *Nassarius reticulatus, Buccinum undatum* and *Crepidula fornicata* (*Gastropoda, Prosobranchia*). *Dev. Biol.* **61**, 104–113.

Drozdov, A. L. (1979). Cortical reaction in eggs of the mussel *Crenomytilus grayanus*. *Sov. J. Dev. Biol.* **10**, 254–257.

Edelman, G. M. (1976). Surface modulation in cell recognition and cell growth. *Science* **192**, 218–226.

Epel, D. (1978). Mechanisms of activation of sperm and egg during fertilization of sea urchin gametes. *Curr. Top. Dev. Biol.* **12**, 185–246.

Epel, D., Steinhardt, R. A., Humphreys, T., and Mazia, D. (1974). An analysis of the partial metabolic derepression of sea urchin eggs by ammonia: The existence of independent pathways. *Dev. Biol.* **40**, 245–255.

Epel, D., Patton, C., Wallace, R. W., and Cheung, W. Y. (1981). Calmodulin activates NAD kinase of sea urchin eggs: An early event of fertilization. *Cell* **23**, 543–549.

Field, I. A. (1922). The biology and economic value of the sea mussel, *Mytilus edulis*. *Fish. Bull.* **38**, 1–159.

Finkel, T., and Wolf, D. P. (1980). Membrane potential, pH and the activation of surf clam oocytes. *Gamete Res.* **3**, 299–304.

Firtel, R. A., and Monroy, A. (1970). Polysomes and RNA synthesis during early development of the surf clam *Spisula solidissima*. *Dev. Biol.* **21**, 87–104.

Franzen, Å. (1955). Comparative morphological investigations into the spermiogenesis among Mollusca. *Zool. Bidr.* **30**, 399–456.

Franzen, Å. (1956) On spermiogenesis, morphology of the spermatozoon, and biology of fertilization among invertebrates. *Zool. Bidr.* **31**, 355–482.

Galtsoff, P. S. (1964). The American oyster, *Crassostrea virginica* Gmelin. *Fish. Bull.* **64**, 324–354.

Galtsoff, P. S., and Philpott, D. E. (1960). Ultrastructure of the spermatozoon of the oyster *Crassostrea virginica*. *J. Ultrastruct. Res.* **3**, 241–253.

Garnault, P. (1888). Sur les phénomènes de la fécondation chez l'*Helix aspersa* et l'*Arion empiricorum*. *Zool. Anz.* **11**, 731–736.

Garnault, P. (1889). Sur les phénomènes de la fécondation chez l'*Helix aspersa* et l'*Arion empiricorum*. *Zool. Anz.* **12**, 10–15.

Giese, A. C., and Pearse, J. S., eds. (1979), "Reproduction of Marine Invertebrates," Vol. 5. Academic Press, New York.

Gillies, R. J., Rosenberg, M., and Deamer, D. W. (1980). Inorganic carbonate release and pH changes during the activation sequence in sea urchin eggs. *J. Cell Biol.* **87**, 136A.

Ginzburg, A. S. (1974). Fertilization of the eggs of bivalve mollusks with different insemination conditions. *Ontogenez* **5**, 341–348.

Gould-Somero, M., Holland, L., and Paul, M. (1977). Cytochalasin B inhibits sperm penetration into eggs of *Urechis caupo* (*Echiura*). *Dev. Biol.* **52**, 11–22.

Guerrier, P. (1970). Les caractères de la segmentation et la dètermination de la polarité dorsoventrale dans le développement de quelques Spiralia III. *Pholas dactylus* et *Spisula subtruncata* (Mollusques Lamellibranches). *J. Embryol. Exp. Morphol.* **23**, 667–692.

Hatanaka, M., Itoh, R., and Imai, F. (1943). The culture of *Venerupis* and *Meretrix*. *Rep. Jpn. Fish. Assoc.* **11**, 218.

Hess, O. (1971). Fresh water gastropoda. *In* "Experimental Embryology of Marine and Fresh-Water Invertebrates" (G. Reverberi, ed.), pp. 215–247. Am. Elsevier, New York.

Hoadley, L. (1930). Polocyte formation and the cleavage of the polar body in *Loligo* and *Chaetopterus*. *Biol. Bull. (Woods Hole, Mass.)* **58**, 256–264.

Horstmann, H. J. (1955). Untersuchungen zur Physiologie der Begattung und Befruchtung der Schlammschnecke *Lymnaea stagnalis* L. *Z. Morphol. Oekol. Tiere* **44**, 222–268.

Huebner, E., and Anderson, E. (1976). Comparative spiralian oogenesis—structural aspects: An overview. *Am. Zool.* **16**, 315–343.

Humphreys, W. J. (1962). Electron microscope studies on eggs of *Mytilus edulis*. *J. Ultrastruct. Res.* **7**, 467–487.

Humphreys, W. J. (1964). Electron microscope studies of the fertilized egg and the two-cell stage of *Mytilus edulis*. *J. Ultrastruct. Res.* **10**, 244–262.

Humphreys, W. J. (1967). The fine structure of cortical granules in eggs and gastrulae of *Mytilus edulis*. *J. Ultrastruct. Res.* **17**, 314–326.

Hylander, B. L., and Summers, R. G. (1977). An ultrastructural analysis of the gametes and early fertilization of two bivalve molluscs, *Chama macerophylla* and *Spisula solidissima* with special reference to gamete binding. *Cell Tissue Res.* **182**, 469–489.

Ii, I., and Rebhun, L. I. (1979). Acid release following activation of surf clam (*Spisula solidissima*) eggs. *Dev. Biol.* **72**, 195–200.

Ikeda, K. (1930). The fertilization cones of the land snail, *Eulota* (*Eulotella*) *similaris stimpsoni* Pfeiffer. *Jpn. J. Zool.* **3**, 89–94.

Ito, S., and Leuchtenberger, C. (1955). The possible role of the DNA content of spermatozoa for the activation process of the egg of the clam, *Spisula solidissima*. *Chromosoma* **7**, 328–339.

Iwata, K. S. (1951). Auto-activation of eggs of *Mactra veneriformis* in sea water. *Annot. Zool. Jpn.* **24**, 187–193.

Jaffe, L. A. (1976). Fast block to polyspermy in sea urchin eggs is electrically mediated. *Nature (London)* **261**, 68–71.

Jordan, H. E. (1910). A cytological study of the egg of *Cumingia* with special reference to the history of the chromosomes and the centrosome. *Arch. Zellforsch.* **4**, 243–253.

Jura, Cz. (1960). Cytological and cytochemical observation on the embryonic development of *Succinea putris* L. (Mollusca) with particular reference to ooplasmic segregation. *Zool. Pol.* **10**, 95–128.

Karp, G. C. (1973). Autoradiographic patterns of [³H]uridine incorporation during the development of the mollusc, *Acmaea scutum*. *J. Embryol. Exp. Morphol.* **29**, 15–25.

Kidder, G. M. (1972a). Gene transcription in mosaic embryos. I. The pattern of RNA synthesis in early development of the coot clam, *Mulinia lateralis*. *J. Exp. Zool.* **180**, 55–74.

Kidder, G. M. (1972b). Gene transcription in mosaic embryos. II. Polyribosomes and messenger RNA in early development of the coot clam, *Mulinia lateralis*. *J. Exp. Zool.* **180**, 75–84.

Kidder, G. M. (1976). RNA synthesis and the ribosomal cistrons in early molluscan development. *Am. Zool.* **16**, 501–520.

Kobayashi, H. (1968a). Cytochemical studies on the second gradient axis of fertilized oyster eggs. III. On the activity of succinate dehydrogenase and acid phosphatase. *Cytologia* **33**, 112–117.

Kobayashi, H. (1968b). Cytochemical studies on the second gradient axis of fertilized oyster eggs. IV. On the activity of ATPase and reduced NAD dehydrogenase. *Cytologia* **33**, 118–124.

Kostanecki, K. (1904). Cytologische Studien an künstlich parthenogenetische sich entwickelnden Eiern von *Mactra*. *Arch. Mikrosk. Anat.* **64**, 1–98.

Kostanecki, K., and Wierzejski, A. (1896). Uber das Verhalten der sogen achromatischen Substanzen im befruchteten Ei. Nach Beobachtungen an *Physa fontinalis*. *Arch. Mikrosk. Anat.* **47**, 309–386.

Krane, S. M., and Crane, R. K. (1960). Changes in levels of triphosphopyridine nucleotide in marine eggs subsequent to fertilization. *Biochim. Biophys. Acta* **43**, 369–373.

Krauss, M. (1950). Lytic agents of the sperm of some marine animals. I. The egg membrane lysin from sperm of the giant keyhole limpet *Megathuria crenulata*. *J. Exp. Zool.* **114**, 239–278.

Kume, M., and Dan, K. (1968). "Invertebrate Embryology." NOLIT Pub. House, Belgrade, Yugoslavia.

Kupelwieser, H. (1909). Entwicklungserregung bei Seeigeleiern durch Molluskensperma. *Arch. Entwickungs mech. Org.* **27**, 434–462.

Lams, H. (1910). Recherches sur l'oeut d'*Arion empiricorum* (Fér.). (Accroissement, maturation, fécondation, segmentation.) *Mem. Acad. R. Belg., Cl. Sci.* **2**, 1–144.

Lewis, C. A., Leighton, D. L., and Vacquier, V. D. (1980). Morphology of abalone spermatozoa before and after the acrosome reaction. *J. Ultrastruct. Res.* **72**, 39–46.

Linville, H. R. (1900). Maturation and fertilization in pulmonate gasteropods. *Bull. Mus. Comp. Zool.* **35**, 213–248.

Longo, F. J. (1972). The effects of cytochalasin B on the events of fertilization in the surf clam, *Spisula solidissima*. I. Polar body formation. *J. Exp. Zool.* **182**, 321–344.

Longo, F. J. (1973). An ultrastructural analysis of polyspermy in the surf clam, *Spisula solidissima*. *J. Exp. Zool.* **183**, 153–180.

Longo, F. J. (1976a). Ultrastructural aspects of fertilization in *Spiralian* eggs. *Am. Zool.* **16**, 375–394.

Longo, F. J. (1976b). Cortical changes in *Spisula* eggs upon insemination. *J. Ultrastruct. Res.* **56**, 226–232.

Longo, F. J. (1976c). Derivation of the membrane comprising the male pronuclear envelope in inseminated sea urchin eggs. *Dev. Biol.* **49**, 347–368.

Longo, F. J. (1976d). The sperm aster in rabbit zygotes: Its structure and function. *J. Cell Biol.* **69**, 539–547.

Longo, F. J. (1977). An ultrastructural study of cross-fertilization (*Arbacia* ♀ × *Mytilus* ♂). *J. Cell Biol.* **73**, 14–26.

Longo, F. J. (1978a). Effects of cytochalasin B on sperm–egg interactions. *Dev. Biol.* **67**, 157–173.

Longo, F. J. (1978b). Surface alterations of fertilized surf clam (*Spisula solidissima*) eggs induced by concanavalin A. *Dev. Biol.* **68**, 422–439.

Longo, F. J. (1980). Organization of microfilaments in sea urchin (*Arbacia punctulata*) eggs at fertilization: Effects of cytochalasin B. *Dev. Biol.* **74**, 422–433.

Longo, F. J. (1981). Regulation of pronuclear development. *In* "Bioregulators of Reproduction" (G. Jagiello and C. Vogel, eds.), pp. 529–557. Academic Press, New York.

Longo, F. J. (1982). Integration of sperm and egg plasma membrane components at fertilization. *Dev. Biol.* **89**, 409–416.

Longo, F. J., and Anderson, E. (1968). The fine structure of pronuclear development and fusion in the sea urchin, *Arbacia punctulata*. *J. Cell Biol.* **39**, 339–368.

Longo, F. J., and Anderson, E. (1969a). Spermiogenesis in the surf clam, *Spisula solidissima*, with special reference to the formation of the acrosomal vesicle. *J. Ultrastruct. Res.* **27**, 435–443.

Longo, F. J., and Anderson, E. (1969b). Cytological aspects of fertilization in the lamellibranch, *Mytilus edulis*. II. Development of the male pronucleus and the association of the maternally and paternally derived chromosomes. *J. Exp. Zool.* **172**, 97–120.

Longo, F. J., and Anderson, E. (1969c). Cytological aspects of fertilization in the lamellibranch, *Mytilus edulis*. I. Polar body formation and development of the female pronucleus. *J. Exp. Zool.* **172**, 69–96.

Longo, F. J., and Anderson, E. (1970a). An ultrastructural analysis of fertilization in the surf clam, *Spisula solidissima*. I. Polar body formation and development of the female pronucleus. *J. Ultrastruct. Res.* **33**, 495–514.

Longo, F. J., and Anderson, E. (1970b). An ultrastructural analysis of fertilization in the surf clam, *Spisula solidissima*. II. Development of the male pronucleus and the association of the maternally and paternally derived chromosomes. *J. Ultrastruct. Res.* **33**, 515–527.

Longo, F. J., and Dornfeld, E. J. (1967). The fine structure of spermatid differentiation in the mussel, *Mytilus edulis*. *J. Ultrastruct. Res.* **20**, 462–480.

Longo, F. J., and Kunkle, M. (1978). Transformations of sperm nuclei upon insemination. *Curr. Top. Dev. Biol.* **12**, 149–184.

Longwell, A. C., and Stiles, S. S. (1968). Fertilization and completion of meiosis in spawned eggs of the American oyster, *Crassostrea virginica* Gmelin. *Caryologia* **21**, 65–73.

Loosanoff, V. L., and Davis, H. C. (1963). Rearing of bivalve molluscs. *Adv. Mar. Biol.* **1**, 1–136.

Luchtel, D. L. (1976). An ultrastructural study of the egg and early cleavage stages of *Lymnaea stagnalis*, a pulmonate mollusc. *Am. Zool.* **16**, 405–419.

McLean, K. (1976). Some aspects of RNA synthesis in oyster development. *Am. Zool.* **16**, 521–528.

McLean, K. W., and Whiteley, A. H. (1974). RNA synthesis during the early development of the Pacific oyster, *Crassostrea gigas*. *Exp. Cell Res.* **87**, 132–138.

Masui, Y., and Clarke, H. J. (1979). Oocyte maturation. *Int. Rev. Cytol.* **57**, 185–282.

Mazia, D. (1961). Mitosis and the physiology of cell division. *In* "The Cell" (J. Brachet and A. E. Mirsky, eds.), Vol. 3, pp. 77–412. Academic Press, New York.

Metz, C. B. (1967). Gamete surface somponents and their role in fertilization. *In* "Fertilization" (C. B. Metz and A. Monroy, eds.), Vol. 1. pp. 163–236. Academic Press, New York.

Meves, F. (1915). Uber den Befruchtungsvorgang bei der Miesmuschel (*Mytilus edulis* L.). *Arch. Mikrosk. Anat.* **87**, 47–62.

Miller, R. L. (1977). Chemotactic behavior of the sperm of chitons (Mollusca: Polyplacophora). *J. Exp. Zool.* **202**, 203–212.

Mirkes, P. E. (1970). Protein synthesis before and after fertilization in the egg of *Ilyanassa obsoleta*. *Exp. Cell Res.* **60**, 115–118.

Monroy, A. (1965), "Chemistry and Physiology of Fertilization." Holt, Rinehart & Winston, New York.

Monroy, A., and Tolis, H. (1964). Uptake of radioactive glucose and amino acids and their utilization for incorporation into protein during maturation and fertilization of the eggs of *Asterias forbesii* and *Spisula solidissima*. *Biol. Bull. (Woods Hole, Mass.)* **126**, 456–466.

Monroy, A., and Tyler, A. (1967). The activation of the egg. *In* "Fertilization" (C. B. Metz and A. Monroy, eds.). Vol. 1, pp. 369–412. Academic Press, New York.

Morgan, T. H. (1936). Further experiments on the formation of the antipolar lobe of *Ilyanassa*. *J. Exp. Zool.* **74**, 381–423.

Morgan, T. H. (1937). The behavior of the maturation spindles in polar fragments of eggs of *Ilyanassa* obtained by centrifuging. *Biol. Bull. (Woods Hole, Mass.)* **72**, 88–98.

Morrill, J. B. (1963). Development of centrifuged *Limnaea stagnalis* eggs with giant polar bodies. *Exp. Cell Res.* **31**, 490–498.

Moy, G. W., and Vacquier, V. D. (1979). Immunoperoxidase localization of bindin during the adhesion of sperm to sea urchin eggs. *Curr. Top. Dev. Biol.* **13**, 31–44.

Niijima, L., and Dan, J. (1965a). The acrosome reaction in *Mytilus edulis*. I. Fine structure of the intact acrosome. *J. Cell Biol.* **25**, 243–248.

Niijima, L., and Dan, J. (1965b). The acrosome reaction in *Mytilus edulis*. II. Stages in the reaction, observed in supernumerary and calcium-treated spermatozoa. *J. Cell Biol.* **25**, 249–259.

Osanai, K. (1969). Relation between cortical change and cytoplasmic movement in the oyster egg. *Annu. Rep. Fac. Educ., Univ. Iwate* **29**, 39–44.

Pasteels, J. J. (1935). Recherches sur le déterminisme de l'entrée en maturation de l'oeut chez divers invertébrés marins. *Arch. Biol.* **46**, 229–262.

Pasteels, J. J. (1963). Sur l'origine de la membrane nucléaire du pronucleus ♂ chez le Mollusque Bivalve *Barnea candida* (Etude au microscope électronique). *Bull. Cl. Sci., Acad. R. Belg.* **49**, 329–336.

Pasteels, J. J. (1965a). Aspects structuraux de la fécondation vus au microscope électronique. *Arch. Biol.* **76**, 463–509.

Pasteels, J. J. (1965b). La fécondation étudiée au microscope électronique. *Bull. Soc. Zool. Fr.* **90**, 195–224.

Pasteels, J. J., and de Harven, E. (1962). Etude au microscope électronique du cortex de l'oeuf de *Barnea candida* (Mollusque Bivalve), et son évolution au moment de la fécondation, de la maturation, et de la segmentation. *Arch. de Biol.* **73**, 465–490.

Pasteels, J. J., and de Harven, E. (1963). Étude au microscope électronique du cytoplasme de l'oeuf vierge et féconde' de *Barnea candida* (Mollusque Bivalve). *Arch. Biol.* **74**, 415–437.

Paul, M. (1975). Release of acid and changes in light-scattering properties following fertilization of *Urechis caupo* eggs. *Dev. Biol.* **43**, 299–312.

Pickett-Heaps, J. D. (1975). Aspects of spindle evolution. *Ann. N.Y. Acad. Sci.* **253**, 352–361.

Pollard, T. D., and Weihing, R. R. (1974). Actin and myosin and cell movement. *CRC, Crit. Rev. Biochem.* **2**, 1–65.

Popham, J. D. (1974). The ultrastructure of the acrosome reaction of the spermatozoon of, and gamete fusion in, the shipworm, *Bankia australis* (Bivalvia: Mollusca). *J. Anat.* **118**, 402–403.

Popham, J. D. (1975). The fine structure of the oocyte of *Bankia australis* (Teredinidae, Bivalvia) before and after fertilization. *Cell Tissue Res.* **157**, 521–534.

Raff, R. A. (1972). Polar lobe formation by embryos of *Ilyanassa obsoleta*. Effects of inhibitors of microtubule and microfilament function. *Exp. Cell Res.* **71**, 455–459.

Raff, R. A., Brandis, J. W., Green, L. H., Kaumeyer, J. F., and Raff, E. C. (1975). Microtubule protein pools in early development. *Ann. N.Y. Acad. Sci.* **253**, 304–317.

Raff, R. A., Newrock, K. M., Secrist, R. D., and Turner, F. R. (1976). Regulation of protein synthesis in embryos of *Ilyanassa obsoleta*. *Am. Zool.* **16**, 529–545.

Raven, C. P. (1959). The formation of the second maturation spindle in the eggs of *Succinea, Physa* and *Planorbis*. *J. Embryol. Exp. Morphol.* **7,** 344–360.

Raven, C. P. (1963). The nature and origin of the cortical morphogenetic field in *Limnaea*. *Dev. Biol.* **7,** 130–143.

Raven, C. P. (1964a). Development. *In* "Physiology of Mollusca" (K. M. Wilbur and C. M. Yonge, eds.), Vol. 1, pp. 165–195. Academic Press, New York.

Raven, C. P. (1964b). The formation of the second maturation spindle in the eggs of various Limnaeidae. *J. Embryol. Exp. Morphol.* **12,** 805–823.

Raven, C. P. (1966). "Morphogenesis: The Analysis of Molluscan Development." Pergamon, Oxford.

Raven, C. P. (1967). The distribution of special cytoplasmic differentiations of the egg during early cleavage in *Limnaea stagnalis*. *Dev. Biol.* **16,** 407–437.

Raven, C. P. (1972). Chemical embryology of Mollusca. *In* "Chemical Zoology" (M. Florkin and B. T. Scheer, eds.), Vol. 7, pp. 155–185. Academic Press, New York.

Raven, C. P., Escher, F. C. M., Herrebout, W. M., and Leussink, J. A. (1958). The formation of the second maturation spindle in the eggs of *Limnaea, Limax,* and *Agriolimax*. *J. Embryol. Exp. Morphol.* **6,** 28–51.

Rebhun, L. I. (1959). Studies of early cleavage in the surf-clam, *Spisula solidissima*, using methylene blue and toluidine blue as vital stains. *Biol. Bull. (Woods Hole, Mass.)* **117,** 518–545.

Rebhun, L. I. (1962a). Electron microscope studies on the vitelline membrane of the surf clam, *Spisula solidissima*. *J. Ultrastruct. Res.* **6,** 107–122.

Rebhun, L. I. (1962b). Dispersal of the vitelline membrane of the eggs of *Spisula solidissima* by alkaline, isotonic NaCl. *J. Ultrastruct. Res.* **6,** 123–134.

Reverberi, G. (1967). Some observations on the ultrastructure of the ovarian *Mytilus* egg. *Acta Embryol. Morphol. Exp.* **19,** 1–14.

Reverberi, G., ed. (1971a) "Experimental Embryology of Marine and Fresh-Water Invertebrates." Am. Elsevier, New York.

Reverberi, G. (1971b). *Mytilus. In* "Experimental Embryology of Marine and Fresh-Water Invertebrates" (G. Reverberi, ed.), pp. 175–187. Am. Elsevier, New York.

Reverberi, G. (1971c). *Dentalium. In* "Experimental Embryology of Marine and Fresh-Water Invertebrates" (G. Reverberi, ed.), pp. 248–264. Am. Elsevier, New York.

Reverberi, G., and Mancuso, V. (1961). The constituents of the egg of *Mytilus* as seen at the electron microscope. *Acta Embryol. Morphol. Exp.* **4,** 102–121.

Rosenthal, E. T., Hunt, T., and Ruderman, J. V. (1980). Selective translation of mRNA controls the pattern of protein synthesis during early development of the surf clam, *Spisula solidissima*. *Cell* **20,** 487–494.

Sachs, M. I. (1971). A cytological analysis of artificial parthenogenesis in the surf clam, *Spisula solidissima*. *J. Ultrastruct. Res.* **36,** 806–823.

Sastry, A. N. (1979). Pelecypoda (Excluding *Ostreidae*). *In* "Reproduction of Marine Invertebrates" (A. C. Giese and J. S. Pearse, eds.), Vol. 5, pp. 113–292. Academic Press, New York.

Sawada, N. (1952). Experimental studies on the maturation division of eggs in *Mactra veneriformis*. I. Inhibitory effect of the body fluid. *Mem. Ehime Univ.* **1,** 63–68.

Sawada, N. (1954a). Experimental studies on the maturation division of eggs in *Mactra veneriformis*. IV. On the effect of certain polysaccharides. *Mem. Ehime Univ., Nat. Sci., Ser. B* **2,** 89–92.

Sawada, N. (1954b). Experimental studies on the maturation division of eggs in *Mactra veneriformis*. V. On the activation by periodate. *Mem. Ehime Univ., Nat. Sci., Ser. B* **2,** 93–100.

Sawada, N. (1960a). Experimental studies on the maturation division of eggs in *Mactra venerifor-mis*. VIII. The extensibility of the surface membrane. *Mem. Ehime Univ., Nat. Sci. Ser. B* **4**, 73–78.

Sawada, N. (1960b). Experimental studies on the maturation division of eggs in *Mactra venerifor-mis*. IX. Some natures of the preparatory reaction for the maturation division. *Mem. Ehime Univ., Nat. Sci., Ser. B* **4**, 79–84.

Sawada, N., and Hosokawa, K. (1959). Experimental studies on the maturation division of eggs in *Mactra veneriformis*. VII. Effects of dinitrophenol and sodium azide on the breakdown of the germinal vesicle. *Mem. Ehime Univ., Nat. Sci., Ser. B* **3**, 243–249.

Sawada, N., and Rebhun, L. I. (1969). The effect of dinitrophenol and other phosphorylation uncouplers on the birefringence of the mitotic apparatus of marine eggs. *Exp. Cell Res.* **55**, 33–38.

Schmekel, L., and Fioroni, P. (1975). Cell differentiation during early development of *Nassarius reticulatus* L. (*Gastropoda, Prosobranchia*). I. Zygote to 16-cell stage. *Cell Tissue Res.* **159**, 503–522.

Schmekel, L., and Fioroni, P. (1976). Cell differentiation during early development of *Nassarius reticulatus* L. (*Gastropoda, Prosobranchia*). II. Morphological changes of nuclei and nucleoli. *Cell Tissue Res.* **168**, 361–371.

Schroeder, T. E. (1975). Dynamics of the contractile ring. *In* "Molecules and Cell Movement" (S. Inoué and R. E. Stephens, eds.), pp. 305–334. Raven Press, New York.

Schuel, H. (1978). Secretory functions of egg cortical granules in fertilization and development. A critical review. *Gamete Res.* **1**, 299–382.

Schuetz, A. W. (1975). Induction of nuclear breakdown and meiosis in *Spisula solidissima* oocytes by calcium ionophore. *J. Exp. Zool.* **191**, 433–440.

Shen, S. S., and Steinhardt, R. A. (1978). Direct measurement of intracellular pH during metabolic derepression of the sea urchin egg. *Nature (London)* **272**, 253–254.

Steinhardt, R. A., Epel, D., Carroll, E. J., and Yanagimachi, R. (1974). Is calcium ionophore a universal activator for unfertilized eggs? *Nature (London)* **252**, 41–43.

Tapaswi, P. K. (1972). RNA synthesis during "oogenesis to the onset of fertilization" in *Limnaea* (Mollusca). *Z. Naturforsch., Teil B: Anorg. Chem., Org. Chem., Biochem., Biophys., Biol.* **27B**, 581–582.

Tapaswi, P. K. (1974). Further investigations on transcription during oogenesis and immediately after activation by sperm in *Limnaea* (Mollusca) eggs. *Acta Embryol. Exp.* **2**, 191–195.

Taylor, G. T., and Anderson, E. (1969). Cytochemical and fine structural analysis of oogenesis in the Gastropod, *Ilyanassa obsoleta*. *J. Morphol.* **129**, 211–248.

Tilney, L. G. (1975). The role of actin in nonmuscle cell motility. *In* "Molecules and Cell Move-ment", (S. Inoué and R. E. Stephens, eds.), pp. 339–388. Raven Press, New York.

Tilney, L. G., and Jaffe, L. A. (1980). Actin, microvilli and the fertilization cone of sea urchin eggs. *J. Cell Biol.* **87**, 771–782.

Tilney, L. G., Hatano, S., Ishikawa, H., and Mooseker, M. S. (1973). The polymerization of actin: Its role in the generation of the acrosomal process of certain echinoderm sperm. *J. Cell Biol.* **59**, 109–126.

Tilney, L. G., Kiehart, P., Sardet, C., and Tilney, M. (1978). Polymerization of actin. IV. Role of Ca^{++} and H^+ in the assembly of actin and in membrane fusion in the acrosomal reaction of echinoderm sperm. *J. Cell Biol.* **77**, 536–550.

Timmermans, L. P. M., Geilenkirchen, W. L. M., and Verdonk, N. H. (1970). Local accumulation of Feulgen-positive granules in the egg cortex of *Dentalium dentale* L. *J. Embryol. Exp. Morphol.* **23**, 245–252.

Tyler, A. (1939). Extraction of an egg membrane lysin from sperm of the giant keyhole limpet (*Megathura crenulata*). *Proc. Natl. Acad. Sci. U.S.A.* **25**, 317–323.

Tyler, A., and Fox, S. W. (1939). Sperm agglutination in the keyhole limpet and the sea urchin. *Science* **90,** 516–517.

Tyler, A., and Fox, S. W. (1940). Evidence for the protein nature of the sperm agglutinins of the keyhole limpet and the sea urchin. *Biol. Bull. (Woods Hole, Mass.* **79,** 153–165.

van den Biggelaar, J. A. M. (1971). Timing of the phases of the cell cycle with tritiated thymidine and Feulgen cytophotometry during the period of synchronous division in *Lymnaea. J. Embryol. Exp. Morphol.* **26,** 351–366.

Verdonk, N. H. (1973). Gene expression in early development of *Lymnaea stagnalis. Dev. Biol.* **35,** 29–35.

von Medem, F. G. (1942). Beiträge zur Frage der Befruchtungsstoffe bei marinen Mollusken. *Biol. Zentralbl.* **62,** 431–446.

von Medem, F. G. (1945). Untersuchungen über die Ei und Spermawirkstoffe bei marine Mollusken. *Zool. Jahrb., Abt. Anat. Ontog. Tiere* **61,** 1–44.

Wada, S. K. (1953). Larviparous oysters from the tropical West Pacific. *Rec. Oceanogr. Works Jpn.* [N. S.] **1,** 66–72.

Wada, S. K., and Wada, R. (1953). On a new pearl oyster from the Pacific coast of Japan, with special reference to the cross-fertilization with another pearl oyster, *Pinctada martensii* (Dunker). *J. Oceanogr. Soc. Jpn.* **8,** 127–138.

Wada, S. K., Collier, J. R., and Dan, J. C. (1956). Studies on the acrosome. V. an egg-membrane lysin from the acrosomes of *Mytilus edulis* spermatozoa. *Exp. Cell Res.* **10,** 168–180.

Weisenberg, R. C., and Rosenfeld, A. C. (1975). In vitro polymerization of microtubules into asters and spindles in homogenates of surf clam eggs. *J. Cell Biol.* **64,** 146–158.

Whitaker, D. M. (1933). On the rate of oxygen consumption by fertilized and unfertilized eggs. *J. Gen. Physiol.* **16,** 475–528.

Wilson, E. (1925). "The Cell in Development and Heredity." Macmillan, New York.

Yamamoto, G., and Nishioka, U. (1943). Artificial fertilization of some pelecypods. *Zool. Mag.* **55,** 372–373.

Yasumasu, I., Tazawa, E., and Fujiwara, A. (1975). Glycolysis in the eggs of the *Echinroid, Urechis unicinctus* and the oyster, *Crassostrea gigas.* Rate-limiting steps and activation at fertilization. *Exp. Cell Res.* **93,** 166–174.

Zagyansky, Y. A., and Jard, S. (1979). Does lectin-receptor complex formation produce zones of restricted mobility within the membrane? *Nature (London)* **280,** 591–593.

Zimmerman, A. M., and Zimmerman, S. (1967). Action of Colcemide in sea urchin eggs. *J. Cell Biol.* **34,** 483–488.

Ziomek, C. A., and Epel, D. (1975) Polyspermy block of *Spisula* eggs is prevented by cytochalasin B. *Science* **189,** 139–141.

Zucker, R. S., Steinhardt, R. A., and Winkler, M. M. (1978). Intracellular calcium release and the mechanism of parthenogenetic activation of the sea urchin egg. *Dev. Biol.* **65,** 285–295.

3

Early Development and the Formation of the Germ Layers

N. H. VERDONK AND J. A. M. van den BIGGELAAR

Zoological Laboratory
State University of Utrecht
Utrecht, The Netherlands

I. Introduction

Except for the cephalopods, development in molluscs is characterized by a regular and constant pattern of cleavage. As a consequence, the history of an individual blastomere can be traced from its origin through the entire development up to the formation of the definitive organs. Therefore, it is not surprising that around the turn of this century, when many embryologists became interested

THE MOLLUSCA, VOL. 3
Development

in the study of molluscan development, several so-called cell-lineage studies were undertaken. Due to these painstaking investigations, cleavage and early development in molluscs are known in greater detail than in most other animals. From this period, cell-lineage studies on the following species may be mentioned: *Neritina* (Blochmann, 1882), *Umbrella* (*Umbraculum*) (Heymons, 1893), *Limax* (Kofoid, 1895; Meisenheimer, 1896), Unionidae (Lillie, 1895), *Crepidula* (Conklin, 1897), *Ischnochiton* (Heath, 1899), *Planorbis* (Holmes, 1900), *Aplysia* (Carazzi, 1900), *Dreissensia* (Meisenheimer, 1901), *Trochus* (Robert, 1902), *Fiona* (Casteel, 1904), *Physa* (Wierzejski, 1905) and *Littorina* (Delsman, 1912, 1914). Since it became evident, that for the interpretation of experimental results, the availability of a cell lineage is a great advantage, further cell-lineage studies have been undertaken: for example, Verdonk (1965) (*Lymnaea*), Camey and Verdonk (1970) (*Biomphalaria*), and van Dongen and Geilenkirchen (1974) (*Dentalium*).

In the first papers dealing with cell lineage of molluscs, various systems were used for the denomination of the blastomeres during cleavage. Since Conklin (1897) published his paper on the embryology of *Crepidula,* however, most embryologists have followed his system of nomenclature. This system enables ready distinction of the successive cleavages and tracing of the origin of the various organs. Moreover, it facilitates a comparison of the cell lineage of different species. Therefore, in all recent literature on the development of molluscs, Conklin's system of nomenclature is used.

II. Cleavage and Cell Lineage

A. The Course of Spiral Cleavage

The type of cleavage in molluscan eggs is a modification of the radial type of holoblastic cleavage, and has been called spiral cleavage by Wilson (1892). In radially cleaving eggs, the first two cleavages are vertical and the third cleavage is horizontal. The three generations of mitotic spindles are perpendicular to each other. In the eight-cell embryo, the four animal cells lie precisely above the vegetal sister cells. In eggs with a spiral cleavage, an eight-cell stage consists of four usually smaller animal micromeres and four larger vegetal macromeres. These two tiers do not lie precisely one above the other; the tier of micromeres has an interblastomeric position on top of the macromeres. During further cleavage, the successive tiers of cells come to lie in an alternating spiral pattern along the egg axis. The chirality of this spiral may be dextral (dexiotropic, clockwise) or sinistral (laeotropic, anticlockwise). If, at third cleavage, an animal micromere is displaced in the direction of the clock relative to its corresponding vegetal macromere, cleavage is called dexiotropic (Fig. 1c). If the micromere is displaced anticlockwise, cleavage is laeotropic (Fig. 1h).

The third cleavage is not the first spiral cleavage. The second cleavage also has a spiral character. The two cleavage spindles are not placed at right angles to the egg axis; their position is not exactly horizontal, nor are they placed precisely parallel to each other, but they make a minute angle (Fig. 1a,f). As a consequence, in each of the two cells, AB and CD, one end of the spindle is slightly displaced in the animal direction and the opposite end in the vegetal direction. This leads to a four-cell stage in which the two cells A and C are situated slightly higher than cells B and D. The blastomeres A and C meet each other at the animal pole in the so-called animal cross-furrow (Fig. 1b). The blastomeres B and D meet each other at the vegetal pole in the vegetal cross-furrow. Usually the blastomeres A, B, C, and D, called the quadrants of the egg, follow each other clockwise (Fig. 1b), but in some species a reverse situation exists (Fig. 1g).

According to the law of alternating cleavage, first formulated by Kofoid (1894), dexiotropic and laeotropic cleavages follow each other. Since in most species third cleavage is dexiotropic, all odd cleavages are dexiotropic and the even ones laeotropic. If, however, third cleavage is laeotropic, the reverse situation exists. This latter type of spiral cleavage is therefore called *reversed cleavage*.

At fourth cleavage, the dexiotropically formed micromeres 1a–1d and the macromeres 1A–1D divide laeotropically (Fig. 1d). The micromeres divide into an upper tier, $1a^1$–$1d^1$, and a lower tier of *primary trochoblasts*, $1a^2$–$1d^2$. The macromeres divide into the second quartet of micromeres, 2a–2d, and the macromeres, 2A–2D.

At fifth cleavage, all four tiers of cells divide dexiotropically. The macromeres 2A–2D give rise to the third quartet of micromeres 3a–3d and the macromeres 3A–3D. The second quartet cells, 2a–2d, divide into the upper tier, $2a^1$–$2d^1$, touching the first quartet of micromeres, and the lower tier, $2a^2$–$2d^2$. The primary trochoblasts, $1a^2$–$1d^2$, divide into an upper and a lower tier of trochoblasts, $1a^{21}$–$1d^{21}$ and $1a^{22}$–$1d^{22}$. The animal cells, $1a^1$–$1d^1$, divide into the four apical cells around the animal pole, $1a^{11}$–$1d^{11}$, and the slightly lower basal cells, $1a^{12}$–$1d^{12}$. By the fifth cleavage, the embryo reaches the 32-cell stage. In a number of species, however, the micromeres of the first quartet do not divide at fifth cleavage, and there the embryo passes from a 16- into a 24-cell embryo.

In many species, the first five cleavages follow each other at regular and relatively short intervals, and the divisions of the different blastomeres are rather synchronous. The interval between fifth and sixth cleavage is generally significantly longer than the preceding intermitotic stages, and is therefore often called a resting stage. Division synchrony is definitively lost at the sixth cleavage round; a few hours difference between the first and last dividing cells is normal. At sixth cleavage, a fourth quartet of micromeres, 4a–4d, is formed laeotropically. The remaining macromeres are indicated as 4A–4D. Each of the other tiers of micromeres also divides laeotropically.

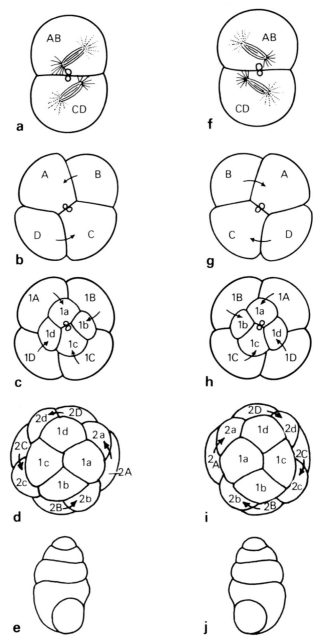

Fig. 1. Diagram comparing the early development of dextral (**a–e**) and sinistral (**f–j**) snails. Dextral snails originate from eggs with a dexiotropic third cleavage and a clockwise arrangement of the quadrants; sinistral snails from eggs with laeotropic third cleavage and an anticlockwise arrangement of the quadrants.

Although cleavage has often not been followed beyond the formation of the fourth quartet of micromeres, we may assume that the vegetal macromeres do not stop dividing, and in several species the formation of a fifth and even a sixth quartet of micromeres has actually been described [e.g., *Crepidula, Umbrella (Umbraculum), Physa,* and *Planorbis*]. As these and later divisions of the macromeres take place immediately before or after the beginning of gastrulation, they will be discussed relative to this period of development.

B. Cleavage with Polar-Lobe Formation

In many molluscan species, first cleavage is accompanied by formation of a protrusion at the vegetal pole of the egg. During division, this protrusion is gradually constricted until it is connected only by a narrow stalk to the remainder of the egg, which meanwhile divides into two equal blastomeres. In this way, a so-called trefoil stage is formed consisting of two nucleated blastomeres and an anucleated polar lobe, connected with either of the two blastomeres. In some species (e.g., *Dentalium*), this connection is so thin that it may easily be overlooked. Subsequently, the polar lobe fuses with the blastomere to which it is still attached. In this way, one obtains a two-cell stage consisting of two blastomeres of unequal size; the bigger one, fused with the polar lobe, is denominated CD; the other, deprived of lobe material, is denominated AB (Fig. 2). At second cleavage, the CD blastomere forms a polar lobe again, which flows into the daughter cells situated in a clockwise direction when viewed from the animal

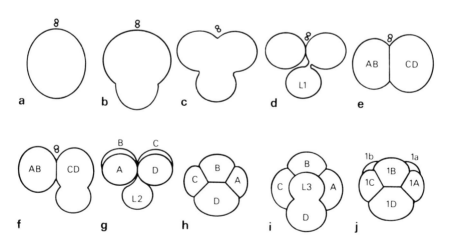

Fig. 2. Diagrammatic representation of polar lobe formation. (**a–c**) Formation of the polar lobe at first cleavage, (**d**) trefoil stage with polar lobe (L1), (**e**) two-cell stage, (**f** and **g**) formation of a polar lobe (L2) at second cleavage, (**h**) four-cell stage seen from the vegetal pole, (**i**) formation of a polar lobe (L3) at third cleavage, (**j**) eight-cell stage seen from the vegetal pole.

pole. This cell is called the D blastomere. Finally, a four-cell stage is reached with three equally sized blastomeres A, B, and C and a bigger D blastomere, following each other in a clockwise direction when viewed from the animal pole. This means that second cleavage is laeotropic as in most equally dividing species. We do not know of any of polar-lobe forming species with an anticlockwise succession of the quadrants.

In some species (e.g., *Dentalium*), a polar lobe is formed again at third cleavage (Fig. 2). In *Littorina*, even during the first five cleavages a polar lobe appears (Moor, 1973).

The polar lobe appears to be of crucial importance for normal development. After removal of the lobe at first cleavage a radially symmetrical embryo is formed, which fails to develop adult structures (Wilson, 1904; Clement, 1952; van Dongen, 1976a,b). For an extensive discussion of the influence of the polar lobe on development the reader is referred to Chapter 6 of this volume.

It is generally assumed that the occurrence of a polar lobe is independent of the systematic position of a species (e.g., Hess, 1962; Fioroni, 1979b). However, some generalizations can be made. Polar lobes have never been observed in eggs of Polyplacophora. Within the Aplacophora, at least one species with a polar lobe has been described (*Epimenia*, Baba, 1951). Of the Scaphopoda, only species of the genus *Dentalium* have been studied so far, all of which have a polar lobe. Among the Bivalvia, cleavage of the Protobranchia has never been studied in detail. From the data of Drew (1901), one might infer that cleavage in *Nucula delphinodonta* is unequal, probably the result of polar-lobe formation. Many Lamellibranchia have a polar lobe, e.g., *Musculus, Mytilus, Crassostrea, Ostrea,* and *Ensis,* but several other species show unequal cleavage without the formation of a lobe: *Cumingia, Spisula, Teredo, Dreissensia, Unio,* and *Anodonta.* No data are available for the Septibranchia.

The situation in the Gastropoda is rather well known, since many species have been studied. The archaeogastropods have not developed polar lobes; cleavage is equal. Most of the mesogastropods have small polar lobes, whereas the neogastropods have relatively large lobes. In the opisthobranchs and the pulmonates, cleavage with polar-lobe formation has never been described. Cleavage is generally equal, except for some opisthobranchs (e.g., *Aplysia*) which have unequal cleavage.

The size of the polar lobe may vary significantly. It may be rather large (Fig. 3a), up to 20–50% of the egg volume, for example, in *Dentalium, Ilyanassa*[1],

---

[1]*Nassarius obsoletus* was removed from the genus *Nassarius* (Dumeril, 1806) by Stimpson and placed into the new genus *Ilyanassa* (Stimpson, 1865) on the basis that *Ilyanassa obsoleta* has a different operculum and lacks the caudal pedal cirri or metapodial tentacles characteristic of the genus *Nassarius*. Although embryologists tend to use the correct term *Ilyanassa obsoleta* (Say), scientists in other fields often still use the now outmoded designation *Nassarius obsoletus.*

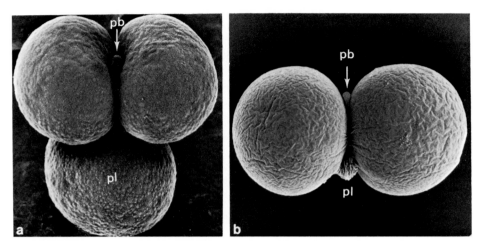

Fig. 3. Scanning electron micrographs of eggs with a polar lobe. (a) *Nassarius reticulatus* with a big polar lobe, (b) *Crepidula fornicata* with a very small polar lobe. pb, polar bodies; pl, polar lobe. (Courtesy of Dr. M. R. Dohmen.)

Nucella or very small (Fig. 3b), less than 1% of the egg volume, as in *Bithynia*, *Crepidula*, and *Littorina*. No correlation seems to exist between the yolk content of the egg and the size of the polar lobe. The yolk-rich egg of *Fulgur* (*Busycon*) has a diameter of 1700 μm and a relatively small polar lobe; the egg of *Crassostrea* has a diameter of 50 μm and a relatively large lobe. In related species, however, the lobe is generally of the same relative size.

The relationship between polar-lobe formation and cleavage has been studied in eggs with large polar lobes. In normal development, polar lobes appear in synchrony with the first cleavages. Starting with a protrusion of the vegetal pole, the constriction of the polar lobe and the incision of the cleavage furrow occur concomitantly, the rate of progress of the cleavage furrow being similar to the rate of constriction of the polar lobe (Conrad, 1973). Cleavage and polar-lobe formation are not obligatorily connected, however, because after isolation of vegetal fragments of uncleaved, *fertilized* eggs of *Dentalium*, those fragments without a nucleus form polar lobes in synchrony with the cleavage of the animal part, which does not form a polar lobe (Wilson, 1904; Verdonk et al., 1971). Similar observations were made by Morgan (1935b, 1936) and Clement (1935) in vegetal halves of *Ilyanassa*, isolated by centrifugation. Polar lobes isolated at the trefoil stage undergo a series of changes more or less synchronous with the cleavage of the egg, periods of amoeboid activity alternating with spherical resting stages. In *Dentalium*, the isolated lobe may form a smaller lobe by a process that closely simulates the formation of a polar lobe by a whole egg. Altogether, three periods of activity were observed by Wilson (1904), although in normal development the lobe is formed only twice. Similar observations were made by Morgan (1933, 1935a,b) on isolated lobes of *Ilyanassa*. Here, three periods of surface activity were observed in lobes isolated at the trefoil stage, whereas in normal development a lobe appears only once more. These periods are not strictly synchronous with the cleavage, but occur with some delay. The above data are best understood on the assumption that some kind of clock mechanism is present in the eggs forming a polar lobe which regulates both cytokinesis and polar-lobe formation, so that in normal development both processes occur synchronously. Deletion experiments indicate that this mechanism is not situated in the nucleus; from centrifugation experiments, it can be concluded that it is bound to the cortex of the egg.

The phenomenon of polar-lobe formation points to a definite organization of the uncleaved egg. A polar lobe is always formed at the vegetal pole of the egg, even under experimental conditions. After centrifugation of eggs of *Ilyanassa* (Morgan, 1933, 1935a; Clement, 1968) or *Dentalium* (Verdonk, 1968), first cleavage is often abnormal and may be oblique or equatorial. Nevertheless, the polar lobe always appears at the vegetal pole, opposite the polar bodies, independent of the direction of cleavage. After removal of a part of the vegetal region from an uncleaved egg of *Dentalium,* before or after fertilization, this egg will not form a polar lobe at first cleavage if the part removed is about equal in size to a polar lobe. If a smaller part of this region is removed, a reduced lobe appears, the reduction being approximately proportional to the removed part (Verdonk et al., 1971). These data indicate that factors responsible for polar-lobe formation are present only in the vegetal polar area of the egg.

Nevertheless, the polar area can also be subject to regulation. If an unfertilized *Dentalium* egg is cut equatorially into an animal and a vegetal half, both halves can be fertilized. The animal half cleaves without forming a polar lobe, whereas the vegetal half forms a polar lobe, the size of which is proportional to the blastomeres (Wilson, 1904; Verdonk et al., 1971). This may be due to the position of the mitotic apparatus. Normally situated in the animal half of the dividing egg, it lies very near to or even in the polar area of a dividing vegetal fragment and may influence the process of cleavage and polar-lobe formation. When a nucleus is not present, as in isolated vegetal fragments of fertilized eggs, no regulation takes place and a polar lobe of normal size is formed at the moment when the corresponding animal part starts cleaving (Verdonk et al., 1971).

In several species, the vegetal region of the egg is characterized by the occurrence of special cytoplasmic structures, which form part of the polar lobe at cleavage. The cytoplasmic structures, described in detail by Dohmen (Chapter 1, this volume) appear to consist of aggregates of electron-dense vesicles. In *Crepidula, Buccinum,* and *Littorina* several small aggregates are found, which are segregated into the polar lobe at first cleavage. In *Bithynia,* one large, cup-shaped aggregate of vesicles is present, the so-called vegetal body (Dohmen and Verdonk, 1974).

A causal relation between polar-lobe formation and the presence of special cytoplasmic structures in the vegetal region of the egg is not likely. By centrifugation of eggs of *Bithynia* prior to first cleavage, the vegetal body can be removed from the vegetal region of the egg (Dohmen and Verdonk, 1979). During first cleavage of centrifuged eggs, a polar lobe was always formed at the normal place, but in about 30% of these eggs the vegetal body was not incorporated in the polar lobe, but in the cytoplasm of one of the two blastomeres.

Similarly, in large polar lobes the mechanism for polar-lobe formation seems to be independent of its cytoplasmic composition, because, as stated before, polar lobes appear at the correct place after centrifugation of eggs prior to first cleavage. Dohmen and Verdonk (1979) did not find any nondisplaceable component in an EM study of centrifuged eggs of *Dentalium* and *Nassarius*. The most obvious conclusion is that the mechanism for polar-lobe formation resides in the cortex or the plasma membrane of the egg.

In the cortical cytoplasm of the *Ilyanassa* egg, a ring of microfilaments is apposed to the plasma membrane at the base of the polar-lobe constriction (Conrad, 1973; Conrad et al., 1973; Conrad and Williams, 1974a). After treatment with cytochalasin B, a drug that interferes with activities in which microfilaments are involved, this ring of microfilaments disappears and a polar lobe is not formed. Colchicine, a drug that depolymerizes microtubules, not only inhibits polar-lobe development but also cytokinesis, if it is applied at the beginning of polar-lobe formation. At later stages, when the lobe is already constricting, colchicine fails to inhibit further development of the polar lobe and its resorption, although it still inhibits cytokinesis (Raff, 1972; Conrad and Williams, 1974a). These data suggest that microtubules might be involved in the initiation of polar-lobe formation. The constriction of the lobe is achieved by a subequatorial band of microfilaments arranged in the cortical cytoplasm of the vegetal hemisphere.

Microfilaments are not restricted, however, to the constriction area of the polar lobe, but appear to be present everywhere in the cortex of spherical eggs (Schmidt et al., 1980). Polar-lobe–like protrusions can be induced by exposing fertilized eggs to isotonic calcium chloride after second maturation division. Then, just as during normal polar-lobe formation, a ring of microfilaments is present in the constriction area of these calcium-induced blebs. Like polar-lobe formation, bleb formation is suppressed by cytochalasin B. This suppression appears to be accompanied by the disappearance of the ring of microfilaments. Colchicine has no effect on the formation of blebs (Conrad and Williams, 1974b). In these experiments, formation of blebs resembling polar lobes is not restricted to the vegetal region of the egg. If calcium is applied at the two-cell stage, both blastomeres, AB and CD, form lobelike structures, whereas in normal development only CD develops a polar lobe. Similarly, after iontophoretic injection of calcium or cyclic AMP, a protuberance resembling a polar lobe can be induced anywhere on the egg surface, but only if calcium is present in the medium (Conrad and Davis, 1977). In normal development, polar lobes are formed also in calcium- and magnesium-free seawater. The same holds true for the induction of large blebs by ionophores, which also occurs in calcium- and magnesium-free artificial seawater (Conrad and Davis, 1980).

These data suggest that cortical microfilaments play a role in polar-lobe formation and also that calcium—probably released from intracellular sources—may be needed for the contraction of these microfilaments, which contain actin or actin-like proteins (Schmidt et al., 1980). They do not account, however, for the formation of the polar lobe at a specific place in the vegetal hemisphere during normal cleavage.

Microtubules may exert a regulatory function in determining the place where the polar lobe will be formed. During cleavage, microtubules extend down from the animal hemisphere into the polar lobe. It is not known whether these microtubules are directly connected to the asters in the animal hemisphere or not (Conrad and Williams, 1974a). If this were the case, then the position of the asters might influence the position of the polar-lobe constriction, which in isolated nucleated vegetal halves is displaced towards the vegetal pole, so that the polar lobe is of the appropriate relative size (see p. 98).

C. Unequal Cleavage

In many bivalves, the plane of first cleavage does not coincide with, but parallels the egg axis. As a consequence, the egg divides into a smaller blastomere AB and a larger blastomere CD. Second cleavage is unequal in CD only, so that at the four-cell stage D is the largest quadrant of the egg; A, B, and C are the same size. In the opisthobranchs *Cavolinia* and *Cymbulia* (Fol, 1875), *Aplysia* (Blochmann, 1883) and *Umbrella (Umbraculum)* (Heymons, 1893), CD is smaller than AB, and D is smaller than the other quadrants, except in the two latter species where C and D are the same size.

In the lamellibranchs with unequal cleavage, the first cleavage spindle is formed in the center of the egg at right angles to the egg axis. Subsequently, the spindle moves toward one side of the egg until one of the asters comes in contact with the egg cortex. As a result, the cleavage plane, formed at right angles to the spindle, divides the egg into two unequal blastomeres (e.g., *Unio*, Lillie, 1901; *Barnea*, Pasteels, 1931). Similar phenomena were observed by Guerrier (1970) in eggs of *Pholas dactylus* and *Spisula subtruncata*. In these eggs, the spindle not only develops in the center of the egg but also at right angles to the plane through the entrance point of the sperm and the egg axis. It then moves either to the right or the left of this plane. As a result the smaller blastomere is situated either to the

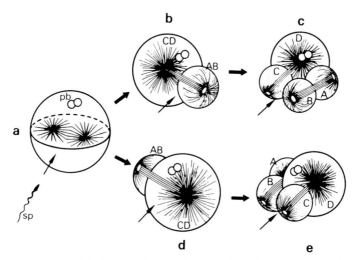

Fig. 4. Orientation of the first two cleavage planes in *Pholas dactylus*. The polar bodies (pb) indicate the position of the animal pole. At first cleavage the spindle moves either to the right (upper row) or to the left (lower row) of the entrance point of the sperm (sp) indicated by a thin arrow. (After Guerrier, 1970.)

right or to the left of the sperm entrance point. At second cleavage, the spindle in CD moves toward the entrance point of the sperm. In this way, the large blastomere D is always situated opposite the entrance point of the sperm (Fig. 4). A similar situation occurs in *Cumingia* (Morgan and Tyler, 1930, 1938). Consequently, in these lamellibranchs second cleavage is either dexiotropic or laeotropic. Whether a relationship between the sperm entrance point and the two first cleavages exists in *Dreissensia* or *Unio* is uncertain. According to Meisenheimer (1901), second cleavage is dexiotropic in *Dreissensia*, whereas Lillie (1895) describes this cleavage as laeotropic in *Unio*. In all species, third cleavage is always dexiotropic and the subsequent cleavages follow the rule of alternating spiral cleavage. When second cleavage is laeotropic the quadrants are arranged in a clockwise direction; when dexiotropic, in an anticlockwise direction (Fig. 4). This arrangement of the quadrants seems to be without consequence for the structure of the embryo, since in *Pholas* and *Spisula*, where both cleavage types occur in equal numbers, all embryos are identical.

D. Reversed Cleavage

The reversed arrangement of the quadrants is found not only in lamellibranchs but also in gastropods. Unlike the lamellibranchs, where third cleavage is always dexiotropic, the gastropods follow the rule of alternating spiral cleavage. Consequently, when second cleavage is dexiotropic instead of laeotropic, the cleavage pattern is the mirror image of the ordinary pattern (Fig. 1). Crampton (1894) observed that a strict relationship exists between the cleavage pattern and the coiling of the shell. Species in which the shell is coiled according to a left-hand spiral have the reversed type of cleavage (e.g., *Ancylus, Biomphalaria, Planorbis, Physa*). Also in dextral species, in which occasionally sinistral races or individuals occur (e.g., *Lymnaea peregra*) the latter originate from eggs with the reversed cleavage type (Ubbels, 1966; Ubbels et al., 1969).

On the basis of data of Boycott and Diver (1923), obtained with self-fertilizing individuals of *L. peregra,* Sturtevant (1923) has theorized that the direction of coiling is determined by one pair of alleles, the gene for dextral coiling being dominant. The direction of coiling of an individual, however, is not dependent on its own genotype but on the genotype of the mother. A difficult problem is the recurrent appearance of dextral snails in sinistral lines (Boycott et al., 1931), which is explained by the action of modifying genes, which interfere with the expression of the recessive gene. No less than six to seven modifiers are supposed to be involved (Diver and Andersson-Kottö, 1938). Freeman and Lundelius (1982) have tried to explain the appearance of dextral individuals in sinistral lines by crossover, which either reassembles two previously dissociated parts of the dextral gene or creates a dextral gene by means of a position effect. Genetic studies on the inheritance of coiling have also been undertaken in the snails *Laciniaria biplicata* (Degner, 1952) and *Partula suturalis* (Murray and Clarke, 1966). In these species, the inheritance of coiling is in accordance with Sturtevant's hypothesis but sinistrality appears to be dominant.

E. Morphogenetic Significance of the Blastomeres

Cell-lineage studies have shown that early development in all spiralian molluscs follows a general scheme whereby a fixed relationship is established between the pattern of the blastomeres and the organization of the embryo.

The first three quartets of micromeres produce the whole ectoderm; the fourth quartet and the macromeres, the endoderm. The pretrochal ectoderm, which will later form the head region with cephalic eyes, tentacles, and cerebral ganglia, originates from the first quartet. Its most apical descendants surrounding the original animal pole may form an apical tuft, which marks the front end of the embryo. The second and third quartet produce the posttrochal ectoderm. An important structure of this latter ectoderm is the *somatic plate,* from which the shell, foot, and mantle cavity develop. In many species, the somatic plate derives from the first somatoblast, 2d. The mesoderm is derived from two different sources. The ectomesoderm originates from cells of the second or third quartet of micromeres. The endomesoderm or primary mesoderm is derived from one cell, the second somatoblast, 4d. Since the somatoblasts (2d and 4d) are situated in the dorsal midline, it follows that in broad outline the D quadrant corresponds to the dorsal side of the future embryo and the B quadrant to the ventral side, and A and C are the lateral quadrants. When second cleavage is laeotropic, the quadrants follow each other clockwise and consequently the A quadrant is situated to the left, the C quadrant to the right of the median plane. After reversed second cleavage the position of the lateral quadrants is reversed too; A to the right and C to the left (see Fig. 1).

The described relationship between the quadrants and the axis of the embryo are true only in broad outline. As Raven (1966) has pointed out, the somatoblasts 2d and 4d are formed in normal (i.e., not reversed) cleavage by laeotropic division from the D quadrant. Consequently, the descendants of the second and fourth quartet of micromeres, also formed by a laeotropic division, are situated along the radii (A, left; B, ventral; C, right; and D, dorsal). The third quartet and the macromeres lie interradially (A, ventral left; B, ventral right; C, dorsal right; D, dorsal left). In the first quartet of micromeres the situation is more complicated. It will be discussed in relation to the origin and development of the molluscan cross (see Section V).

III. The Blastula

A. Cleavage Cavity and Blastocoel

In freshwater molluscs and land pulmonates, a cleavage cavity appears between the blastomeres during the intervals between the successive cleavages. This phenomenon was first described in detail by Kofoid (1895) in *Limax*. Soon after the completion of first cleavage, when the blastomeres flatten against each other, small lenticular clefts appear between the apposed walls of the blastomeres. These spaces coalesce into a large intercellular cavity, filled with a clear fluid. This cavity grows to such an extent that the blastomeres are connected only at the peripheral margin. The maximum development of the cavity is followed by a sudden expulsion of its contents either at the animal or the vegetal pole, after which the egg resumes its original size. The whole process may repeat itself once or twice before second cleavage. During the four-cell stage, the cleavage cavity reappears in the center between the blastomeres and is emptied once or twice. This regular succession of formation and extrusion of the cleavage cavity continues until gastrulation. From the eight-cell stage, however, the cleavage cavity is no longer situated in the center of the egg but in the animal half between the macromeres and micromeres. During the 24-cell stage of *Lymnaea* and *Physa,* the cleavage cavity disappears almost completely, some lenticular clefts between the micromeres excepted. For some time, an intimate contact is formed between the macromeres (3D) and the micromeres of the first quartet (van den Biggelaar, 1976). As will be described in Chapter 5, this contact is of great importance for establishing the bilateral symmetry and dorsoventrality in the embryo.

The recurrent cleavage cavity has been described for several land and freshwater snails, for example, *Limax* (Kofoid, 1895; Meisenheimer, 1896), *Physa* (Wierzejski, 1905), the prosobranch *Bithynia tentaculata* and various species of pulmonates (Comandon and de Fonbrune, 1935), *Agriolimax (Deroceras)* (Carrick, 1939), and *Lymnaea stagnalis* (Raven, 1946). It also occurs in freshwater lamellibranchs such as *Dreissensia* (Meisenheimer, 1901) and *Anodonta* (Herbers, 1913).

According to Raven (1966), this recurrent cleavage cavity in freshwater and land molluscs serves as a mechanism for osmotic regulation. As these eggs are surrounded by an hypotonic medium, they need a water-excreting mechanism even during the earliest stages of development.

In eggs of marine molluscs, which are in osmotic equilibrium with the surrounding seawater, a cleavage cavity is absent during the earliest stages of development. At later stages, a more or less distinct cavity may be present, situated in the animal half, although it is generally smaller than in freshwater molluscs, for example, *Trochus* (Robert, 1902), *Patella* (van den Biggelaar, 1977), and *Dentalium* (van Dongen, 1977). The transition of the cleavage cavity into the blastocoel is not accompanied by real changes, it is only a matter of denomination.

In eggs, surrounded by capsule fluid, the uptake of albumen may start during the cleavage stages; in *Limax*, as early as the 16-cell stage (Meisenheimer, 1896). In *Lymnaea*, the ingestion of capsule fluid, starting also at about the 16-cell stage (Arnolds, unpublished), takes place by pinocytosis. A ring-shaped elevation of the cell membrane appears, surrounding a central depression. Subsequently, the ring wall becomes higher and converges from all sides, fusing in the middle (Fig. 5). In this way, a vacuole with capsule fluid is formed, which detaches from the surface and migrates into the interior (Elbers and Bluemink, 1960). Thorotrast or ferritin particles added to egg-capsule fluid appear to be taken up by the cells of a blastula and are found in the albumen vacuoles (Elbers and Bluemink, 1960; Bluemink, 1967).

B. Types of Blastulas

In embryos in which a wide cleavage cavity is present, the blastula takes the form of a *coeloblastula*. As the vegetal macromeres are larger than the animal micromeres, the blastocoel is situated excentrically in the animal half of the embryo. This type of blastula is found in the Polyplacophora, Scaphopoda, in all freshwater and most marine Lamellibranchiata, and in the Archaeogastropoda: *Haliotis, Patella,* and *Trochus*. In most other gastropods, the blastula has the character of a placula. Shortly before gastrulation, the animal and vegetal side of the blastula flatten. As the animal and vegetal cells approach each other, the blastocoel is reduced to a narrow slit: *Paludina (Viviparus), Littorina, Bithynia, Lymnaea, Planorbis, Physa,* and *Limax*.

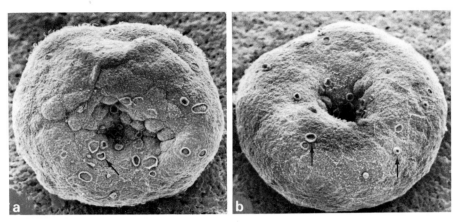

Fig. 5. Scanning electron micrographs of two gastrula stages of *Lymnaea stagnalis*. Note the pinocytotic activity at the vegetal side. Ring-shaped elevations (arrows) of the cell membrane, enclosing small amounts of capsule fluid, converge and fuse in the center. (**a**) Young gastrula. A wide depression is formed at the vegetal side. (**b**) Older gastrula with a reduced blastopore. (Courtesy of Dr. M. R. Dohmen.)

In molluscs with very yolk-rich eggs, the yolk is concentrated during cleavage in the vegetal macromeres. The very small micromeres are situated as a cap on top of the large macromeres. This special type of blastula, called *sterroblastula,* is found in some marine lamellibranchs, such as *Ensis, Spisula, Ostrea (Crassostrea), Mytilus, Pecten, Teredo,* and *Venus;* and in the gastropods *Neritina, Crepidula, Ilyanassa, Fusus, Urosalpinx, Purpura,* and *Fulgur (Busycon).*

IV. Gastrulation

A. Formation of the Archenteron

In molluscs, gastrulation starts at an early moment in development, when the embryo consists of relatively few cells, for example, about 70 cells in *Paludina,* 120 cells in *Lymnaea,* 150 cells in *Littorina,* and 200 cells in *Physa.*

The course of gastrulation is dependent on the type of blastula formed during cleavage. In species with a coeloblastula, the archenteron is formed by *invagination.* The macromeres and their descendants, situated in the center of the vegetal half, change in shape. They reduce their external surface whereas the inner part widens so that they become wedge shaped. In this way, a pit is formed at the vegetal pole of the embryo, which subseqoently deepens into the archenteron (Fig. 5). In *Lymnaea,* pseudopodia are formed at the top of the endomeres which make connections with the overlying animal cells. These pseudopodia may assist the inward movement of the archenteron (Raven, 1946). As the macromeres are the first cells to invaginate, they form the top of the archenteron. They are followed by endomeres, separated from the macromeres after the formation of the fourth quartet (4a–4c), which are the last cells to invaginate.

The mesentoblast 4d withdraws from the surface either before or during gastrulation and sinks individually into the blastocoel. In *Dentalium*, however, the mesentoblasts $4d^1$ and $4d^2$ are invaginated together with the endomeres and remain in the lining of the archenteron until they divide into enteroblasts and mesoblasts. The latter become situated outside the primitive gut (van Dongen, 1977).

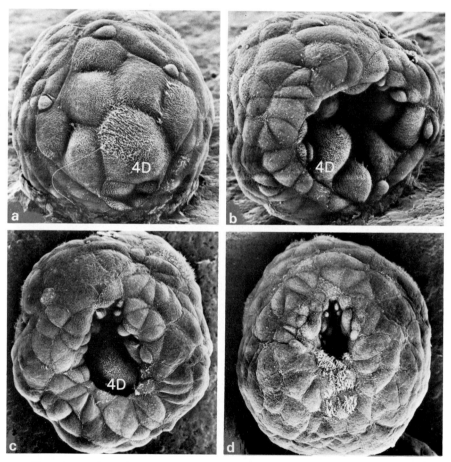

Fig. 6. Gastrulation by epiboly in *Crepidula fornicata*. Note the presence of surface structures especially on the macromere 4D. (**a**) Young gastrula at the beginning of epiboly, (**b**) gastrula with a wide circular blastopore, (**c**) advanced gastrula with a more slitlike blastopore, (**d**) old gastrula, almost completely overgrown by ectoderm. (Courtesy of Dr. M. R. Dohmen.)

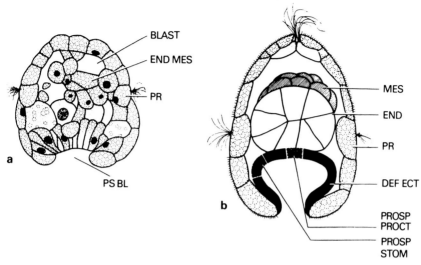

Fig. 7. Gastrulation and the formation of a larval test in *Neomenia* (Aplacophora). (**a**) Section through an embryo at the close of gastrulation. The blastocoelic cavity is almost completely filled by immigrated presumptive endo- and mesoderm cells. (**b**) Diagrammatic section through an embryo at the hatching stage. The pseudoarchenteron is surrounded by the definitive ectoderm, shown black. The endoderm is interposed between ectoderm and mesoderm. The whole complex of germ layers is surrounded by the test, partly composed of cells of the prototroch. BLAST, blastocoelic cavity; DEF ECT, definitive ectoderm; END, endoderm; END MES, endomesoblastic cells; MES, mesoderm; PR, prototrochal tier of test cells; PROSP PROCT, prospective proctodaeum; PROSP STOM, prospective stomodaeum; PS BL, pseudoblastopore. (After Thompson, 1960.)

In those forms which have a sterroblastula, gastrulation is by *epiboly*. The micromeres, representing the ectoblast, grow over the large endomeres and the margin of the blastopore is formed by derivatives of the second and third quartet. Several examples include *Crepidula* (Fig. 6), *Neritina, Ilyanassa,* and *Fulgur (Busycon)*. In some cases, the epibolic movement is followed by an invagination of the endomeres, as seen in the lamellibranchs *Ensis, Mytilus, Ostrea (Crassostrea),* and *Teredo* and in the gastropods *Umbrella (Umbraculum), Clione, Cymbulia,* and *Rhodope.*

In species where the blastula has the shape of a placula, gastrulation starts by incurving of the whole vegetal field, followed by invagination (e.g., *Paludina (Viviparus), Pomatias*) or by epiboly (e.g., *Littorina, Bithynia*).

A very special form of gastrulation has been described by Thompson (1960) for the solenogaster *Neomenia*. At first, a shallow pit is formed at the vegetal side of the embryo. From this pseudoblastopore, prospective endo- and mesoderm cells immigrate into the interior of the embryo, which is covered by cells forming the larval test (Fig. 7). The cells lining the pseudoblastopore after completion of this immigration later form the adult ectoderm.

B. The Blastopore

At the beginning of gastrulation, the blastopore is very wide, its circular margin marking the border between ectoderm and endoderm. When gastrulation proceeds, the blastopore narrows and assumes the form of a slit by the approaching lateral lips (Fig. 6). It then closes from back to front until only a small opening remains, the margin of which is lined by the stomatoblasts. In some species, such as *Ischnochiton, Dentalium, Paludina (Viviparus), Littorina, Aplysia,* and *Limax,* the blastopore remains open. In other species it closes completely so that the archenteron loses its connection with the outside. This occurs in *Anodonta, Sphaerium, Dreissensia, Neritina, Ocinebra (Purpura), Fulgur,* and *Lymnaea,* among others.

At its first appearance the blastopore is situated at the vegetal pole. During gastrulation it starts moving anteriorly over the ventral side of the embryo. In this displacement the closure of the blastopore from back to front is only of minor importance. It is mainly due to the outgrowth of the dorsal region, where the somatic plate is situated, which forms the shell gland and the anlage of the foot.

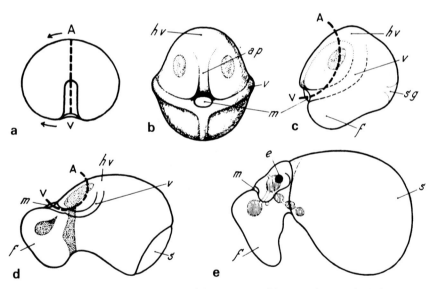

Fig. 8. Diagrammatic representation of the curvature of the animal–vegetal axis by expansion of the dorsal side in *Lymnaea stagnalis.* (**a**) Gastrula with the blastoporus directly opposite the animal pole. Arrows indicate later displacement of the animal pole and the blastopore. (**b**) Trochophore, ventral view. (**c**) The same stage, seen from the left side. (**d**) Early veliger. (**e**) Old veliger. The broken line indicates the animal–vegetal axis (A - - V). ap, apical plate; e, eye; f, foot; hv, head vesicle; m, mouth; s, shell; sg, shell gland; v, velum. (After Raven, 1949.)

As a result, not only the blastopore is ventrally displaced but also the pretrochal region, the future head region, is shifted forward. As a result of this extension of the dorsal region, the main axis of the embryo (running through the animal pole and the middle of the blastopore) becomes bent, and the blastopore becomes situated in front of the prototroch or velum (Fig. 8).

At this point, the stomodaeum is formed by an invagination of the *stomatoblasts*, which in *Ischnochiton* originate from $2a^{222}$–$2d^{222}$ and $3a^2$–$3d^2$ (Heath, 1899), and in *Dentalium* from $2b^{22}$ (van Dongen and Geilenkirchen, 1974). As these cells either form the lining of the still open blastopore or are situated at the point where the final closure of the blastopore took place, it is generally assumed that the blastopore becomes the mouth of the molluscan embryo.

There is one exception to this rule. In the prosobranch *Paludina (Viviparus)* the blastopore becomes the anus, as was first established by Blochmann (1883) and later confirmed by von Erlanger (1891), Tönniges (1896), Otto and Tönniges (1906), Dautert (1929), and Fernando (1931). The stomodaeum originates by an invagination of the ectoderm at the ventral side just below the velum. According to Fioroni (1979a), at the beginning of gastrulation, the position of the stomodaeum in all molluscs corresponds to the anterior margin; the anus corresponds to the posterior margin of the blastopore. The blastopore closes from back to front and develops into the mouth except in *Paludina (Viviparus)*. The anus originates later in development at a point indicated in some species (e.g., *Patella, Crepidula, Littorina, Physa,* and *Lymnaea*) by the anal cells. In the opistobranchs *Aplysia* (Blochmann, 1883) and *Umbrella (Umbraculum)* (Heymons, 1893), the anal cells can be distinguished at the early gastrula stage, when they are situated indeed at the posterior margin of the blastopore.

V. The Germ Layers and Their Derivatives

A. Ectoderm

In molluscs, the ectoderm is divided into a pretrochal and a posttrochal region by a band of ciliated cells, the so-called *prototroch* cells. The *pretrochal region,* the future head region, originates from the first quartet of micromeres (1a–1d), formed at third cleavage. In all molluscs, except the lamellibranchs, this first quartet forms a typical structure, known as the *molluscan cross*. At fourth cleavage, the first quartet divides laeotropically into the apical cells $1a^1$–$1d^1$ and the primary *trochoblasts* or *turret cells* $1a^2$–$1d^2$. At the next cleavage of the cells $1a^1$–$1d^1$, the apical daughter cells, $1a^{11}$–$1d^{11}$, become situated interradially, whereas the outer daughter cells, $1a^{12}$–$1d^{12}$, take a radial position. The latter join the upper cells of the second quartet $2a^{11}$–$2d^{11}$, which become the *tip cells*.

The first indication of the cross is now visible (see Fig. 9a). In the center and surrounding the animal pole are the four apical cells, $1a^{11}-1d^{11}$. The arms of the cross consist of four basal cells, $1a^{12}-1d^{12}$, and the four tip cells, $2a^{11}-2d^{11}$. The arms are situated along the radii; the b arm (ventral) and the d arm (dorsal) in the median plane; the a arm (left) and the c arm (right) in the transverse plane. They are kept apart by the trochoblasts $1a^2-1d^2$, which may divide once or twice. By subsequent divisions of the cells $1a^{12}-1d^{12}$, the arms elongate in a radial direction (Fig. 9a,b,c). In some species [e.g., *Patella, Trochus, Fulgur (Busycon), Umbrella (Umbraculum),* and *Rhodope*], the arms of the cross may soon split into several rows of cells and the form of the cross is obliterated. In other species the splitting of the arms is delayed and the cross figure is maintained much longer (e.g., *Crepidula, Littorina, Lymnaea, Planorbis, Biomphalaria,* and *Physa*). In the center of the cross, the cells $1a^{11}-1d^{11}$ divide once more forming the *apical rosette cells,* $1a^{111}-1d^{111}$, surrounding the animal pole and the *peripheral rosette cells,* $1a^{112}-1d^{112}$, occupying the reentering angles between the arms of the cross (Fig. 9b,c).

Originally the cross is radially symmetrical. The first sign of bilateral symmetry always becomes apparent in the dorsal arm of the cross, whereas the ventral arm later deviates from the lateral arms, which develop symmetrically. Generally, the basal cell of the dorsal arm, $1d^{121}$, divides later and in some species, such as *Lymnaea* (Verdonk, 1965), in a transverse direction so that the daughter cells, $1d^{1211}$ and $1d^{1212}$, become situated side by side at the base of the arm, whereas in the other arms the corresponding cells become situated one after the other in the direction of the arm (Fig. 9c). In gastropods, the division of cells in the dorsal arm stops soon after its formation and the cells form the central part of the *head vesicle*. The divisions continue in a very limited number of cells of the cross; the basal cells ($1a^{1211}$ and $1c^{1211}$), the inner median cells ($1a^{1212}$ and $1c^{1212}$) of the lateral arms, the dorsal peripheral rosette cells ($1c^{1112}$ and $1d^{1112}$) and the inner median cell of the ventral arm ($1b^{1212}$) (Fig. 9c,d). From these cells, the *cephalic plates* are formed, which give rise to eyes, tentacles, and cerebral ganglia. The cephalic plates are separated by the apical plate, consisting of six or seven ciliated cells in the pulmonates *Lymnaea, Biomphalaria, Planorbis,* and *Physa*. These cells originate from the apical rosette cells ($1a^{111}-1d^{111}$); the ventral peripheral rosette cells ($1a^{112}-1b^{112}$) and the basal cell of the ventral arm ($1b^{1211}$). In Polyplacophora, Aplacophora, marine Bivalvia, *Dreissensia,* Scaphopoda, and the gastropods *Patella* and *Acmaea,* an *apical tuft* is present situated on a pronounced protrusion, the *apical organ*. This structure originates from the apical rosette cells surrounding the original animal pole. In *Dentalium,* only a part of these cells, namely $1c^{1111}$ and $1d^{1111}$ bear the apical tuft (van Dongen and Geilenkirchen, 1974).

The prototroch consists of one or more rows of ciliated cells surrounding the embryo. In its formation, the *primary trochoblasts* play an important role. They

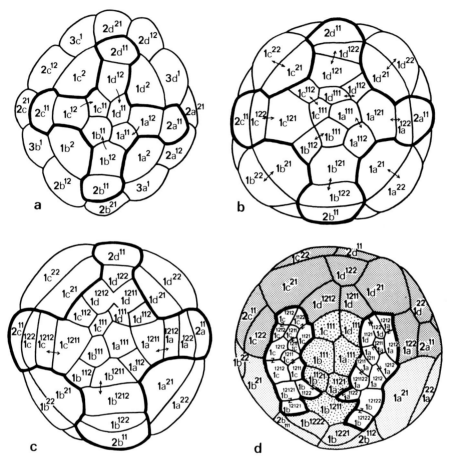

Fig. 9. Four stages in the development of the molluscan cross in *Lymnaea stagnalis*. (a) Radially symmetrical configuration of the cross (heavily outlined). (b) Elongation of the four symmetrical arms of the cross in radial direction by the division of the cells $1a^{12}$–$1d^{12}$. The primary trochoblasts $1a^2$–$1d^2$ have divided. They keep the four arms of the cross apart. (c) Bilaterally symmetrical configuration of the cross. The cells $1a^{121}$, $1b^{121}$ and $1c^{121}$ divide in radial direction whereas the cell $1d^{121}$ divides in a direction at right angles to the dorsal arm of the cross. (d) Early trochophore. The cross transforms into the head region of the embryo with the apical plate (lightly stippled); the cephalic plates (shown white and heavily outlined); the head vesicle (densely stippled) and the prototroch (less densely stippled).

are descendants of the cells $1a^2-1d^2$, which divide once or twice. In this way, four interradial groups of two or four primary trochoblasts are formed, which may develop cilia as early as during the blastula stage, for example, in *Patella* (van den Biggelaar, 1977). Except dorsally, the gaps between these primary trochoblasts are filled up by *secondary trochoblasts* descending from the tip cells of the ventral and lateral arms $(2a^{11}-2c^{11})$. The dorsal gap is closed by a shifting of the margins of the prototroch, which fuse at the dorsal side (*Trochus*, Robert, 1902; *Ischnochiton*, Heath, 1899). In some species (e.g., *Patella*, Wilson, 1904; *Dentalium*, van Dongen and Geilenkirchen, 1974) accessory trochoblasts derived from $1a^{12}-1d^{12}$ are added.

In the pulmonates, the prototroch is scarcely developed, consisting of 21–23 cells in *Physa* (Wierzejski, 1905) and *Planorbis* (Holmes, 1900). In *Lymnaea*, the prototroch consists of only eight cells (Verdonk, 1965). It does not surround the whole embryo but runs in the form of a horseshoe from the mouth to the lateral sides of the embryo, where it meets the *head vesicle*, a prominent structure in pulmonates consisting of 12–14 large flattened cells, which occupy the dorsal side of the head (Fig. 9d). The cells are derived from the dorsal arm of the cross, the dorsal primary trochoblasts and the tip cells and outer median cells of the lateral arms (Verdonk, 1965; Camey and Verdonk, 1970).

In the Aplacophora and Protobranchia, the prototroch develops into a larval test covering nearly the whole body. It consists of a series of tiers of regularly shaped cells. In the Protobranchia, three rows of cells bear long cilia, whereas the apical and posterior row have shorter cilia (Drew, 1901). In the Aplacophora, only one tier develops powerful cilia (e.g., *Neomenia;* Thompson, 1960). The definite ectoderm develops from cells that wander in from the region of the blastopore.

The *posttrochal ectoderm* originates from the cells of the second and third quartet of micromeres. At the dorsal side, the *somatic plate* is formed, which by rapid cell multiplication produces an important part of the trunk region. From its anterior region, bordering the prototroch or the head vesicle, the shell gland develops. The posterior area of the somatic plate, situated between the shell gland and the blastopore, becomes the *ventral plate* from which the foot is formed. It is generally assumed that the whole somatic plate originates from the first somatoblast, 2d. In all probability, this is true for the lamellibranchs (Lillie, 1895; Guerrier, 1970). In *Dentalium*, however, the somatic plate is formed from the cells 2d and 3d; the shell gland originates from $2d^1$, whereas the foot develops from a paired anlage, the left part from $3d^2$ and the right from $2d^2$ (van Dongen and Geilenkirchen, 1974). In gastropods, the first somatoblast 2d is not a very conspicuous cell, as in lamellibranchs and scaphopods, and its history is therefore difficult to follow by cell-lineage studies. Conklin's (1897) statement that in *Crepidula*, the shell gland and the foot originate from 2d as in the lamellibranchs, has to be questioned. Experimental deletion of specific cells in gastropods has shown that the origin of shell and foot in these molluscs is more complicated. In *Ilyanassa,* Cather (1967) has demonstrated that the shell originates not only from 2d but also from 2c, whereas Clement (1971) found that the foot does not originate from 2d but from 3c and 3d. Also, in the gastropod *Bithynia,* shell and foot are not only formed from the D quadrant but also from the C quadrant (Verdonk and Cather, 1973; Cather *et al.,* 1976). This might indicate that in gastropods the C quadrant is of greater importance for the construction of the posttrochal region than in other molluscs (Verdonk, 1979).

The *stomodaeum* in molluscs is an ectodermal structure that forms the mouth cavity, pharynx, and esophagus. The boundary between the ectoderm and endoderm is situated at the place where the esophagus debouches into the stomach.

At the place where the hindgut makes contact with the ectoderm, an anal plate may be present, which invaginates as an ectodermal *proctodaeum,* as seen in *Dreissensia* (Meisenheimer, 1901) and *Sphaerium* (Okada, 1936). In gastropods, the proctodaeum is very short.

B. Endoderm

The endoderm originates from the macromeres 4A–4D and the micromeres 4a–4c; 4d, as the mesentoblast, forms both endoderm and mesoderm. Both the micromeres and the macromeres may divide before gastrulation starts.

The roof of the archenteron is formed by the macromeres, the sides by cells of the fifth quartet of micromeres; the cells of the fourth quartet adjoin the stomodaeum. During the development of the primitive gut, the endoderm is subdivided in large-celled and small-celled endoderm. As the latter is situated in the tip of the archenteron, it may be concluded that the small-celled endoderm is derived from the macromeres (Raven, 1966). The large-celled endoderm is probably derived from the fifth quartet of micromeres, which in some species (e.g., *Umbrella (Umbraculum), Fulgur (Busycon)* consists of larger cells than the macromeres. These cells start to bulge out into two lateral lobes, which form the primordium of the larval liver. The small-celled endoderm then grows forward in the direction of the esophagus and forms the wall of the stomach (e.g., in *Lymnaea* and *Physa*). The part of the mesenteron adjoining the stomodaeum is formed by descendants of the fourth quartet, which are situated here from the end of gastrulation (Conklin, 1897).

It is generally accepted that the hindgut is derived from enteroblasts, which are split off from the mesentoblast 4d. In *Crepidula, Fulgur (Busycon)* (Conklin, 1897, 1907), and *Fiona* (Casteel, 1904), these cells become situated in the posterior region of the enteron, from which the hindgut grows out. In *Physa* (Wierzejski, 1905), the enteroblasts are situated between the posterior wall of the enteron and the ectoderm. They arrange themselves into a solid strand of cells, which secondarily connects with the hindgut. A lumen later appears.

Meisenheimer's (1896) view that in *Limax* the whole hindgut is derived from the ectoderm has to be questioned, especially since in the related species *Agriolimax (Deroceras)* (Carrick, 1939) it develops in the normal way. Clement (1960) has shown convincingly that in *Ilyanassa,* the hindgut originates from 4d, as after removal of this cell the hindgut is missing.

C. Mesoderm

According to its origin, the mesoderm in molluscs can be subdivided into endomesoderm and ectomesoderm. In the literature, the endomesoderm is often designated as primary or adult mesoderm, whereas the ectomesoderm is called secondary or larval mesoderm. As these indications might suggest that the endomesoderm is most important for mesoderm formation in molluscs and contributes primarily to the development of adult structures, whereas the ectomesoderm would form larval structures only, these indications are inadequate. Both the endomesoderm and the ectomesoderm may form larval and adult structures.

The endomesoderm originates from the second somatoblast 4d or M, formed from the macromere 3D after fifth cleavage. Usually this macromere starts dividing long before the other macromeres (3A, 3B, and 3C). In *Dentalium,* however, 3D divides slightly in advance (Guerrier et al., 1978), whereas in *Patella* (van den Biggelaar, 1977) and *Haliotis* (van den Biggelaar, unpublished observations), the division of 3D is slightly retarded. Except in very yolk-rich eggs, such as those of *Crepidula, Ilyanassa,* and *Fulgur (Busycon),* the mesentoblast 4d is larger than the macromere 4D. According to Smith (1935), Crofts (1938), and Creek (1951), the endomesoderm in *Patella, Haliotis,* and *Pomatias* does not originate from 4d but from the macromere 4D. Because one of us has found that, both in *Patella* (van den Biggelaar, 1977) and *Haliotis* (unpublished), the mesodermbands are derived from 4d, one may assume that in molluscs (except in *Paludina (Viviparus),* see later) the endomesoderm originates from 4d.

Depending on the number of divisions that have taken place in the first, second, and third quartet of micromeres before the macromere 3D starts dividing, the number of cells in the embryo at the time of mesentoblast formation varies from 24 in most gastropods to 72 in *Ischnochiton*. After its formation the mesentoblast divides equally into the cells $4d^1$ (MEr) and $4d^2$ (MEl), situated to the right and to the left, respectively, of the dorsal midline. Usually these cells start to sink into the blastocoel. In *Chiton (Middendorffia?) polii* (Kowalewsky, 1883) and *Dentalium* (van Dongen, 1977), the mesentoblasts remain at the surface until gastrulation, when they move in together with the endoderm.

Both mesentoblasts split off some enteroblasts before they become true mesodermal teloblasts. These teloblasts, situated on the left and the right dorsal side

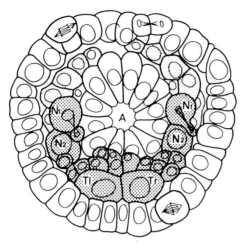

Fig. 10. Diagrammatic section through an embryo of *Physa fontinalis* showing the position of the mesoderm. The endomesoderm (densely stippled) and the ectomesoderm (lightly stippled) form a nearly continuous ring surrounding the endoderm. A, archenteron; N_1 and N_2, nephroblasts; Tl, left teloblast; Tr, right teloblast. (After Wierzejski, 1905.)

behind the archenteron, subsequently start budding off small cells in front and in this way produce two *mesoderm bands,* which grow in length both by addition of new cells at their posterior ends and by subdivision of cells already formed. In species with well-developed mesoderm they may embrace the archenteron in the shape of a horseshoe (Fig. 10).

Due to the inward migration of the mesentoblast, the development of the mesoderm bands takes place inside the embryo and is therefore difficult to follow. Only in a few cases has the history of the mesentoblasts been traced so far that the fate of the cells became clear. In the pulmonates *Planorbis* (Rabl, 1879; Holmes, 1900) and in *Physa* (Wierzejski, 1905), the cells in the anterior part of each mesoderm band (N_1 and N_2) are the nephroblasts, which form the protonephridia (Fig. 10). In the lamellibranch *Sphaerium* (Okada, 1936, 1939), apart from the larval kidney each mesentoblast (M_r and M_l) also forms a primordial gonocyte. In *Sphaerium striatinum,* the latter is already set apart after the mesentoblasts have split off three mesoblastic cells (Woods, 1931). Also the larval musculature would originate, at least partly, from the mesentoblast (e.g., the adductor muscle in *Unio;* Lillie, 1895). According to Crofts (1955), the larval retractor muscle in the gastropods *Haliotis, Patella, Patina,* and *Calliostoma* is derived from 4d. In lobeless embryos of *Ilyanassa,* however, in which a mesentoblast is not formed, large blocks of granular muscle cells are found, which resemble the retractor muscle of normal larvae (Atkinson, 1971). After removal of the mesentoblast 4d in *Ilyanassa,* a larval retractor muscle may be present (Clement, 1960). It is generally assumed that adult structures such as heart, kidney, and gonads also originate from 4d, but direct proof is lacking.

The *ectomesoderm* originates from cells of the second or third quartet of micromeres. In the lamellibranchs and scaphopods, the stem cells belong to the second quartet, for example, in *Sphaerium* $2a^2$ and $2c^2$ (Okada, 1936), in *Unio* $2a^2$ (Lillie, 1895), in *Dentalium* $2a^{212}$ and $2c^{122}$ (van Dongen, 1977). In the Gastropoda, they originate from the third quartet. Even in species that are not closely related, such as *Physa* (Wierzejski, 1905), *Fiona* (Casteel, 1904), and *Littorina* (Delsman, 1914), the ectomesoderm is derived from exactly the same stem cells: $3a^{2111}$, $3a^{2211}$, $3b^{2111}$, and $3b^{2211}$. Conklin' (1897) view that the ectomesoderm in *Crepidula* is derived from 2a, 2b, and 2c has to be questioned, as it is based only on the position of the first ectomesoblasts at the beginning of gastrulation.

After their formation, the cells of the ectomesoderm retract from the surface and sink into the blastocoel, where they multiply by repeated divisions. Their descendants may join the cells of the endomesoderm, as seen in *Dentalium* and *Physa*. In the latter species, a nearly continuous ring is formed around the archenteron (cf. Fig. 10). As the ectomesoderm and the endomesoderm grow so closely together, it is obvious that an exact indication of the contribution of each source of mesoderm to the formation of embryonic and adult structures is hardly possible.

With respect to mesoderm formation, the gastropod *Paludina (Viviparus)* stands quite apart, as 4d does not behave as a mesentoblast, but as an ordinary endoblast of the fourth quartet. Mesoderm bands, which in all other molluscs originate from this cell, are not formed in this species. The mesoderm does not appear before the end of gastrulation, when at the ventral side of the embryo a layer of cells is formed between the ectoderm and the endoderm.

According to von Erlanger (1891, 1894) and Fernando (1931), the mesoderm in *Paludina (Viviparus)* originates from the endoderm. The ventral wall of the archenteron bulges out and forms a coelomic sac, which separates from the endoderm and extends in the animal and lateral direction at both sides of the archenteron. Finally, the whole structure falls apart into separate cells, which fill the body cavity.

A quite different description of the origin of the mesoderm in *Paludina (Viviparus)* is given by Tönniges (1896), Otto and Tönniges (1906), and Dautert (1929). According to these authors, the mesoderm originates from the ectoderm. At a late gastrula stage, cells of the ectoderm at the ventral side of the embryo start protruding into the cavity between the ectoderm and the archenteron. They move out of the ectoderm and spread in the animal and lateral direction at both sides of the archenteron. According to Dautert, the mesoderm in *Paludina (Viviparus)* corresponds to the ectomesoderm of the other molluscs, whereas the endomesoderm would be missing. Although his view is based on sections of many embryos, the absence of relevant pictures in his paper leaves room for some doubt.

VI. Concluding Remarks

The basic research on the early development of molluscs was carried out at the turn of the century, when it became more and more evident that the developmental significance of the cleavage cells was strongly influenced by the inheritance of a special part of the ovum. This idea was corroborated by striking examples of ooplasmic segregation combined with a remarkable determinate cleavage pattern. A fixed relationship seemed to exist between the successive cleavage products and their contribution to the organization of the embryo. That is the reason why a number of cell-lineage studies have been performed on a variety of molluscan eggs. For the basic information on the early development and the formation of the germ layers, one still depends on the monumental works of that period, that is, Lillie's paper on "The embryology of the Unionidae" (1895) and Conklin's "Embryology of *Crepidula*" (1897).

In this chapter, we have discussed the main data of the older literature in the light of more recent cell-lineage and other developmental studies.

Interest in the study of early molluscan development mainly depends upon its qualification as a model system for cell differentiation. Therefore, the study of the embryology of the different species has not been determined by their systematic position, and, as a consequence, large gaps exist in our knowledge about the development of a number of molluscan genera. The primitive groups have either not been studied at all (Monoplacophora) or only scantily (Polyplacophora, Aplacophora). With respect to the two major classes (Bivalvia and Gastropoda), our knowledge is more complete, but even here important data are missing. Among the bivalves, data on the early development of the Protobranchia are almost entirely missing, whereas the literature on the gastropods mainly concerns the Prosobranchia and Pulmonata without adequate studies of the Opisthobranchia. Unfortunately, many recent papers provide only a general description of development, often restricted to larval development, whereas cleavage and early development are discussed superficially or not at all. For an understanding of the evolutionary lines within the phylum of the Mollusca, comparative data on early development are indispensable because during ontogenesis of the molluscs, the basic body plan is laid down very early.

An extensive study of unexplored molluscan species will undoubtedly deepen our insight in the phylogeny of the Mollusca, and simultaneously indicate a number of species that are even more applicable as model systems for the study of cytoplasmic localization, regulation of cell division, cellular interactions, and cell differentiation.

References

Atkinson, J. W. (1971). Organogenesis in normal and lobeless embryos of the marine prosobranch *Ilyanassa obsoleta*. *J. Morphol.* **133**, 339–352.

Baba, K. (1951). General sketch of the development in a solenogastre, *Epimenia verrucosa* (Nierstrasz). *Misc. Rep. Res. Inst. Nat. Resour. (Tokyo)* **19–21**, 38–46.

Blochmann, F. (1882). Ueber die Entwicklung der *Neritina fluviatilis* Müll. *Z. Wiss. Zool.* **36**, 125–174.

Blochmann, F. (1883). Beiträge zur Kenntnis der Entwicklung der Gastropoden. *Z. Wiss. Zool.* **38**, 392–410.

Bluemink, J. G. (1967). The subcellular structure of the blastula of *Limnaea stagnalis* (Mollusca) and the mobilization of the nutrient reserve. Ph.D. Thesis, University of Utrecht, Utrecht, The Netherlands.

Boycott, A. E., and Diver, C. (1923). On the inheritance of sinistrality in *Limnaea peregra*. *Proc. R. Soc. London, Ser. B* **95**, 207–213.

Boycott, A. E., Diver, C., Garstang, S. L., and Turner, F. M. (1931). The inheritance of sinistrality in *Limnaea peregra* (Mollusca, Pulmonata). *Philos. Trans. R. Soc. London, Ser. B* **219**, 51–131.

Camey, T., and Verdonk, N. H. (1970). The early development of the snail *Biomphalaria glabrata* (Say) and the origin of the head organs. *Neth. J. Zool.* **20**, 93–121.

Carazzi, D. (1900). L'embriologia dell'*Aplysia limacina* L. *Anat. Anz.* **17**, 77–102.

Carrick, R. (1939). The life-history and development of *Agriolimax agrestis* L., the gray field slug. *Trans. R. Soc. Edinburgh* **59**, 563–597.

Casteel, D. B. (1904). The cell-lineage and early larval development of *Fiona marina*, a nudibranch mollusc. *Proc. Acad. Nat. Sci. Philadelphia* **56**, 325–405.

Cather, J. N. (1967). Cellular interactions in the development of the shell gland of the gastropod *Ilyanassa*. *J. Exp. Zool.* **166**, 205–224.

Cather, J. N., Verdonk, N. H., and Dohmen, M. R. (1976). Role of the vegetal body in the regulation of development in *Bithynia tentaculata* (Prosobranchia, Gastropoda). *Am. Zool.* **16**, 455–468.

Clement, A. C. (1935). The formation of giant polar bodies in centrifuged eggs of *Ilyanassa*. *Biol. Bull. (Woods Hole, Mass.)* **69**, 403–414.

Clement, A. C. (1952). Experimental studies on germinal localization in *Ilyanassa*. I. The role of the polar lobe in determination of the cleavage pattern and its influence in later development. *J. Exp. Zool.* **121**, 593–625.

Clement, A. C. (1960). Development of the *Ilyanassa* embryo after removal of the mesentoblast cell. *Biol. Bull. (Woods Hole, Mass.)* **119**, 310.

Clement, A. C. (1968). Development of the vegetal half of the *Ilyanassa* egg after removal of most of the yolk by centrifugal force, compared with the development of animal halves of similar composition. *Dev. Biol.* **17**, 165–186.

Clement, A. C. (1971). *Ilyanassa. In* "Experimental Embryology of Marine and Freshwater Invertebrates" (G. Reverberi, ed.), pp. 188–213. North-Holland Publ., Amsterdam.

Comandon, J., and de Fonbrune, P. (1935). Recherches effectuées aux premiers stades du développement d'oeufs de Gastéropodes et d'un ver à l'aide de la cinématographie. *Arch. Anat. Microsc.* **31**, 79–100.

Conklin, E. G. (1897). The embryology of *Crepidula*. *J. Morphol.* **13**, 1–226.

Conklin, E. G. (1907). The embryology of *Fulgur*. A study on the influence of yolk on development. *Proc. Acad. Nat. Sci. Philadelphia* **59**, 320–359.

Conrad, G. W. (1973). Control of polar lobe formation in fertilized eggs of *Ilyanassa obsoleta* Stimpson. *Am. Zool.* **13**, 961–980.

Conrad, G. W., and Davis, S. E. (1977). Microionthophoretic injection of calcium ions or of cyclic AMP causes rapid shape changes in fertilized eggs of *Ilyanassa obsoleta*. *Dev. Biol.* **61**, 184–201.

Conrad, G. W., and Davis, S. E. (1980). Polar lobe formation and cytokinesis in fertilized eggs of *Ilyanassa obsoleta*. III. Large blebs formation caused by Sr^{2+}, ionophores X537A and A23187, and compound 48/80. *Dev. Biol.* **74**, 152–172.

Conrad, G. W., and Williams, D. C. (1974a). Polar lobe formation and cytokinesis in fertilized eggs of *Ilyanassa obsoleta*. I. Ultrastructure and effects of cytochalasin B and colchicine. *Dev. Biol.* **36**, 363–378.

Conrad, G. W., and Williams, D. C. (1974b). Polar lobe formation and cytokinesis in fertilized eggs of *Ilyanassa obsoleta*. II. Large bleb formation caused by high concentrations of exogenous calcium ions. *Dev. Biol.* **37**, 280–294.

Conrad, G. W., Williams, D. C., Turner, F. R., Newrock, K. M., and Raff, R. A. (1973). Microfilaments in the polar lobe constriction of fertilized eggs of *Ilyanassa obsoleta*. *J. Cell Biol.* **59**, 228–233.

Crampton, H. E. (1894). Reversal of cleavage in a sinistral gastropod. *Ann. N.Y. Acad. Sci.* **8**, 167–169.

Creek, G. A. (1951). The reproductive system and embryology of the snail *Pomatias elegans* (Müller). *Proc. Zool. Soc. London* **121**, 599–640.

Crofts, D. R. (1938). The development of *Haliotis tuberculata*, with special reference to organogenesis during torsion. *Philos. Trans. R. Soc. London, Ser. B 2*, **28**, 219–268.

Crofts, D. R. (1955). Muscle morphogenesis in primitive gastropods and its relation to torsion. *Proc. Zool. Soc. London* **125**, 711–750.

Dautert, E. (1929). Die Bildung der Keimblätter bei *Paludina*. *Zool. Jahrb., Abt. Anat. Ontog. Tiere* **50**, 433–496.

Degner, E. (1952). Der Erbgang der Inversion bei *Laciniaria biplicata*. M.T.G. (Gastr., Pulm.). *Mitt. Hamburg Zool. Mus. Inst.* **51**, 3–61.

Delsman, H. C. (1912). Ontwikkelingsgeschiedenis van *Littorina obtusata*. Ph.D. Thesis, University of Amsterdam, Amsterdam, The Netherlands.

Delsman, H. C. (1914). Entwicklungsgeschichte von *Littorina obtusata*. *Tijdschr. Ned. Dierkd. Ver.* [2] **13**, 170–340.

Diver, C., and Andersson-Kottö, I. (1938). Sinistrality in *Limnaea peregra* (Mollusca, Pulmonata). The problem of mixed broods. *J. Genet.* **35**, 447–525.

Dohmen, M. R., and Verdonk, N. H. (1974). The structure of a morphogenetic cytoplasm, present in the polar lobe of *Bithynia tentaculata* (Gastropoda, Prosobranchia). *J. Embryol. Exp. Morphol.* **31**, 423–433.

Dohmen, M. R., and Verdonk, N. H. (1979). The ultrastructure and role of the polar lobe in development of molluscs. *In* "Determinants of Spatial Organization" (S. Subtelny and I. R. Konigsberg, eds.), pp. 3–27. Academic Press, New York.

Drew, G. A. (1901). The life-history of *Nucula delphinodonta* (Mighels). *Q. J. Microsc. Sci.* **44**, 313–391.

Elbers, P. F., and Bluemink, J. G. (1960). Pinocytosis in the developing egg of *Limnaea stagnalis*. *Exp. Cell Res.* **21**, 619–622.

Fernando, W. (1931). The origin of the mesoderm in the gastropod *Viviparus* (= *Paludina*). *Proc. R. Soc. London, Ser. B* **107**, 381–390.

Fioroni, P. (1979a). Phylogenetische Abänderungen der Gastrula bei Mollusken. In "Ontogenese und Phylogenie" (P. Siewing, ed.), pp. 82–100. Parey, Berlin.

Fioroni, P. (1979b). Zur Struktur der Pollappen und der Dottermakromeren—eine vergleichende Übersicht. *Zool. Jahrb., Abt. Anat. Ontog. Tiere* **102**, 395–430.

Fol, H. (1875). Etudes sur le développement des Mollusques. I. Sur le développement des Ptéropodes. *Arch. Zool. Exp. Gen.* **4**, 1–214.

Freeman, G., and Lundelius, J. W. (1982). The developmental genetics of dextrality and sinistrality in the gastropod *Lymnaea peregra*. *Wilhelm Roux's Arch. Dev. Biol.* **191**, 69–83.

Guerrier, P. (1970). Les caractères de la segmentation et la détermination de la polarité dorsoventrale dans le développement de quelques Spiralia. III. *Pholas dactylus* et *Spisula subtruncata* (Mollusques, Lamellibranches). *J. Embryol. Exp. Morphol.* **23**, 667–692.

Guerrier, P., van den Biggelaar, J. A. M., van Dongen, C. A. M., and Verdonk, N. H. (1978). Significance of the polar lobe for the determination of dorsoventral polarity in *Dentalium vulgare* (da Costa). *Dev. Biol.* **63**, 233–242.

Heath, H. (1899). The development of *Ischnochiton*. *Zool. Jahrb., Abt. Anat. Ontog. Tiere* **12**, 567–656.

Herbers, K. (1913). Entwicklungsgeschichte von *Anodonta cellensis* Schröt. *Z. Wiss. Zool.* **108**, 1–174.

Hess, O. (1956). Beobachtungen der Normogenese des Süsswasser-Prosobranchiers *Bithynia tentaculata* L. *Biol. Zentralbl.* **75**, 664–682.

Hess, O. (1962). Entwicklungsphysiologie der Mollusken. *Fortschr. Zool.* **14**, 130–163.

Heymons, R. (1893). Zur Entwicklungsgeschichte von *Umbrella mediterranea*. *Z. Wiss. Zool.* **56**, 245–298.

Holmes, S. J. (1900). The early development of *Planorbis*, *J. Morphol.* **16**, 369–458.

Kofoid, C. A. (1894). On some laws of cleavage in *Limax*. *Proc. Am. Acad. Arts Sci.* **29**, 180–203.

Kofoid, C. A. (1895). On the early development of *Limax*. *Bull. Mus. Comp. Zool.* **27**, 35–118.

Kowalewsky, A. (1883). Embryogénie du *Chiton polii* (Philippi). *Ann. Mus. Hist. Nat. Marseille* **1**, No. 5.

Lillie, F. R. (1895). The embryology of the Unionidae. *J. Morphol.* **10**, 1–100.

Lillie, F. R. (1901). The organization of the egg of *Unio*, based on a study of its maturation, fertilization and cleavage. *J. Morphol.* **17**, 227–292.

Meisenheimer, J. (1896). Entwicklungsgeschichte von *Limax maximus* L. I. Furchung und Keimblätterbildung. *Z. Wiss. Zool.* **62**, 415–468.

Meisenheimer, J. (1901). Entwicklungsgeschichte von *Dreissensia polymorpha*. Pall. *Z. Wiss. Zool.* **69**, 1–137.

Moor, B. (1973). Zur frühen Furchung des Eies von *Littorina littorea* L. (Gastropoda Prosobranchia). *Zool. Jahrb., Abt. Anat. Ontog. Tiere* **91**, 546–573.

Morgan, T. H. (1933). The formation of the antipolar lobe in *Ilyanassa*. *J. Exp. Zool.* **64**, 433–467.

Morgan, T. H. (1935a). Centrifuging the eggs of *Ilyanassa* in reverse. *Biol. Bull. (Woods Hole, Mass.)* **68**, 268–279.

Morgan, T. H. (1935b). The rhythmic changes in form of the isolated antipolar lobe of *Ilyanassa*. *Biol. Bull. (Woods Hole, Mass.)* **68**, 296–299.

Morgan, T. H. (1936). Further experiments on the formation of the antipolar lobe of *Ilyanassa*. *J. Exp. Zool.* **74**, 381–425.

Morgan, T. H., and Tyler, A. (1930). The point of entrance of the spermatozoön in relation to the orientation of the embryo in eggs with spiral cleavage. *Biol. Bull. (Woods Hole, Mass.)* **58**, 59–73.

Morgan, T. H., and Tyler, A. (1938). The relation between entrance point of the spermatozoön and bilaterality of the egg of *Chaetopterus*. *Biol. Bull. (Woods Hole, Mass.)* **74**, 401–402.

Murray, J., and Clarke, B. (1966). The inheritance of polymorphic shell characters in *Partula* (Gastropoda). *Genetics* **54**, 1261–1277.

Okada, K. (1936). Some notes on *Sphaerium japonicum biwaense* Mori, a fresh water bivalve. IV. Gastrula and fetal larva. *Sci. Rep. Tohoku Imp. Univ., Ser. 4* **11**, 49–68.

Okada, K. (1939). The development of the primary mesoderm in *Sphaerium japonicum biwaense* Mori. *Sci. Rep. Tohoku Imp. Univ., Ser. 4* **14**, 25–47.

Otto, H., and Tönniges, C. (1906). Untersuchungen über die Entwicklung von *Paludina vivipara*. *Z. Wiss. Zool.* **80**, 411–514.

Pasteels, J. (1931). Recherches sur le déterminisme du mode de segmentation des Mollusques Lamellibranches (actions des rayons ultraviolets sur l'oeuf de *Barnea cand.*). *Arch. Biol.* **42**, 389–413.

Rabl, C. (1879). Über die Entwicklung der Tellerschnecke. *Morph. Jahrb.* **5**, 562–660.

Raff, R. A. (1972). Polar lobe formation by embryos of *Ilyanassa obsoleta*. Effects of inhibitors of microtubule and microfilament function. *Exp. Cell Res.* **71**, 455–459.

Raven, C. P. (1946). The development of the egg of *Limnaea stagnalis* L. from the first cleavage till the trochophore stage with special reference to its "chemical embryology." *Arch. Néerl. Zool.* **7**, 353–434.

Raven, C. P. (1949). On the structure of cyclopic, synophthalmic and anophthalmic embryos, obtained by the action of lithium in *Limnaea stagnalis*. *Arch. Néerl. Zool.* **8**, 323–353.

Raven, C. P. (1966). "Morphogenesis, the Analysis of Molluscan Development," 2nd ed. Pergamon, Oxford.

Robert, A. (1902). Recherches sur le développement des Troques. *Arch. Zool. Exp. Gen.* [3] **10**, 269–525.

Schmidt, B. A., Kelly, P. T., Mary, M. C., Davis, S. E., and Conrad, G. W. (1980). Characterization of actin from fertilized eggs of *Ilyanassa obsoleta* during polar lobe formation and cytokinesis. *Dev. Biol.* **76**, 126–140.

Smith, F. G. W. (1935). The development of *Patella vulgata*. *Philos. Trans. R. Soc. London, Ser. B* **225**, 95–125.

Sturtevant, A. H. (1923). Inheritance of direction of coiling in *Lymnaea*. *Science* **58**, 269–270.

Thompson, T. E. (1960). The development of *Neomenia carinata* Tullberg (Mollusca, Aplacophora). *Proc. R. Soc. London, Ser. B* **153**, 263–278.

Tönniges, C. (1896). Über die Bildung des Mesoderms bei *Paludina vivipara*. *Z. Wiss. Zool.* **61**, 541–605.

Ubbels, G. A. (1966). Morphological and cytochemical aspects of oogenesis in *Limnaea stagnalis*. *Arch. Neerl. Zool.* **16**, 544–547.

Ubbels, G. A., Bezem, J. J., and Raven, C. P. (1969). Analysis of follicle cell patterns in dextral and sinistral *Limnaea peregra*. *J. Embryol. Exp. Morphol.* **21**, 445–466.

van den Biggelaar, J. A. M. (1976). Development of dorsoventral polarity preceding the formation of the mesentoblast in *Lymnaea stagnalis*. *Proc. K. Ned. Akad. Wet., Ser. C* **79**, 112–126.

van den Biggelaar, J. A. M. (1977). Development of dorsoventral polarity and mesentoblast determination in *Patella vulgata*. *J. Morphol.* **154**, 157–186.

van Dongen, C. A. M. (1976a). The development of *Dentalium* with special reference to the significance of the polar lobe. V and VI. Differentiation of the cell pattern in lobeless embryos of *Dentalium vulgare* (da Costa) during late larval development. *Proc. K. Ned. Akad. Wet., Ser. C* **79**, 245–266.

van Dongen, C. A. M. (1976b). The development of *Dentalium* with special reference to the significance of the polar lobe. VII. Organogenesis and histogenesis in lobeless embryos of *Dentalium vulgare* (da Costa) as compared to normal development. *Proc. K. Ned. Akad. Wet., Ser. C* **79**, 454–465.

van Dongen, C. A. M. (1977). Mesoderm formation during normal development of *Dentalium dentale*. *Proc. K. Ned. Akad. Wet., Ser. C* **80**, 372–376.

van Dongen, C. A. M., and Geilenkirchen, W. L. M. (1974). The development of *Dentalium* with special reference to the significance of the polar lobe. I, II and III. Division chronology and development of the cell pattern in *Dentalium dentale* (Scaphopoda). *Proc. K. Ned. Akad. Wet.*, *Ser. C* **77**, 57–100.

Verdonk, N. H. (1965). Morphogenesis of the head region in *Limnaea stagnalis* L. Thesis, University of Utrecht, Utrecht, The Netherlands.

Verdonk, N. H. (1968). The effect of removing the polar lobe in centrifuged eggs of *Dentalium. J. Embryol. Exp. Morphol.* **19**, 33–42.

Verdonk, N. H. (1979). Symmetry and asymmetry in the embryonic development of molluscs. *In* "Pathways in Malacology" (S. van der Spoel, A. C. van Bruggen, and J. Lever, eds.), pp. 25–45. Bohn, Scheltema & Holkema, Utrecht.

Verdonk, N. H., and Cather, J. N. (1973). The development of isolated blastomeres in *Bithynia tentaculata* (Prosobranchia, Gastropoda). *J. Exp. Zool.* **186**, 47–61.

Verdonk, N. H., Geilenkirchen, W. L. M., and Timmermans, L. P. M. (1971). The localization of morphogenetic factors in uncleaved eggs of *Dentalium. J. Embryol. Exp. Morphol.* **25**, 57–63.

von Erlanger, R. (1891). Zur Entwicklung von *Paludina vivipara. Morphol. Jahrb.* **17**, 337–374.

von Erlanger, R. (1894). Zum Bildung des Mesoderms bei der *Paludina vivipara. Morphol. Jahrb.* **22**, 113–118.

Wierzejski, A. (1905). Embryologie von *Physa fontinalis* L. *Z. Wiss. Zool.* **83**, 502–706.

Wilson, E. B. (1892). The cell lineage of *Nereis. J. Morphol.* **6**, 361–481.

Wilson, E. B. (1904). Experimental studies on germinal localization. I. The germ-regions in the egg of *Dentalium. J. Exp. Zool.* **1**, 1–72.

Woods, F. H. (1931). History of the germ cells in *Sphaerium striatinum*. (Lam.). *J. Morphol.* **51**, 545–595.

4

Organogenesis[1]

BEATRICE MOOR

CH-4000 Basel
Switzerland

I. Introduction

This synopsis of molluscan organogenesis primarily stresses research performed since the second edition of Raven's *Morphogenesis* (1966). However, an important part of our morphological knowledge is still based on the detailed investigations carried out earlier, especially the monumental work of Korschelt and Heider (1936).

Some remarks on the present status of morphological research on molluscan

[1]This study was performed during a stay at the Zoological Institute of the University of Münster/Westfalen, Federal Republic of Germany, granted by the Swiss Academy of Sciences.

THE MOLLUSCA, VOL. 3
Development

ontogenesis may explain the concept on which this account is based. First, a synopsis that emphasizes general traits of molluscan development may obscure characteristic differences between the classes. This happens all the more so, because our knowledge about the different classes is rather unequal. This handicap is also apparent at lower systematic levels. Second, an account of organogenesis predominantly deals with development of adult organs. Because molluscan development is essentially indirect (Fioroni, 1971) one must bear in mind that the larval body (or the corresponding stages in intracapsular development) represents, so to speak, the environment where differentiation of the adult structures takes place. On the other hand, an analysis of larval structures cannot disregard the adult organization (Thiriot-Quiévreux, 1971a, 1974). Last, comparative embryology also plays an important role in providing basic data for the evaluation of theories of phylogenetic research (Jägersten, 1972). Recent success in the cultivation of planktotrophic larvae allows investigation of species hitherto inaccessible (Harris, 1975; Perron and Turner, 1977).

In order to gain a better insight into the amazing variety of molluscan development, organogenesis of each major systematic group is treated separately.

II. Polyplacophora

Development of the generally yolk-rich eggs shows little morphological variation. However, our knowledge is restricted to those forms that spend a longer or shorter period as free-swimming lecithotrophic larvae before settling (Thorson, 1946). Except for *Schizoplax brandti,* in which Kussakin (1960, quoted by Smith, 1966) observed a more direct development related to the suppression of the free-swimming stage, nothing is known about possible morphological correlations with variations of hatching stage (Dell, 1962; Smith, 1966). Actual metamorphosis *(sensu stricto)*, has not been observed in placophorans. Their development is characterized by a very early onset of the differentiation of the definitive organs in the trochophora-like larva (Christiansen, 1954).

The primordial mouth and foregut are produced by invagination of the ectoderm surrounding the blastopore, which is somewhat slitlike during its shift from the polar position onto the prospective ventral half (Metcalf, 1893; Heath, 1899). Through the invagination of the primordial foregut, the mouth reaches a position immediately behind the prototroch, basically corresponding to the definitive subtermainal area.

Most data concerning the inner organs are derived from Hammarsten and Runnström's (1925) description of the development of *Acanthochiton discrepans.* In the development of the alimentary canal, the anlage of the ectodermal part of the foregut first becomes recognizable. The anlage of the radular sac soon grows to a large size; on the ventral side the rudiment of the subradular organ differentiates, and from the dorsal wall of the buccal cavity the paired salivary glands grow out. During this period of differentiation, the mouth temporarily closes. According to Hammarsten and Runnström (1925) the ciliated esophagus is endodermal in origin and, as a result, so are the sugar glands, which arise as lateral diverticula of the esophagus. Asymmetric organization is characteristic of the midgut and intestine. The intestine becomes rather long, forming many loops. As in gastropods, the left digestive gland diverticulum is larger than the right one.

In *Acanthochiton*, the heart develops dorsal to the posterior part of the gut. There is a primordial connection of the pericardium with the anlagen of the kidneys, from which each of the ureters grows. A lesser distal part may be formed by the ectoderm. The anlage of the genital coelom differentiates as a median solid outgrowth of the pericardium. On both sides, Hammarsten and Runneström (1925) observed two cells with nuclei that differed in size from those of the surrounding mesenchyme. Because of their initial independence from the pericardial outgrowth, they may be called true primordial germ cells (see Section V,J). Paired gonadal anlagen, which only secondarily form a single median organ, were observed by Higley and Heath (1912) in *Trachydermon* and *Nuttallina*, whereas other species are characterized by paired gonads in the adult stage (Hammarsten and Runnström, 1925).

With respect to the organogenesis of the nervous system, extensively analyzed by Hammarsten and Runnström (1925), we will confine our remarks to its origin. In an early stage, cells proliferate from the area close to the apical tuft (Heath, 1899; Hammarsten and Runnström, 1925). The form of the anlagen initially resembles the cross shape of the proliferation area. The paired anlagen are soon connected by a commissure lying immediately beneath the apical tuft. According to Hammarsten and Runnström, who criticize Kowalevsky's (1883a) description of an additional proliferation from the posttrochal body wall, the nervous system is entirely pretrochal in origin, in contrast to gastropods, bivalves, and scaphopods. Although Hammarsten and Runnström (1925) also emphasize that the buccal ganglia develop from a common anlage, they discuss the possibility of having overlooked the stage at which the epithelium of the pharynx could have made some contribution to the main anlage. As will be discussed later, the aplacophoran nervous system may also have some posttrochal components. Thus, only a detailed developmental analysis will reveal whether the molluscan nervous system in general derives from both pretrochal and posttrochal proliferation sites or not.

Toward the end of the larval period, paired ocelli develop at the left and right side behind the prototroch. They persist through the juvenile phase, remaining recognizable for as long as the transparent juvenile tissue allows light to penetrate (Heath, 1904). Rosen et al. (1979), therefore, called them the "so-called larval eyes." These authors confirm the findings of Heath (1904) (made in several species, e.g., *Katharina tunicata,* in the same geographical area) that the ocelli are located within the epithelium. Fischer (1980), on the other hand, verifies for *Lepidochitona cinerea* Kowalevsky's (1883a) statement about *Chiton polii* (= *Middendorffia caprearum*). Fischer notes that the ocelli are sunken beneath the epithelium, although he does not find the structureal changes in the adjacent epithelium to be as described by Kowalevsky. Despite the differences in the epithelium, the ocelli are generally similar (Fig. 1). Several pigment cells form a cup around a rhabdome originating from two or three sensory cells (in *Katharina* from a single one). The cytological features of the sensory cells in the various species are similar, although the system of folded membranes seems to be absent in *Katharina.* In a comparison of the larval ocelli with the visual cells of the aesthetes (Fischer, 1978, 1979) and with the shell eyes (Boyle, 1969; Haas and Kriesten, 1978), structural differences are evident. For instance, no densely packed membraneous structures (special agranular ER) are formed; the folded membranes of the larval ocelli originate from the Golgi apparatus. This suggests that there is probably no structural relationship between the juvenile ocelli and the photoreceptors of the adult (Fischer, 1980).

In the species investigated, shell-plate formation starts soon after hatching. In

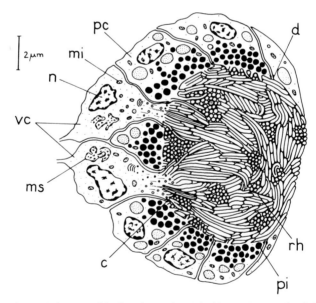

Fig. 1. Schematic diagram of the larval eye of *Lepidochitona cinerea.* The rhabdome is not represented to scale: the microvilli involved in its formation are, in reality, more slender and more numerous. See text for explanation. c, cilium; d, desmosome; mi, mitochondrium; ms, system of folded membranes; n, nucleus; pc, pigment cell; pi, pigment granulum; rh, rhabdome; vc, visual cell. (Modified after Fischer, 1980.)

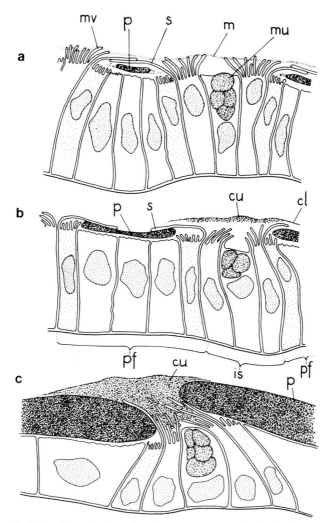

Fig. 2. Shell plate formation in *Ischnochiton rissoa*. (**a**) (5 h after hatching, at 14–16 °C). Primordial shell plate formation; the crystallization chamber is completely covered by the stragulum. The cells that bear the stragulum are slightly more electron dense than the other cells of the plate field. The entire epithelium is covered by a thin mucuslike layer. This formation is restricted to the dorsal side of the larva. (**b**) (9 h). As the shell plate grows the stragulum only covers the marginal parts of it. The mucoid substance has changed into a condensed layer restricted to the intersegmental ridges and the marginal parts of the shell plates. (**c**) (20 h). The growth of the shell plates is restricted to the anterior margin of the plate. The intersegmental space has been filled with cuticular substance which extends onto the marginal regions of the plates. cl, condensed layer; cu, cuticle; is, intersegmental ridge; m, mucuslike layer; mu, mucus(?) vesicles; mv, microvilli; p, shell plate; pf, plate field; s, stragulum. (After Kniprath, 1980e; from Wilhelm Roux's Archives by permission of Springer-Verlag, Heidelberg.)

the epithelium of the prospective intersegmental ridges and notum, cells with large vacuoles, probably containing mucus, differentiate. In the prospective plate-forming tissue, the *plate fields* (Kniprath, 1980e; in analogy to the *shell field* in Conchifera), the marginal cells develop flat extensions from the apical membrane covering the plate field (Haas et al., 1979, 1980, Kniprath, 1979b, 1980e) (Fig. 2). An extension of this sort is therefore called a *stragulum* (cover-

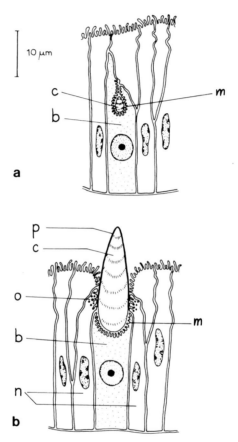

Fig. 3. Stages of development of a cylindrical spine on the notum of *Lepidochitona cinerea*. The calcareous part of the spine is formed by a single cell. (**a**) The initial stage shows a deep invagination forming an extracellular chamber into which the calcium carbonate is precipitated. (**b**) The apex of the growing spine, which is exposed to the surrounding medium, is covered with an organic pellicle secreted by the neighboring cells. In this way the functions of secretion of inorganic material and of organic substance are separated. This may point to a similarity between the formation of the spine and the formation of the shell of the conchifera. b, basal, calcium carbonate-secreting cell; c, calcareous part of the spine; m, microvilli; n, neighboring cell; o, vesicles filled with organic material forming the organic pellicle of the spine; p, organic pellicle of the spine. (After Haas, 1981.)

ing) by Kniprath. The stragulum seals the plate field against the surrounding medium, thus allowing deposition of lime without any external disturbance. Development of the shell plates follows a gradient, the most advanced differentiation being found in the middorsal region of shell plates 2–5. From there, differentiation progresses laterad, rostrad (shell plate 1), and caudad (shell plates 6–8). In *Middendorffia,* normal development produces shell-plate primordia consisting of a continuous layer of calcium. The precipitation of lime in several disjointed sites that secondarily coalesce (Kowalevsky, 1883a; Christiansen, 1954; Haas et al., 1979, 1980) is pathological, at least in *Middendorffia.* (Its possible phylogenetic significance was interpreted by von Salvini-Plawen, 1972.) It is probably induced by a nonoptimal developmental temperature (Kniprath, 1979b, 1980e).

With respect to theories of the possible evolutionary relationship between the placophoran hardparts and the conchiferan shell, it is worth mentioning that the stragulum performs the same function as the periostracum of the Conchifera, although periostracum formation during later development is still a matter of disagreement (Haas and Kriesten, 1974; Haas, 1976, 1981; Kniprath, 1980a, 1981). However, the formation of the cylindrical spine in the notum of *Lepidochitona* (analyzed by Haas, 1976; revised in some essential points in 1981) exhibits a morphological pattern (Fig. 3) comparable to the organization of the early shell field (see Section V,B,1). Nevertheless, Haas (1976, 1981) and Haas and Kriesten (1977) hold the view that this is probably a parallel or analogous differentiation rather than a homology. In this respect, the structural relationship between the aesthetes, the epidermal papillae of the perinotum (Haas and Kriesten, 1975, 1977; Fischer et al., 1980), and the morphogenetic process of the incorporation of the aesthetes into the shell plates should be analyzed.

III. Aplacophora

The striking similarities between the larvae of the Protobranchia and Aplacophora (*Dondersia = Nematomenia banyulensis;* Pruvot, 1890) was already noted by Drew (1899a,b; 1901), who first studied protobranch ontogenesis. However, Drew pointed out that lack of knowledge of the internal development of the aplacophoran larvae precludes serious comparisons (1899a,b). The few studies of aplacophoran development (Pruvot, 1892; Heath, 1918; Thompson, 1961a) support this view. Therefore, one should be cautious about generalizing theories such as those of von Salvini-Plawen (1969, 1972, 1973, 1980). Furthermore, it was not the aim of Drew, who first used the term *test-cell-larva* to create a distinct type of larva, as the homology of the test with the typical velum was beyond discussion (Drew, 1901). In his view this was confirmed by the development of the filibranch *Pecten tenuicostatus* (Drew, 1906); in this species, prior to the formation of a typical velum, the prospective velar area resembles the test of the Protobranchia. However, the statement of Thompson (1961a; see also Hadfield, 1979) that the special problems of comparative embryology will remain uncertain until cell lineage has been performed to clarify the basic questions of homology is worth repeating.

The relationship between the test and the definitive ectoderm is an important problem. For *Neomenia,* Thompson (1961a) states that an epithelial continuity exists between the test and the anlage of the definitive ectoderm. In *Pecten tenuicostatus,* Drew (1906) emphasizes the continuity between the testlike prospective velar area and the remaining ectoderm of the larval body, whereas in the Protobranchia he was not able to trace an epithelial continuity from the test to the so-called new ectoderm (Drew, 1899a,b, 1901). In *Yoldia limatula,* however, he mentions that "just inside the test, between it and the new ectoderm, there are frequently a few scattered nuclei lying in a very thin film of protoplasm" (Drew, 1899a, p. 23). It may be that this is an epithelial structure that was not recognized as such and may extend from the edge of the test to the new ectoderm.

In the Protobranchia (Drew, 1901) and the aplacophorans *Dondersia* and *Proneomenia* (Pruvot, 1890, 1892) the test is cast off at metamorphosis, whereas in *Neomenia* it is overgrown by definitive ectoderm and incorporated into the head region. The test cells still containing yolk serve as food supply for the postlarval body (Thompson, 1961a).

Another problem concerning the relationship between the test and the definitive body is the formation of the nervous system. In *Neomenia,* Thompson (1961a) observed that, during the larval stage in which the test is functional, the major part of the definitive nervous tissue proliferates from three pairs of slightly depressed areas of small yolk-free cells, all situated on the prospective ventral side, anterior to the prototrochal tier of test cells. The cerebral ganglia arise from the two anterior pairs, whereas the pedal ganglia and the ventral (pedal) nerve cords originate from the posterior pair. In early postlarval development, after the test cells have been incorporated into the anterior part of the body, two other components of the nervous system come into existence, each originating from a pair of prolieration sites:

1. The ectodermal postlarval foregut forms two outgrowths, tentatively interpreted by Thompson as rudiments of the buccal ganglia, which finally fuse with the cerebral ganglia.

2. In the definitive ectoderm, immediately anterior to the mouth, two areas of proliferation are observed which form cells that finally fuse with the cerebral ganglia.

As noted by Thompson, the interpretation is handicapped by the fact that basic questions in homology remain unsolved without verification by cell-lineage studies. Because we do not know the structure with which the test is really homologous, the following discussion is somewhat speculative. The nervous tissue originating from the proliferation areas in the test is undoubtedly pretrochal in origin. The foregut is posttrochal, hence the part of the nervous system that develops from it is also posttrochal. However, the relationship of the portion anterior to the mouth remains uncertain, because we do not know the cellular homology of the test.

It is nearly impossible to compare these results for *Neomenia* with the findings of Baba (1938) about *Epimenia,* a species that does not form a test-cell-larva (Baba, 1940, 1951). When the proliferating nervous tissue is said to derive from the ''cephalic segment,'' one might presume that it is pretrochal in origin. But Baba (1938, p. 32) defines the sites of proliferation as lying ''laterally and in front of the stomodaeal pit.'' As the mouth lies in the ''middle segment,'' just posterior to the prototroch, it could be that, despite Baba's statement about the general origin from the cephalic segment in *Epimenia,* the nervous system would have the double origin that seems to be characteristic of *Neomenia.*

Exclusive pretrochal origin of the nervous system, typical of polyplaco-phorans, probably represents the primitive condition. From that point of view, organogenesis of the nervous system in *Neomenia* (and *Epimenia?*) would exhibit some signs of a more evolved state, since the combined origin from pre- and posttrochal proliferation sites is typical of the Gastropoda, Bivalvia (including Protobranchia), and Scaphopoda. Finally, in *Neomenia,* a secondary fusion of separately differentiated anlagen forms a more complex unit. This type of organogenesis of the nervous system is known to be a common feature in Gastropoda and Bivalvia and is often emphasized in considerations of evolutionary position (see the Opisthobranchia).

It is evident that more detailed knowledge of the ontogenetic patterns in these forms could provide valuable contributions to the discussion (for references, see von Salvini-Plawen, 1969, 1972) of suggested progressive and/or retrogressive evolutionary tendencies that become manifest in the organization of the Aplaco-phora.

IV. Scaphopoda

Information on organogenesis in scaphopods is rather scarce. A survey of development in *Dentalium* has been given by Lacaze-Duthiers (1857) and Kowalevsky (1883b). The cell lineage and some points of early organogenesis have been described in detail by van Dongen and Geilenkirchen (1974a,b,c, 1975) and van Dongen (1976a,b,c, 1977).

The scaphopod larva is characterized by a conspicuous prototroch, which appears at the late gastrula stage (van Dongen and Geilenkirchen, 1974a,c). The trochoblasts are arranged in three rows covering a considerable part of the embryo's surface. Even when the posttrochal region grows, the ciliary apparatus

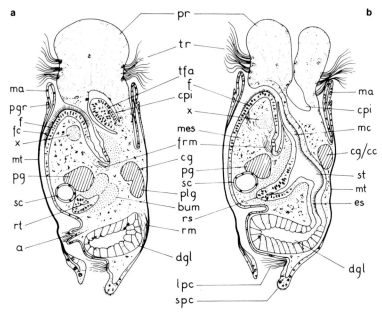

Fig. 4. *Dentalium vulgare,* larva about 7 days old (at 20°C) showing the definitive body plan. (**a**) paramedian, (**b**) slightly oblique median sections; shell omitted. The radular sac already has a rather conspicuous dimension that is typical of the adult animal (Morton, 1959). a, anus; bum, anlage of the buccal musculature; cc, cerebral commissure; cg, cerebral ganglion; cpi, cerebral pit; dgl, digestive gland; f, foot; fc, cilia of the foot epithelium; frm, foot retractor muscle; lpc, long posterior cilia; ma, mantle edge; mc, mouth cavity; mes, mesenchyme; mt, mantle; es, esophagus; pg, pedal ganglion; pgr, periostracal groove; plg, pleural (pallial) ganglion; pr, pretrochal region; rm, retractor muscle; rs, radular sac; rt, radular tooth; sc, statocyst; spc, small posterior cilia; st, stomodaeum; tfa, tentacle field anlage; tr, prototrochal cilia; x, enigmatic organ (mucus gland?). (125×.) (After van Dongen, 1976c.)

remains a conspicuous feature (de Lacaze-Duthiers, 1857; Wilson, 1904). This is still the case when the late larva is ready to shed the prototroch (Fig. 4).

Organogenesis within the posttrochal region begins in the early trochophora stage. On the dorsal side, the primordium of the shell gland becomes distinguishable, as does the anlage of the foot on the ventral side (van Dongen and Geilenkirchen, 1974b,c). Invagination of the stomatoblasts (descendants of blastomere $2b^{22}$) forms the anlage of the stomodaeum. The narrow pit caused by this invagination represents the future mouth (van Dongen, 1976c). As in all molluscs, the future shell-bearing epithelium passes through an invagination stage. After evagination, the cells of the central region of the shell field become flattened, and the mantle spreads, mostly laterally. Finally, the mantle edges meet midventrally and fuse (Kowalevsky, 1883b), while the periostracal groove at the anterior mantle edge becomes circular (Fig. 4). In this way the condition for the elaboration of the tusk-shaped shell is established at the late trochophore stage (deLacaze-Duthiers, 1857).

Organogenesis of the head region has been reinvestigated by van Dongen (1976c). In an early stage, two *cephalic plates* become recognizable by the reduced size of their cells. They soon form a pair of slight depressions, which grow inward posteriorly and build up a pair of tubular invaginations that finally extend beyond the prototroch (van Dongen, 1976c) (Fig. 4). As these invaginations are partly involved in the formation of the nervous system, they have been compared to the cerebral tubes of the gastropod embryo (Korschelt and Heider, 1936). Nevertheless basic differences do exist: it is not the pit itself (or a portion of it) that becomes a part of the cerebral ganglia (which would justify the comparison). Van Dongen (1976c) reports that the anlagen of the cerebral ganglia are formed by cells separating by proliferation from the inner end of each tube, which is the same as found in the Gastropoda and Bivalvia. Moreover, the dorsal wall of the *cerebral pits* becomes multilayered. Two thickenings protrude into the lumen of the cerebral pits (Fig. 4a) and are covered by a columnar epithelium. These protrusions are the anlagen of the tentacle (captacula) fields (van Dongen, 1976c). Further information about the development of the nervous system is almost completely lacking. It seems that all of the remaining parts of the nervous system develop from posttrochal ectoderm as in gastropods and bivalves. Kowalevsky's (1883b) description only refers to the statocysts formed by invagination and to the pedal ganglia, which originate from thickenings in the epithelium of the foot.

New studies in mesoderm formation reveal that it occurs "in full agreement with current conceptions on mesoderm formation in molluscs" (van Dongen, 1977). In *Dentalium dentale* (Geilenkirchen et al., 1971) and in *D. antillarum* (Timmermans et al., 1970), symbiotic bacteria are associated with the egg. The bacteria, initially attached to the vegetal pole of the egg, become associated with cell 4d during formation of the primary mesoderm. They become situated on the exterior surface of the mesentoblasts during gastrulation, and eventually in the follicle of growing oocytes (Geilenkirchen et al., 1971). Their presence may be interpreted as "indicative of an organogenetic relationship between the archenteron and the gonad" (van Dongen, 1977). The secondary mesoderm, in *D. dentale* derived from the second micromeres $2a^{212}$ and $2c^{122}$ (van Dongen, 1977), most likely forms the retractor muscles of the larva (C. A. M. van Dongen, personal communication, 1981).

As the early larva has a transitory ciliary apparatus it may be compared to the test-cell-larva of the Protobranchia and Aplacophora where transitory cells initially cover almost the entire embryo (Drew, 1899a,b, 1901; Korschelt and Heider, 1936; von Salvini-Plawen, 1972, 1973, 1980). One may consider the scaphopod feature as intermediate between the test-cell-larva and the trocho-phore of gastropods and bivalves, having only two relatively small rows of ciliary cells.

Although the early embryo is predominantly characterized by transitory struc-tures, development of the definitive organs starts precociously, and progresses directly. This might be due to a division of the young larva into two parts: a posterior region from which the definitive organs arise, and an anterior region of almost exclusively transitory structures. In an early stage, the pretrochal region is gradually retracted inside the prototrochal belt (van Dongen, 1976c). Moreover, the cephalic plates do not perform their organogenetic activity *in situ*. On the contrary, they form deep invaginations by which they come to lie beyond the transitory anterior part. Definitive organogenesis is thus essentially restricted to the posterior region.

V. Gastropoda and Bivalvia

A. Organogenetic Aspects of the Velum and Its Reduction

In the Bivalvia, the velum is most commonly described as a disk-shaped organ (Erdmann, 1935), whereas in the Gastropoda it is generally characterized by lobe-shaped outgrowths dorsolateral to the mouth (Fretter, 1967). However, in relatively large bivalve larvae the velum has a tendency to be lobed (Lebour, 1938, *Teredo norvegica;* Allen, 1961, *Pandora inaequivalvis;* Culliney and Turner, 1976, *Xylophaga atlantica*). The extreme case occurs in *Planktomya henseni* (Allen and Scheltema, 1972) which apparently has a long planktonic period. In gastropods, the corresponding functional needs lead to augmentation of the lobes: four lobes are common in prosobranchs with a long planktotrophic phase. Four lobes are also found in *Cymbulia peroni,* unusual among the Opisthobranchia which generally have a relatively small bilobed velum (Thiriot-Quiévreux, 1970). Even six lobes can occur as in veligers of the holoplanktonic prosobranchs (Richter, 1968; Thiriot-Quiévreux, 1975). Giant veligers, de-scribed by Dawydoff (1940), may have a twelve-lobed velum.

Some aspects of the morphogenesis of the head organs are discussed to illus-trate the variety and complexity of the processes connected with velar reduction at metamorphosis.

In aplysiids (Kriegstein, 1977b; Switzer-Dunlap and Hadfield, 1977; Switzer-Dunlap, 1978), a gradual resorption reduces the velar lobes after ciliary loss and dissociation of the ciliary cells from the velar edge. The remaining two rounded lobes seem to be the rudiments of the cephalic tentacles. Despite the absence of velar lobes, trochoblasts do differentiate in the Basommatophora (Wierzejski, 1905; Verdonk, 1965). In *Lymnaea,* the row of ciliary cells runs across the anlage of the lip tentacles ventral to the cephalic plates (Verdonk, 1965). Even the stylommatophoran *Bradybaena* shows an interrupted row of homologous, vestigially differentiated velar cells (Moor, 1977). In the prosobranch *Mangelia nebula,* Fretter (1972) notes that blobs of orange pigment, scattered in the velar lobes, pass into the cephalic hemocoel, where they concentrate and form posttentacular patches. The velar lobes themselves are thought to be cast off. What is called *head lobes* in the Calyptraeids *Crucibulum spinosum* and *Calyptraea chinensis* develops between the tentacles and the mouth during the veliger phase. They will border the mouth and eventually form the snout (Fretter, 1972).

In bivalves, more or less extensive portions of the tissue of the intra- or prevelar area are probably involved in the formation of the upper pair of labial palps (Cole, 1938; Creek, 1960; Oldfield, 1964; Bayne, 1965; D'Asaro, 1967). Their large size in the settling larvae of *Ostrea edulis* and *Pandora inaequivalvis* is interpreted as reflecting their functional importance, because in these species filtering by the gills does not yet suffice for nutrition (Yonge, 1926; Allen, 1961). D'Asaro (1967), on the other hand, notes that in settling *Chione cancellata* only rudimentary palps occur; the gills are well-developed, and the dorsal edges of the inner demibranchs lead directly into the mouth.

Loss of the velum in the Bivalvia is accompanied by rapid disintegration of the velar retractor muscles (Creek, 1960; D'Asaro, 1967). In many prosobranchs, the dorsal components of the larval retractor muscles will persist as cephalic retractors after metamorphosis. In those forms that develop a proboscis, the ventral components of the larval retractor muscles become associated with the proboscis sheath (Fretter, 1972).

B. Mantle and Shell

1. Early Development of the Shell Field

In early developmental stages, when gastrulation is not yet complete, the first sign of the shell field (Kniprath, 1977, 1979b, 1981) is an enlargement of the cells in an area at the prospective dorsal (posttrochal) side of the embryo (Ziegler, 1885; Drew, 1906; Moor, 1977; Kniprath, 1981). The wide range in morphology of this developmental stage does not support Raven's (1952) suggestion that inductive processes are involved in the development of the shell (an extensive review of the problem is given by Kniprath, 1981).

The most striking feature of the early development of the shell field is the transitory invagination of the central area forming the shell or preconchylian gland (Kowalevsky, 1883b; Horst, 1884, quoted by Jackson, 1888; Fullarton, 1890). In gastropods, the shell gland is a narrow pit with a circular opening, whereas in bivalves, it forms "a transverse groove . . . attenuating at its ends" (Horst, 1883–1884, quoted by Waller, 1981; Fullarton, 1890). The invagination

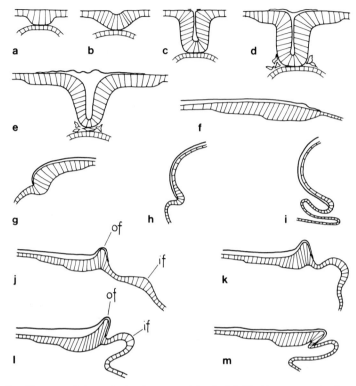

Fig. 5. Diagrams of the early development of the shell field of *Lymnaea stagnalis* (**a**) Anlage of the shell field, (**b–d**) invagination of the shell field (= shell gland stage), (**e**) evagination of the shell gland, (**f**) evagination is complete; the central area becomes thinner, (**g–i**) morphogenesis of the columellar mantle edge, (**j–m**) morphogenesis of the free mantle edge. if, inner fold; of, outer fold. (After Kniprath, 1977, from Wilhelm Roux's Archives by permission of Springer Verlag, Heidelberg.)

stage is a general feature with only a few exceptions (*Acteonia (Cenia) cocksi,* Pelseneer, 1899; Chia, 1971; *Okadaia elegans,* Baba, 1937). The functional significance already postulated by Ziegler (1885), is as follows: only the cells in the marginal zone of the shell field are able to form the periostracum. By means of the invagination, the periostracum-secreting cells are brought close together during the initial phase of secretion, thus preventing the formation of a hole in the periostracum (Kniprath, 1977, 1979a,b, 1981) (Fig. 5).

2. Morphogenesis of Mantle and Shell

Subsequently, the shell gland evaginates and the shell field spreads. This is caused not only by a flattening of the cells in the central area of the mantle roof, but also by cell divisions. The marginal zone of the mantle roof and the mantle edge, that is, the periostracum-forming tissue and the prospective mantle folds, grow only by mitotic activity (Kniprath, 1977, 1981). No shifting of tissue from one zone into another occurs in the embryonic phase or during postembryonic development (Kniprath, 1975; *Lymnaea;* 1978: *Mytilus*); differentiation of the various areas of the shell-forming tissue is definitive. The gastropod shell field has a more or less round shape and thus gives rise to a somewhat saucer- to cup-shaped shell primordium. In bivalves, the evagination and subsequent spreading lead to a dumbbell-shaped primordium (Waller, 1981) (Fig. 6). "When the shell gland actually becomes the mantle is a matter of definition made difficult by the fact that the change is gradual" (Waller, 1981, p.4). Calcification of the bivalve shell starts at an early stage of development (Erdmann, 1935; LaBarbera, 1974; Kniprath, 1980d). The midline remains uncalcified and becomes the early ligament. The underlying mantle (mantle isthmus) and adjacent parts of the fused mantle edge will be responsible for the further elaboration of ligamental structures (Owen et al., 1953; Beedham, 1958b).

Early shell calcification also occurs in most gastropods. However, in those prosobranchs that swallow a large amount of nutritive eggs, the first-formed shell is not calcified. It is elastic and is extended in the course of the storage of ingested eggs. In *Buccinum* and *Xancus,* the periostracum shrinks when the mass of stored food decreases; in *Buccinum,* a new apical shell forms under the wrinkled periostracum. In *Xancus,* on the other hand, the uncalcified part of the larval shell is eventually shed, and the apical opening is sealed off by the secretion of shell material (Bandel, 1975), which resembles shell repair without participation of the mantle edge. A delay in calcification is also reported for those larvae which spend a long time in the plankton (Scheltema, 1971).

As compared with the Gastropoda, the formation of the larval shell of the Bivalvia exhibits a rather uniform pattern (Rees, 1950). Our knowledge about details of early shell morphogenesis is restricted to planktotrophic species (Le-Pennec, 1974; LePennec and Masson, 1976). Bernard (1896) first distinguished

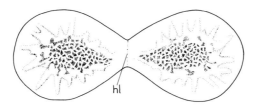

Fig. 6. Diagram of early prodissoconch I of *Ostrea edulis* shortly after evagination of the shell gland and before folding along the presumptive hinge. Pitted zones (stippled) represent the two initial centers of calcification beneath the single dumbbell-shaped periostracal layer. hl, line of the future hinge. (195×.) (After Waller, 1981, by permission of the Smithsonian Institution Press from Smithsonian Contributions to Zoology, No. 238, Smithsonian Institution Press, Washington, D.C.)

two different parts of the premetamorphic shell or prodissoconch (Jackson, 1888), today known as prodissoconch I and II (Werner, 1939). At present it is not known whether these terms are also applicable to the morphological variations observed by Ockelmann (1965) in the lecithotrophic larvae (with and without pelagic period). There is a striking difference in shell sculpture between prodissoconch I and II. Although the surface of prodissoconch I has some microsculpture (Werner, 1939; Carriker and Palmer, 1979; Waller, 1981) (Fig. 6), it is rather smooth in comparison with prodissoconch II, which is generally characterized by striae running parallel to the margin of the shell. The suggestion that prodissoconch I is the product of the shell gland and prodissoconch II of the mantle (Ockelmann, 1965; Jablonski and Lutz, 1980) appears to be incorrect (Waller, 1981). This also applies to the analogous definition of the protoconch in Gastropoda by Fretter and Graham (1962) and Fretter and Pilkington (1971). The transition from shell gland to mantle takes place in a very early stage of prodissoconch I formation. In contrast, the mantle edge does not become retractible from the margin of the valves in response to irritation until growth of prodissoconch II has advanced considerably (Waller, 1981).

The size of prodissoconch I is correlated with the egg size (Ockelmann, 1965). According to Waller (1981), formation of prodissoconch II starts as soon as the two valves surround the body and close against each other along their free margins. Waller's interpretation is based on criteria of functional morphology. On the other hand, for prosobranchs, Robertson (1971) distinguishes an embryonic and postembryonic (= larval) shell, based on the different environments in which the shells are formed. Concerning these two types, Thiriot-Quiévreux (1972) uses the terms protoconch I and II. Therefore, Robertson's (1974) terminological parallelism of protoconch I/II and prodissoconch I/II remains doubtful (Jablonski and Lutz, 1980). For the prosobranch shell, this terminology remains unsatisfactory, because a comparative study of shell formation reveals a higher complexity of shell formation patterns (Bandel, 1982).

A striking feature of embryonic shell morphogenesis is found in certain shell-less Opisthobranchia. Whereas most shell-less species form a larval shell in which the width of the aperture corresponds to the greatest diameter of the whorl (shell Type 1; Thompson, 1961b), a few species among the most evolved nudibranchs (Thompson, 1961b) form a so-called inflated shell, having a whorl that is much wider than the aperture (shell Type 2; Thompson, 1961b, 1976; Pelseneer, 1911; Williams, 1971) (Fig. 7). The latter type of shell is formed even in forms with planktotrophic larvae. Shell Type 1 shows additional larval shell growth. In *Aplysia californica* (Kriegstein, 1977b), where the shell persists, and in *Doridella obscura* (Perron and Turner, 1977), where the larval shell is cast off, larval shell growth is restricted to the initial phases of planktonic life. Type 2 shells are calcified to an extent probably similar to Type 1 shells, and the insertion of the larval retractor muscles resembles the structure formed by other molluscs (Bonar, 1978a).

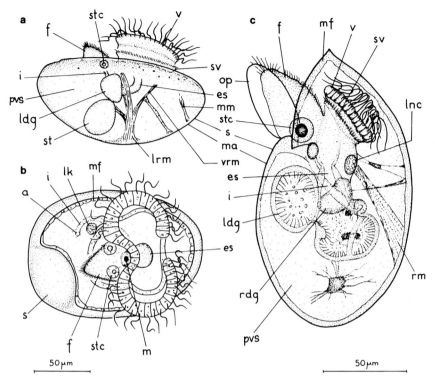

Fig. 7. Shell Type 2 in Opisthobranch veligers. (**a** and **b**) *Dendronotus frondosus*, early veliger, (**a**) lateral, (**b**) ventral view. (After Williams, 1971.), (**c**) *Hancockia burni*, free swimming veliger. (After Thompson, 1972.) Both species belong to the Nudibranchia Dendronotacea. The "inflated" feature of the shell is apparent from the relative dimension of the perivisceral space. Mark the slight sinistrality of the shell in the ventral view, which, in fact, is a hyperstrophy (Thompson, 1958; Horikoshi, 1967). a, anus; f, foot; i, intestine; ldg, left diverticle of the digestive gland; lk, larval kidney; lnc, left nephrocyst; lrm, left retractor muscle; m, mouth; ma, mantle; mf, mantle fold; mm, mantle muscle; es, esophagus; op, operculum; pvs, perivisceral space; rdg, right diverticulum of the digestive gland; rm, larval retractor muscle; s, shell; st, stomach; stc, statocyst; sv, subvelum; v, velum; vrm, velar retractor muscle.

3. The Mantle Edge

In most Gastropoda, two folds enclose the periostracal groove; the outer one is involved in the formation of the periostracum (Kniprath, 1970, 1971, 1972). Although in *Lymnaea* the two folds develop synchronously shortly after evagination of the shell gland (Kniprath, 1977, 1981) (Fig. 5), there are prosobranch and stylommatophoran species (e.g., *Marisa*, Kniprath, 1979a,b, 1980b, and *Achatina*, Brisson, 1968) where the prospective inner fold precociously overgrows the anlage of the outer fold and may also overgrow the marginal parts of the evaginating shell gland (Kniprath, 1979a,b, 1980b,f). In slugs, fusion of the tissue of the fold constitutes an epithelial sac wherein the rudimentary shell develops. Shell formation does not really occur inside the shell gland, because only the proximal wall is homologous with the shell-forming tissue of ecto-cochleate (external shelled) forms (Kniprath, 1979a,b, 1980c).

Although adult bivalves have three (Yonge, 1957; Beedham, 1958a) or, in a few cases, four marginal folds (Hillman, 1964), the settling larva (pediveliger)

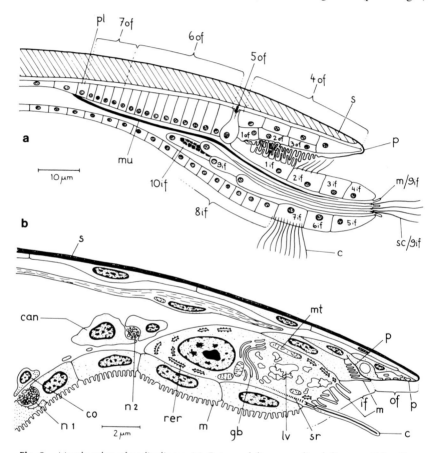

Fig. 8. Mantle edge of pediveligers. (**a**) *Ostrea edulis*, generalized diagram. (After Cran-field, 1974; modified with permission of the author; from the Journal of the Marine Biological Association by permission of the Cambridge University Press, Cambridge.) (**b**) *Pecten maximus*, mantle edge in the anterior region of the mantle. (From Cragg, 1976.) These figures indicate the structural diversity already present in the larvae competent to settle. The numbers in (**a**) indicate the different cell types that are discernable. c, cilium; can, cell associated with n2; co, com-panion cell to n1; gb, Golgi body; if, inner fold; lv, electron lucent vesicle; m, microvilli; mt, mitochondrium; mu, musculature; n1, n2, nerves; of, outer fold; p, periostracum; pl, pallial line; rer, rough endoplasmic reticulum; s, shell; sc, sensory cilium; sr, striated root.

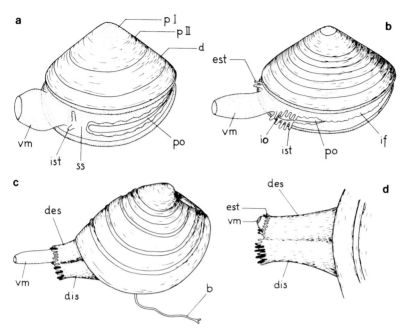

Fig. 9. Stages of the postlarval development of the siphons in *Mercenaria mercenaria*. For explanation, see text. b, byssus; d, dissoconch; des, definitive exhalant siphon; dis, definitive inhalant siphon; est, tentacles of the exhalant siphon; if, inner fold; io, inhalant opening; ist, tentacles of the inhalant siphon; po, pedal opening; pI, prodissoconch I; pII, prodissoconch II; ss, siphonal septum; vm, valvular membrane. Magnifications: (**a**) 50×, (**b**) 15×, (**c**) 5×, (**d**) 1×. (After Carriker, 1961.)

generally has two folds (Ansell, 1962; Cranfield, 1974; Cragg, 1976) (Fig. 8). Three mantle folds in the planktonic veliger stage are reported for *Planktomya henseni* (Allen and Scheltema, 1972).

In the boring clams *Xylotrya gouldi* and *Zirphaea crispata*, the pediveliger forms the exhalant siphon (Sigerfoos, 1908; Werner, 1939), which always develops first (Kändler, 1927; Quayle, 1952; Ansell, 1962). In *Venus striatula*, the first process of mantle fusion, the formation of the siphonal septum, occurs in a late larval stage. There are few detailed studies on the morphological aspects of mantle fusion (Belding, 1912; Quayle, 1952; Carriker, 1961; Ansell, 1962). In *Mercenaria mercenaria* (Carriker, 1961), the exhalant siphon forms by a rather long sleevelike extension of the inner mantle fold (valvular membrane) (Fig. 9a). Formation of the inhalant siphon starts with a splitting of a part of the formerly fused tissue, which marks the siphon septum prior to the growth of the valvular membrane (Fig. 9b). The valvular membrane further elongates (Fig. 9c) during development of the definitive exhalant and inhalant siphons from the middle fold. It eventually becomes reduced to a short, funnel-shaped outlet, when the definitive siphons extend well beyond the margin of the shell valves (Fig. 9d). As Caddy (1969) concludes from a study of *Macoma balthica,* this transition from a well-adapted planktonic organism to a well-adapted benthic animal involves specific anatomic and behavioral adaptations.

C. Mantle Cavity

It is commonly assumed that in bivalves the mantle cavity is formed by invagination (Raven, 1966). According to Waller (1981), formation of the mantle cavity of *Ostrea edulis* does not begin before growth of prodissoconch II (see Section V,B,2) has begun. Apparently the mantle cavity is formed by an outgrowth of the mantle and enlarges secondarily by invagination. This agrees with the earlier observations of Erdmann (1935). The gills will develop in that secondarily formed region of the mantle cavity. Erdmann (1935) therefore called this region the gill cavity, referring to Hatschek (1880) who first mentioned it.

In the Gastropoda, the outgrowth of the mantle edge contributes to the development of the mantle cavity (Demian and Yousif, 1973d). As in the Bivalvia, enlargement of the mantle cavity is intimately associated with shell growth; the growing shell supports the growing mantle. Additional extension by invagination may occur (Raven, 1966). For a discussion of the relationship between the mantle cavity and the gills, the reader is referred to the extensive review of Raven (1966).

The external morphology of early stages in development of the filibranch type has been studied by Waller (1981) in*Ostrea* by scanning electron microscopy. The gill primordium forms as an epithelial ridge along the dorsal side of the gill cavity. This is preceded by the formation of a cluster of cilia on each side near the posterior edge of the mantle. The cilia represent the precursors of the gill bridge, which at a later stage is formed by an epithelial cross-fusion (probably preceded by cross-fusion of the cilia). Filament development begins in the anterior part of the gill primordium. Minute transverse ridges crossing the gill ridge at right angles represent the primordia of the gill filaments.

The anlage of the pulmonate lung appears as an ectodermal invagination even before development of the mantle cavity (Meisenheimer, 1898; Heyder, 1909; Régondaud, 1961a, 1964; Brisson, 1968; Moor, 1977). During development, the lung widens and penetrates further into the body. When the rudimentary mantle cavity develops, the lung will open into it. The anus is also located in the mantle cavity, from which the anlage of the ureter arises as a tubular outgrowth. The osphradium, present in aquatic pulmonates (Arni, 1973), also originates from it (Régondaud, 1964). The extension of the rudimentary mantle cavity between the mantle fold and the body has been designated the supranuchal cavity by Régondaud (1964).

D. Formation of the Dorsal Body Wall in Shell-less Opisthobranchia

Metamorphosis of the shell-less Opisthobranchia involves a shedding of the veliger shell. Such a change into a sluglike body plan involves not only the rearrangement of the inner organs, but also the formation of a new body wall, for the tissue which has elaborated the larval shell is also transitory. In *Aeolidiella alderi*, whose shell is of Type 1 (see previous discussion), Tardy (1970b) describes cellular multiplication in the zone immediately beneath the mantle edge which had formed the larval shell (Thompson, 1958, 1962). This causes a

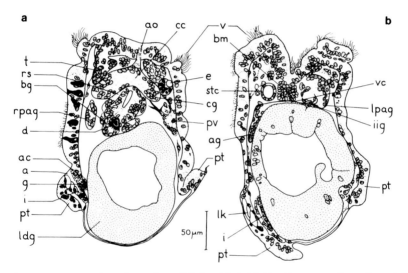

Fig. 10. Horizontal sections of *Aeolidiella alderi*. (**a**) Hatching stage. Spreading of the pallial thickening, which will give rise to the notum, begins some time before hatching at the right side of the larva, whereas at the left side, the thickened tissue is still in its original position at the hatching stage. (**b**) Several hours later, the covering of the visceral mass by the pallial thickening has advanced markedly. a, anus; ac, anal cell; ag, abdominal ganglion; ao, apical organ; b, brain; bg, buccal ganglion; bm, buccal mass; cc, cerebral commissure; cg, cerebropleural ganglion; cn, cnidosac; d, radular tooth; dt, digestive tissue (stippled); e, eye; g, gonadal anlage; i, intestine; iig, infraintestinal ganglion; ldg, left lobe of the digestive gland; lk, larval kidney, lpag, left parietal ganglion; n, notum; p, anlage of the palp; pt, pallial thickening ("bourrelet palléal" of Tardy); pv, pulsating vesicle; rdg, right lobe of the digestive gland; rh, rhinophoral anlage; rpag, right parietal ganglion; rs, radular sac; st, tissue of the prospective stomach; stc, statocyst; svp, spot of violet pigment; swp, spot of white pigment; t, ectodermal invagination involved in the formation of the central nervous system ("telencephalization" of Tardy) (see Section V, F); vc, visceral commissure; I_1, II_1, etc., first papillae of the series I, II, etc. (After Tardy, 1970b; from the Annales des Sciences Naturelles, Zoologie et Biologie Animale, by permission of Masson Editeur, Paris.)

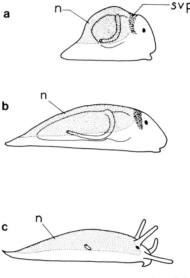

Fig. 11. Morphogenesis of the notum (stippled area) of *Aeolidiella alderi*. (Animals not to scale.) For key to abbreviations, see Fig. 10 legend. (After Tardy, 1970b; from the Annales des Sciences Naturelles, Zoologie et Biologie Animale, by permission of Masson Editeur, Paris.)

thickening from which the dorsal epidermis forms (Fig. 10). The developing notum (Fig. 11) appears to be pallial in origin.

Proliferation of the mantle tissue is probably also involved in the formation of a new dorsal body wall in *Aegires punctilucens*, which shows the unusual feature of shedding its Type 1 shell and forming the notum during the veliger stage (Thiriot-Quiévreux, 1977).

Bonar (1976) concludes from his observations on *Phestilla sibogae*, which bears a Type 2 shell, that the dorsal epidermis might originate from epipodial tissue. These arguments are based exclusively on an analysis of the actual course of shell shedding, accompanied and followed by a relatively rapid spreading of the epipodial epidermis (Bonar and Hadfield, 1974; Hadfield, 1978). Nearly nothing is known about the organogenesis that follows shell shedding, that is, about the formation of the notum. The cerata, which originate from the notum during juvenile development (Fig. 12), are probably innervated by the pleural ganglia (Russell, 1929). Contradictory suggestions of pedal innervation need reinvestigation (Tardy, 1970b; Bonar, 1976).

During the postmetamorphic stages of other species, the dorsal area may be covered by newly formed epidermis, such as in *Berthella plumula*. This species does not discard its shell, and exhibits the kind of pallial thickening that causes proliferation of mantle over the shell (Tardy, 1970b).

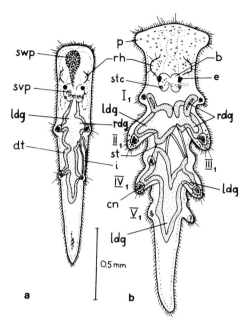

Fig. 12. Later larval morphogenesis in *Aeolidiella alderi*. (a) The pseudovermis stage, characterized by the presence of four rudimentary papillae (Tardy, 1962). (b) Further development shows that the first four papillae belong to series II and IV, whereas series I, III, V, etc. become successively manifest during further development. The right lobe of the digestive gland remains small, only protruding into the papillae of series I and II of the right side, whereas the left lobe will produce the diverticula extending into all the remaining papillae. For key to abbreviations, see Fig. 10 legend. (After Tardy, 1970b; from the Annales des Sciences Naturelles, Zoologie et Biologie Animale, by permission of Masson Editeur, Paris.)

E. Foot

In most gastropods, the anlage of the foot becomes visible as a projection behind the mouth at an early embryonic stage. In free larval stages, differentiation of the sole contributes a median band of cells with cilia to the *cephalopedal ciliary apparatus* of the veliger larva. These cilia create a rejection current which removes particles too large for ingestion (Thompson, 1959; Fretter, 1967; Fretter and Montgomery, 1968). In the Bivalvia, on the other hand, the foot develops at a relatively late larval stage, the pediveliger (Creek, 1960; Ansell, 1962; Sastry, 1965).

The premetamorphic organogenesis of the foot is characterized by an important contribution from the ectoderm. In addition to a considerable portion of the pedal mesenchyme, the pedal ganglia and statocysts are also formed from the epithelium of the foot (Honegger, 1974; Giese, 1978; Raven, 1966). In most prosobranchs and opisthobranchs, the posterior dorsal side bears an operculum, which is transitory in several groups, especially in the Opisthobranchia. Various patterns of strongly developed cilia are present. Conspicuous features in both gastropods and bivalves are the pedal mucus glands and, in the latter, the byssus apparatus. These glands occupy a large part of the interior of the foot (Atkinson, 1971; Cranfield, 1973a; Bonar and Hadfield, 1974; Lane and Nott, 1975). They are composed of enlarged flask-shaped cells with narrow necks, within the epithelium. The nuclei have a basal position. The abundant rough endoplasmic reticulum is involved in the elaboration of the secretory product (Cranfield, 1973a; Lane and Nott, 1975).

In *Mytilus edulis* nine types of glands are identified by Lane and Nott (1975) (Fig. 13); they may be divided into three main groups. Five types of glands form the byssus apparatus. They are associated with a pair of lateral pouches having a posterior duct which opens at the ventral surface of the foot near the heel. From its opening, a groove extends along the median line of the sole. Two pedal glands

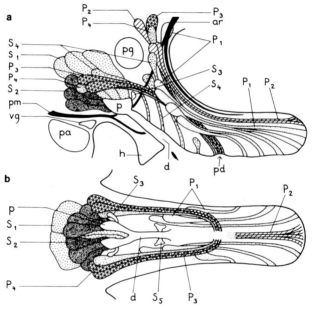

Fig. 13. Diagrams showing the arrangement of glands in the foot of the pediveliger of *Mytilus edulis.* (**a**) Sagittal section, (**b**) horizontal section. Cells of the nine different glands are indicated as P_1, P_2, P_3, P_4, S_1, S_2, S_3, S_4, and S_5. ar, anterior retractor muscle; d, posterior duct; h, heel; p, lateral pouch; pa, posterior adductor muscle; pd, pedal depression; pg, pedal ganglion; pm, posterior retractor muscle; vg, visceral ganglion. (After Lane and Nott, 1975; from the Journal of the Marine Biological Association by permission of Cambridge University Press, Cambridge.)

open into the depression within this groove, at some distance anterior to the opening of the posterior duct. They are probably functional only at metamorphosis, attaching the secondary byssus to the substratum (Cranfield, 1975). Two pedal glands open at the tip and onto the ventral and ventrolateral surfaces of the foot, secreting mucus for ciliary locomotion in the initial phase of settlement (Quayle, 1952; Carriker, 1961; Ansell, 1962; Cranfield, 1973b,c; Gruffydd et al., 1975).

Mytilus retains a functional byssus apparatus—the five gland types invariably persist in the adult stage—and remains able to break the byssus to change the place of attachment (Lane and Nott, 1975). *Ostrea,* which resorbs the whole foot after the larva has become cemented to the substratum (Cole, 1938; Cranfield, 1973b), resembles *Mytilus* also in the differentiation of nine types of glands (Cranfield, 1973a). In *Pecten,* where the byssus apparatus is a transitory structure used only at metamorphosis, five types of glands are recorded (Gruffydd et al., 1975). Surveying the presence of a byssus apparatus in adult bivalves, Yonge (1962) interpreted this phenomenon as the retention of an organ originally fitted for function during metamorphosis only.

In the nudibranch *Phestilla sibogae,* well-developed pedal glands are recorded by Bonar and Hadfield (1974). Besides the relatively small propodial mucus glands, not yet developed in the premetamorphic stage, large metapodial glands are present. Secretion by these glands firmly anchors the larva to the substratum during shell shedding (see Section V,D) performed by vigorous muscle activity (Bonar, 1976).

The podocyst is an organ *sui generis,* an embryonic extension originating from the posterior tip of the foot, which occurs only in the stylommatophoran pulmonates, but is lacking in the Succineidae (Cather and Tompa, 1972). Other than functioning in circulation and gas exchange (see Section V,H) it also participates in the uptake of albumen. Cather and Tompa (1972) found that the ciliated podocyst bears an extensive coat of microvilli, and thus is especially fitted for the uptake of albumen by pinocytosis, as demonstrated by the use of tracers. But there is also some evidence that the cells of the podocyst are involved in the removal of catabolic products (Quattrini and Sacchi, 1971).

F. Nervous System and Sense Organs

Attention paid to the organogenesis of the nervous system has remained almost exclusively confined to the more obvious structures such as ganglia and their connections. Nearly nothing is known about the *primitive nervous plexus* (Werner, 1955), which coordinates the behavior of the larva, particularly the movements of the velum (Carter, 1926, 1928). Carter proposes a connection of the neural elements of the velar lobes with the cerebral ganglia; this is denied by Werner (1955) and Fretter (1967). The apical organ, probably present in all free-

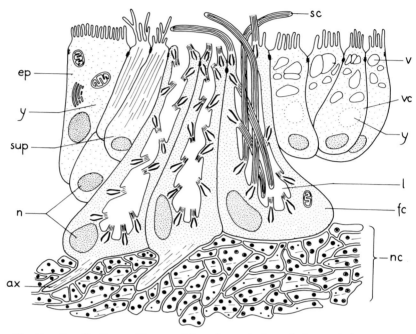

Fig. 14. *Phestilla sibogae*. Diagram of the cell types in the cephalic or apical sense organ of the veliger. For clarity, most of the cilia in the flask cells are "amputated." Not to scale. ax, axon; ep, typical epidermal cell; fc, flask cell; l, lumen of a flask cell; n, nucleus; nc, nerve cells of the cerebral commissure; sc, sensory cilia; sup, supporting cell; v, vacuoles; vc, vacuolated cells; y, yolk. (After Bonar, 1978b.)

swimming gastropod larvae (Bonar, 1978c), has been investigated by Bonar (1978b) in the nudibranch *Phestilla sibogae* (Fig. 14). For bivalve veligers, Boyle and Turner (1976) suggested that disagreement about the structure of the homologous organ (whether there are flagella present of not) is partly due to technical difficulties.

The cerebral ganglia of gastropods and bivalves, the different forms of tentacular ganglia in gastropods, and the accessory parts of the brainlike nervous center in certain opisthobranchs and pulmonates, arise from the pre- or intravelar area. All the remaining parts (except the pleural ganglia in some forms) have a postvelar (posttrochal) origin. In prosobranchs and pulmonates, the cerebral ganglia originate from a pair of epithelial areas (The cerebral, cephalic, or sense plates) close to the cells that form the median apical plate (Wierzejski, 1905; Régondaud, 1964; Verdonk, 1965; Honegger, 1974; Giese, 1978). During further development tentacles will grow out from the cephalic plates. In pulmonates, a pair of invaginations, the so-called cerebral tubes (Meisenheimer, 1898), arise on each side posterolateral to the tentacular anlage; they will give rise to the procerebrum (in Stylommatophora) or lateral lobes (in Basommatophora) and the follicle gland (Lever, 1958; Lever et al., 1959; van Mol, 1967).

In the nudibranch *Aeolidiella alderi* (Tardy, 1970b, 1974) two deep invaginations grow inward from the intravelar area (Thompson, 1958, 1962). They form parts homologous with the cerebral ganglia of organisms with a less concentrated nervous system (e.g., aplysiids), the components of the pleural ganglia (according to Tardy they are not discernable from the cerebral parts), the eyes, the optic ganglia, the rhinophoral ganglia, and the ganglia of the palps. The latter will fuse with the brain, whereas the optic and rhinophoral ganglia remain separate. Finally, the invagination separates from the body wall and is transformed into the ganglionic complex. Tardy (1970b, 1974) calls this phenomenon *telencephalization* (cf. Fig. 10a) in order to distinguish it from the *cerebralization* (the concentration of the ganglia) which also happens in late development of *Aeolidiella*. However, for an evaluation of Tardy's comparison to the pulmonate pattern further studies are needed.

In *Aplysia californica* no such invaginations are present, and a concentration of the ganglia does not occur; in fact, they will move farther away from each other during juvenile development (Kriegstein, 1977a,b). Organogenesis of the mediodorsal bodies is not yet elucidated. If the suggestion of Boer et al. (1968; Boer and Joosse, 1975) is correct, even mesodermal parts are incorporated into the cephalic nervous centers.

A similar diversity exists in the Bivalvia. For instance, in *Dreissensia,* the cerebral ganglia and their connections arise simultaneously from the ectoderm. The pleural ganglia, which will fuse with the cerebral ganglia, proliferate separately from the posttrochal epithelium (Meisenheimer, 1901a). In *Lasaea rubra,* on the other hand, Oldfield (1964) reports that the cerebral and pleural components arise together from a deep invagination, necessitated by the presence of a large yolky cephalic mass.

Whereas in *Ampullarius* the pleural ganglia, proliferating from the pedal ectoderm, approach the pedal ganglia (Honegger, 1974), in *Littorina saxatilis* they come to lie near the cerebral ganglia. In this species the clearly discernable postvelar proliferation is not immediately followed by the formation of a distinct anlage. The cells remain diffusely arranged (Guyomarc'h-Cousin, 1974). A similar observation is reported by Moritz (1939) for *Crepidula adunca*. In *Lymnaea*, the pleural ganglia also arise immediately behind the velum (Raven, 1975).

The pedal ganglia proliferate from the epithelium of the foot anlage. They become closely associated with the statocysts, which are, however, innervated by the cerebral ganglia. The early stages of the formation of the anlage of the visceral commissure and its ganglionic components have not been thoroughly investigated. Several distinct points of proliferation exist in *Aeolidiella* (Tardy, 1970b) and in *Bradybaena* (Moor, 1977). The proliferation sites are restricted to the area where the rudimentary mantle cavity will form. Ganglionic fusion is also observed in this part of the nervous system (Brisson, 1964a; Régondaud, 1961b, 1964).

A striking transitory feature, restricted to the early phase of neural organogenesis in the Stylommatophora, is the presence of the so-called integumental sense organs (Meisenheimer, 1898). They differentiate exclusively in areas of the ectoderm from which the anlagen of the ganglia have previously separated. Though they may resemble small sense buds (Raven, 1966), a cytological investigation reveals that they lack the structure of true sense buds. Rather, they are dynamic structures from which cells proliferate before they seem to be fully differentiated and finally move away from the epithelium (Moor, 1982).

Several bivalve veligers are characterized by a pair of pallial eyes (Pelseneer, 1900, 1908; Erdmann, 1935; Bayne, 1964), generally situated near the gills or gill rudiments. In *Ostrea edulis,* the early stages of development show a differentiation of a strong microvillous tuft (Waller, 1981) on a thickened area (Erdmann, 1935) in the region where the epithelium will invaginate to form the future eye cup. This resembles the development of the eyes in *Haliotis* (Crofts, 1938) and in *Gibbula cineraria* (Underwood, 1972), where differentiation of the eye pigment precedes the morphogenetic movements of the epithelium. The eye rudiments develop on the projecting anlagen of the optic tentacles, and the formation of the eye cup starts later. The course of eye development in *Helix aspersa* (Eakin and Brandenburger, 1967) is quite different; no cytological differentiation occurs before the basic morphogenetic processes. Regarding further development, Zunke's (1978, 1979) results in *Succinea* are in agreement with those found in *Helix,* which supports the typically stylommatophoran condition of *Succinea* that was questioned by Rigby (1965).

Although in the Gastropoda the statocysts usually appear rather early (e.g., in *Lamellaria*), even before the formation of the pedal ganglia (Fioroni and Meister, 1976), their appearance in bivalve larvae seems to be postponed until the late veliger stage (Sastry, 1965; Cragg and Nott, 1977; Frenkiel and Mouëza, 1979). In the Bivalvia they are generally formed by invagination. This holds also for the prosobranchs *Marisa* (Demian and Yousif, 1973a, 1975) and *Ampullarius* (Honegger, 1974), whereas in *Buccinum* vesicle formation follows separation of the anlage from the epithelium (Giese, 1978). A similar variety in early organogenesis seems to be present in the pulmonates (Raven, 1975). Complete separation from the body wall is typical of the gastropods. It also occurs in many bivalves.

In *Ostrea edulis* and *Pecten maximus,* however, the statocysts retain a funnel-shaped connection with the exterior (the static duct; Werner, 1939; Cragg and Nott, 1977; Waller, 1981). Cilia are present not only in the cells of the statocyst but also in the cells lining the static duct (Cragg and Nott, 1977). The direction of the cilia in the duct and the flexion of the cilia bordering the exterior orifice suggest that their beat supports an inwardly directed flow (Cragg and Nott, 1977;

Waller, 1981). In this way small objects could be taken up, functioning as statoconia. This is in agreement with the observation of Cragg and Nott that the statoconia show a great variety in shape, size, and composition. Their size never prevents them from passing through the duct. Therefore, the old term of *excretory duct* (Barber, 1968) should be abandoned. In each statocyst, one nonciliated cell is present containing a variety of granules that somewhat resemble statoconia. However, the function of these nonciliated cells is not yet known (Cragg and Nott, 1977).

G. Alimentary Tract

1. Foregut

In pulmonates, gastrulation takes place by invagination. Initially a communication exists between the archenteron and the foregut (Fig. 16a). The latter originates by multiplication and invagination of the stomodaeal cells lying close to the blastopore. In the prosobranchs *Marisa* (Demian and Yousif, 1973a) and *Ampullarius* (Honegger, 1974), which show a development similar to the pulmonates (small ova and the presence of albumen as a nutrient), the stomodaeal invagination is a blind tube (Fig. 15a). It develops after closure of the blastopore, and fuses secondarily with the endoderm to establish a continuous lumen. This behavior is probably typical of all prosobranchs, although the time of fusion may vary considerably. It is also true for *Viviparus*, in which the blastopore is not closed, and represents the future site of the anus (Dautert, 1929; Fioroni, 1980).

Opisthobranchs behave the same way as prosobranchs (Thompson, 1958, 1962, 1967; Tardy, 1970b). In all forms cilia develop in the foregut except in areas that will form the radular sac and the jaw (Demian and Yousif, 1973a,b; Moor, 1977) (Fig. 15). The radular sac develops as an invagination of the prospective ventral wall of the buccal cavity. In pulmonates (e.g., *Bradybaena*) its development starts as soon as the prospective buccal cavity forms a wide depression (Moor, 1977). In *Helix* (Fol, 1880) and *Physa* (Wierzejski, 1905), the anlage of the radular sac seems to appear outside the mouth, which is interpreted by Riedl (1960) as a retardation of the formation of the buccal cavity. In *Rhodope*, this part of the foregut is reduced, and the radular sac is a transient structure. In the adult, the buccal ganglia and the opening of the salivary glands lie near the mouth (Riedl, 1960).

Fretter (1969, 1972) describes the complex organogenesis of the foregut in monotocardians, in which the adult bears a proboscis. In *Nassarius incrassatus* (acrembolic type) the prospective anterior esophagus develops as a diverticulum beneath the larval foregut of the veliger. The radular sac is formed from its blind end. The salivary glands develop alongside the anterior part of the esophagus. Their ducts do not communicate with the buccal cavity prior to metamorphosis.

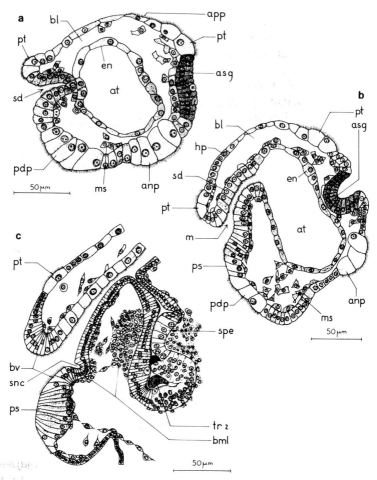

Fig. 15. *Marisa cornuarietis.* Sagittal sections. (**a**) 4-day-old embryo, (**b**) at 6 days, and (**c**) at 14 days. The average period of embryonic development leading to a crawling hatching stage is 20 days at 15–20°C. anp, anal cell-plate; app, apical cell-plate; asg, anlage of the shell gland; at, archenteron; bl, blastocoel; bml, buccal muscles; bv, buccal vestible; en, endoderm; hp, head plate; m, mouth; ms, mesenchyme cells; pdp, pedal cell plate; ps, peristome; pt, prototroch; sd, stomodaeum; snc, sublingual cavity; spe, supraradular epithelium; tr2, tooth of the second row. (After Demian and Yousif, 1973a (**a** and **b**), and 1973b (**c**).)

The cells that build up the proboscis, the salivary glands and their ducts, the muscles, and the cartilage of the odontophore proliferate either from the blind end of the early anlage of the anterior esophagus or from the gut wall near the origin of the diverticulum. The pleurembolic proboscis also develops beneath the larval foregut.

Kerth (1979) studied the morphogenesis of the radula in several Basommatophora. The uniform tooth pattern in pulmonates is attained secondarily. Initially two rows of lateral teeth are formed. As the radular sac and the girdle of the odontoblasts grow, further rows are added laterally before the radula gets its typical uniformity by development of the central teeth. The juvenile growth of the pulmonate radula consists of the following processes:

1. An increase in the number of transverse rows by a rate of formation at the proximal end of the radula exceeding the rate of loss at the distal end

2. Lengthening of the girdle of odontoblasts by differentiation of new odontoblasts at its ends so that the number of longitudinal rows of teeth is also increased

3. Growth of the odontoblasts giving rise to larger teeth (Kerth and Hänsch, 1977). Runham (1975) observed that odontoblasts bear exceedingly long microvilli, the arrangement and length of which are responsible for the characteristic morphogenesis of the teeth. They become incorporated in the hardening matrix of the developing teeth and finally become detached from the apex of the odontoblasts.

New teeth form at the proximal end of the radula, and disintegrate at the distal end. In *Viviparus fasciatus,* Kerth et al. (1981) observed this process as early as the late embryonic period. Up to 30 rows of teeth may be lost before hatching. The process starts with disintegration of the radular membrane, liberating single teeth; all radular substances seem to be entirely dissolved. The enzymes involved in this process are apparently secreted by the embryo itself; there are no signs of bacterial contributions (Kerth et al., 1981).

2. Midgut and Hindgut

If gastrulation occurs by invagination, the archenteron represents the primordial lumen of both midgut and hindgut (Demian and Yousif, 1973a,b; Honegger, 1974); in the pulmonates it represents only the midgut. If gastrulation takes place by epiboly, a lumen is formed secondarily; the anlagen of the different parts of the migdut and hindgut are first made apparent by their histological differentiaton. Often the yolk is concentrated in a few giant cells, the yolk macromeres, which will never divide (Fioroni, 1965a,b, 1966b; Fioroni and Meister, 1976; Giese, 1978). In the embryos of *Marisa* (Demian and Yousif, 1973a,b) and *Ampullarius* (Honegger, 1974), which ingest albumen, most of the cells lining the archenteron have large vacuoles. The small-celled endoderm, which will form the definitive structures, is restricted to the posteroventral part, with an open communication to the anlage of the hindgut. In pulmonates such as *Bradybaena* (Moor, 1977), the small-celled endoderm has a dorsal position, and the anlage of the intestine is independent of it (Fig. 16). Thus, despite a similarity in the type of nutrition used, the difference in the pulmonate and prosobranch patterns is evident. In *Buccinum* (Portmann and Sandmeier, 1965; Giese, 1978), however, a swelling of the archenteron, caused by the uptake of large amounts of nutritive eggs, does not mask the prosobranch pattern. The hindgut develops a vesicular dilation, which plays an important role in the process of yolk assimilation. The midgut retains its saclike shape for a relatively long time, so that the

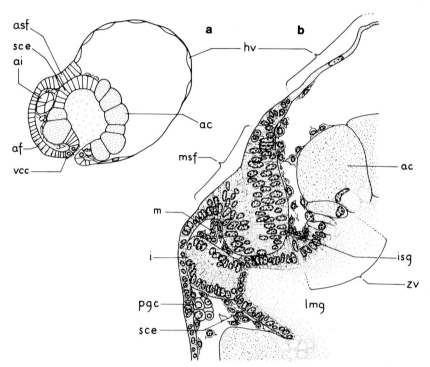

Fig. 16. *Bradybaena fruticum.* Sagittal sections. (**a**) 5-day-old embryo, (**b**) at 8 days (development at 18–22°C; the crawling stage hatches after 28–29 days). The most prominent features are the independent anlagen of the intestine and of the primary germ cells; the latter are recognizable for the first time at the age of 6 days. ac, giant (primary) albumen cells of the midgut; af, anlage of the foot; ai, anlage of the intestine; asf, anlage of the shell field; hv, head vesicle; i, intestine; isg, invaginated part of the shell field (= shell gland); lmg, lumen of the midgut; m, mesenchyme; msf, marginal zone of the shell field which does not invaginate; pgc, primordial germ cells; sce, small celled endoderm; vcc, ventral ciliary cells of the esophagus; zv, zone of vacuolization giving rise to the secondary albumen cells which will never reach the dimensions of the primary ones (ac). (After Moor, 1977.)

assumption of the definitive pattern is delayed until a relatively late stage. The development of the stomach and of the two digestive gland diverticula starts with a histological differentiation of the respective areas.

In planktotrophic veligers of prosobranchs, the stomach, provided with a gastric shield, passes into a style sac, from which the intestine originates. In living larvae, Fretter and Montgomery (1968) have observed an intestinal groove and typhlosole in the wall of the style sac, whereas the histological study of Thiriot-Quiévreux (1974) points to the presence of a protostyle-like structure in the sac lumen. In herbivorous forms this will be transformed into the crystalline style. In adult carnivorous species these structures are lacking; they are lost at metamorphosis.

In the opisthobranch *Aeolidiella* (Fig. 10a,b), the yolk is in the large cells of the anlagen of the stomach and in the left (greater) and right (smaller) diverticulum of the digestive gland, whereas the anlage of the hindgut originally consists of small cells (Tardy, 1970b). In *Aeolidiella* (Fig. 12) as in most gastropods, the left diverticulum will become much larger than the right one. In *Onchidella* a third opening comes into existence between the left diverticulum and the stomach (Fretter, 1943). In pulmonates, the organogenesis of the left and right diverticula seems to be somewhat different. The diverticula differ in the number and size of the cells involved in the storage and digestion of albumen. The left diverticulum becomes much larger, particularly in the Stylommatophora, where it protrudes anteriorly into the head vesicle (Brisson, 1964b, 1968; Weiss, 1968; Moor, 1977). The diverticula also differ with respect to the fate of the cells that bear the large albumen vacuoles. Their relationship to the differentiation of the definitive digestive tissue (of the diverticula) has yet to be elucidated (Weiss, 1968; Arni, 1973, 1974, 1975; Moor, 1977).

For the Bivalvia, consistent new data on the alimentary tract are not available. Development of the foregut is characterized by a lack of any recapitulative differentiation corresponding to the absence of a radula in the adult. The marked structural changes that occur in the transition from a pelagic to a benthic life do not affect the alimentary tract, since the diet does not fundamentally change. The style sac is already present in the larva and takes part in the digestion of the food (Nelson, 1918). Because it persists in the adult, it is typically a larvo-adult organ (Yonge, 1926). Extensive reviews are given by Raven (1966) and Wada (1968); the accounts of Sastry (1979) and Andrews (1979) present valuable reference lists.

H. Excretory and Circulatory Systems

Paired protonephridia are found in the Bivalvia and Pulmonata. In the bivalves and basommatophoran gastropods, they consist of two to four cells (Meisenheimer, 1899, 1901a; Wierzejski, 1905). In *Lymnaea,* the efferent duct that connects the terminal cell with the exterior is formed by three cells, each of a different shape (Meisenheimer, 1899). However, the function of these cells is not yet known (Brandenburg, 1966). In the Stylommatophora, the efferent duct is built up by numerous cells (Meisenheimer, 1898; Carrick, 1938). The terminal cells—a single in each protonephridium in the Bivalvia and Basommatophora; a few in each organ in the Stylommatophora—are variously formed. However, a general feature comparable with the organization of the choanocytes of the Porifera is evident (Wierzejsky, 1905; Kümmel and Brandenburg, 1961; Brandenburg, 1966). That part of the terminal cell membrane that joins to the efferent duct and forms its proximal part shows a structure of parallel rods comparable with a weir; the cells have, therefore, been named cyrtocytes or cyrtomastigocytes (Kümmel and Brandenburg, 1961; Brandenburg, 1975).

The long cilia, originating from the terminal cells, form a flame that beats in the proximal part of the duct. It is assumed that the terminal cells play an important role in the filtering process. Brandenburg (1966) observed a membranous structure that surrounds the protonephridium; it seems to be continuous with the basement membrane of the ectoderm (Brandenburg, 1966). However, this does not support the ectodermal origin of the protonephridia that has been postulated by Meisenheimer (1898, 1899, 1901a); the observations of Holmes (1900), Wierzejski (1905), Carrick (1938), and Moor (1977) suggest a mesodermal origin for the protonephridia.

In prosobranchs, the larval kidneys generally consist of a pair of enlarged cells, lateral or dorsolateral in the neck region behind the velum, protruding over the epithelium. Conspicuous vacuoles and various granules are present in these larval kidneys, which are most prominent in forms that develop within an egg capsule (Portmann, 1930; Fioroni, 1966a,b; D'Asaro, 1969; Giese, 1978). In most opisthobranchs, two kinds of transitory structures are recorded, presumed to be excretory: (1) the nephrocysts, paired unicellular organs, and (2) the secondary kidney, an unpaired multicellular structure (Saunders and Poole, 1910; Raven, 1966; Bonar, 1978c). In most species the nephrocysts lie beneath the epithelium, although an ectodermal origin seems more probable (Saunders and Poole, 1910). In *Aeolidiella,* the nephrocysts or primitive kidneys remain within the epithelium, and increase in size during embryonic development (Tardy, 1970b). In *Phestilla,* the nephrocysts are beneath the epithelium in close contact with the larval retractor muscles, initially containing large amounts of yolk, especially in the vicinity of the nephrocysts (Bonar and Hadfield, 1974).

The larval or secondary kidney lies close to the anus; in *Phestilla sibogae* it almost completely surrounds it (Bonar and Hadfield, 1974). It seems to be an organ *sui generis,* not homologous with any other excretory structure in gastropod larvae (Saunders and Poole, 1910). It is often darkly pigmented (Bonar, 1978c). Its presence in veligers of the Architectonicidae represents one of the arguments for the discussion of their systematic position (Robertson, 1973). In

Aeolidiella, Tardy (1970b) observed another structure that might have an excretory function; it is only found at the right side and resembles a basommatophoran protonephridium.

In addition, various kinds of specifically differentiated ectodermal cells occur, the functional significances of which are still unknown. In particular, the so-called anal cells should be mentioned, because they are very prominent in prosobranchs and opisthobranchs. They have been also observed in Basommatophora, but not in the Stylommatophora.

An enigmatic structure is a pair of so-called pulsating vesicles, in the hemocoelic space dorsolaterally in the nuchal region of *Aeolidiella* (Tardy, 1970b). A larval heart, comparable to that of many prosobranchs (Werner, 1955; D'Asaro, 1969; Giese, 1978), is not a common feature in opisthobranchs (see Bonar, 1978c). Whereas the larval heart is a transitory structure in the Prosobranchia, in the Stylommatophora it becomes integrated into the definitive circulatory system. It originates on the right side somewhat below the lung. During the lung's invagination, the larval (or embryonic) heart elongates considerably and comes beneath the lung. It maintains its own pulsation during the first phase in which the definitive heart starts to function. Eventually it forms a portion of the anterior aorta. Although the embryonic stages of *Achatina* and *Bradybaena* differ greatly, the pattern of their embryonic circulation is essentially the same (Brisson, 1968; Moor, 1977). Blood flows through paired pedal vessels into the podocyst and is then directed anteriorly through the median dorsal lacune of the foot. In some tropical forms, such as *Achatina,* the podocyst assumes giant proportions (Brisson, 1964b, 1968). In early stages before differentiation of the podocyst, the head vesicle is an important site of oxygen uptake.

The origin of the anlagen of the pericardium, heart, and kidney is traced back to cells proliferating from the two teloblasts derived from the 4d blastomere, such as in *Sphaerium japonicum* (Okada, 1939) and in *Marisa cornuarietis* (Demian and Yousif, 1973c). In both gastropods and bivalves the timing between the appearance of the solid anlagen and the formation of the cavities within them, as well as the timing between the formation of the anlagen and their fusion in bilaterally symmetrical forms, show great variability (Meisenheimer, 1901b; Okada, 1939; Oldfield, 1964). In the ampullariids *Marisa* and *Ampullarius,* a right and a left anlage appear, but only the right (posttorsionally left) will persist, whereas the somewhat smaller left one disintegrates. In *Marisa,* the auricle and ventricle develop by invagination from the pericardial wall on the posterodorsal and left sides. These invaginations grow into the pericardial cavity and finally fuse.

In *Ampullarius,* in which the entire heart arises from one invagination site, Honegger (1974) observed irregular beating of the heart before histological differentiation was completed. At the same time a green concretion appeared in the cells of the anlage of the kidney. In later embryonic stages, when the kidney enlarges, the newly formed tissue lacks such inclusions. Towards the end of the embryonic period the concretion is eliminated, while the cells assume the adult structure. Thus, a part of the definitive (posterior) kidney functions temporarily as a larval kidney of the accumulating type. In the Stylommatophora (e.g., *Bradybaena*), concretion is stored within vacuoles of the cells of the efferent duct of the protonephridia. Their accumulating capacity becomes exhausted when the definitive heart begins to beat. This is also accompanied by the appearance of concretions in the definitive kidney. In *Marisa,* the anterior kidney—a homolg of the ureter of the Mesogastropoda (Demian and Yousif, 1973c)—is formed by an ectodermal invagination before development of the mantle cavity, whereas in *Ampullarius* both processes occur simultaneously.

I. Musculature and Connective Tissue

1. Musculature

The recent literature is concerned mainly with the relationship between larval and adult muscles. Among the shell-bearing gastropods apparently only *Patella* has a velar musculature that is purely transitory (Smith, 1935), whereas in other forms it undergoes functional changes during metamorphosis. The pedal retractor represents the main part of the columellar muscle, which in most forms is a larvo-adult structure (Fretter and Graham, 1962; Fretter, 1969, 1972). In the opisthobranch *Retusa obtusa,* which retains the shell, the larval retractor muscles degenerate, and the adult columellar muscle arises independently of the larval structure (Smith, 1967). In species without a shell, such as *Phestilla,* the large pedal and velar retractor muscles undergo rapid autolysis immediately after shell loss (Bonar, 1976, 1978c), but elements of the subepithelial cephalopedal muscle complex are retained throughout metamorphosis (Thompson, 1958, 1962).

In most bivalves the velar retractors seem to be transitory. The larval adductors begin to differentiate before prodissoconch I is complete, and finally become the definitive muscles. The anterior adductor appears prior to the posterior (e.g., *Cardium edule,* Creek, 1960; *Pandora inaequivalvis,* Allen, 1961). In *Lasaea rubra,* the anterior and posterior muscles appear simultaneously (Oldfield, 1964). In such monomyarian species as *Ostrea* and *Aequipecten* the single adult adductor derives from the posterior larval adductor muscle (Jackson, 1888; Sastry, 1965). Sastry observed that the posterior adductor develops during the pediveliger stage. It then moves more toward the center, and the anterior muscle disappears.

2. Nuchal cells

In pulmonates (Raven, 1966; Régondaud, 1972; Moor, 1977) and in freshwater prosobranchs (Honegger, 1974), mesenchyme cells situated in the head vesicle differentiate into the so-called nuchal cells. Specific differences exist with respect to their genesis, structure, localization, and fate (Bloch, 1938; Régondaud, 1972; Cumin, 1972; Arni, 1973; Honegger, 1974; Jones and Bowen, 1979). For instance, in *Lymnaea*, the nuchal cells are concentrated in the vicinity of the protonephridia (Régondaud, 1972; Cumin, 1972). In *Bradybaena*, they are distributed rather regularly over the surface of the rostral portion of the left diverticulum of the digestive gland (Moor, 1977). In contrast to these obvious differences, a striking similarity exists at the ultrastructural level (Régondaud, 1972; B. Moor, unpublished observations, 1980).

The nucleus has an eccentric position and, in *Bradybaena*, its surface is enlarged by numerous folds. The nuchal cells show a rather voluminous Golgi zone and an enormously developed endoplasmic reticulum. Presumably these cells are important for embryonic metabolism, since their structure suggests synthetic activities (Régondaud, 1972; Minganti, 1950).

In *Ampullarius* and *Bradybaena*, where the nuchal cells have identical positions, marked differences exist with respect to their mode of differentiation. In *Bradybaena*, cells that have started to differentiate do not divide and do not show the dimorphism of *Ampullarius*. This dimorphism correlates with two different modes of arrangement of the cells in the hemolymph space, either in compact groups or as single cells. Only the latter show a structure that is taken for a vacuole by Honegger (1974). Presumably only these cells are structurally homologous with the nuchal cells of the pulmonates. Toward the end of the embryonic period all nuchal cells are transformed into cells resembling special components in the connective tissue of adult ampullariids (Honegger, 1974). Structural changes are also observed in *Lymnaea* (Régondaud, 1972).

The most important argument in favor of continuous structural changes is probably the occurrence of grooves and pores in the cell membrane in *Lymnaea*, *Bradybaena*, and *Deroceras*. These structures have also been observed in postembryonic *Deroceras reticulatum* (Jones and Bowen, 1979). This suggests that the embryonic nuchal cells are probably related to the pore cells described in adult gastropods (Jones and Bowen, 1979). The contradictory statements of partial loss by extrusion in pulmonates should be reinvestigated. The observation of the different behavior of pore cells in postembryonic life (Jones and Bowen, 1979; Richardot, 1979) could stimulate further study of the obviously complex role and structural evolution of the nuchal cells.

J. Reproductive Organs

Morphogenesis of the reproductive organs may be initiated in three different ways. When the gonad primordium is formed relatively late, it either proliferates from the pericardium (*Viviparus viviparus*, Drummond, 1902; Otto and Tönniges, 1906; Griffond, 1977; *Venus striatula*, Ansell, 1961; *Dreissensia poly-*

morpha, Meisenheimer, 1901a), or it becomes first recognizable as an entirely independent group of cells in the primary body cavity (*Aeolidiella alderi, Amphorina doriae,* Tardy, 1970a,b 1971a,b; *Cymbulia peroni,* Thiriot-Quiévreux, 1970; several species of Atlantidae and Carinariidae, Thiriot-Quiévreux, 1969, 1971b, 1975; *Ampullarius canaliculatus,* Honegger, 1974). The gonad primordium may also become recognizable at rather early developmental stages by differentiation of *primordial germ cells* (PGC). This is reported for the bivalves *Cyclas cornea* (Meisenheimer, 1901b), *Sphaerium japonicum* (Okada, 1936, 1939), and *Lasaea rubra* (Oldfield, 1964), for the basommatophorans *Lymnaea stagnalis, Physa acuta, Biomphalaria straminea, Bulimus truncatus, B. globosus, Acroloxus lacustris,* and *Ancylus fluviatilis* (Brisson and Régondaud, 1971, 1977), and in the stylommatophoran *Bradybaena fruticum* (Moor, 1977) (Fig. 16b).

Ultrastructural examinations of *Acroloxus* (Brisson, 1973), and *Lymnaea* (Brisson and Besse, 1975) reveal cytoplasmic structures typical of PGC: a nucleus with finely dispersed chromatin, cytoplasm with many mitochondria, and a rudimentary endoplasmic reticulum. At first sight this type of gonad development, starting with the appearance of PGC seems typical of pulmonates, but Larambergue (1949, quoted by Brisson and Régondaud, 1977) notes two PGC in the hatching larva of the prosobranch *Calyptraea sinensis.*

Whether gonad development starts with a group of undifferentiated cells or with the appearance of PGC, the Gastropoda always form a single asymmetrically situated anlage, whereas the Bivalvia (Lubet et al., 1976) show a bilaterally symmetrical development of either paired anlagen or an originally single anlage giving rise to paired gonads (*Dreissensia polymorpha,* Meisenheimer, 1901b). Initially paired gonads may secondarily fuse by anastomoses of some of their tubules (*Mya arenaria,* Coe and Turner, 1938; *Venus striatula,* Ansell, 1961). Literature on gonad development in the Bivalvia has been reviewed by Sastry (1979).

In prosobranchs, the gonoduct is partly mesodermal and partly ectodermal in origin. According to Guyomarc'h-Cousin (1976), formation of the mesodermal part of the efferent duct in *Littorina saxatilis* has no relationship to the gonad, which apparently arises independently from the pericardium. In larvae of female Atlantidae, Thiriot-Quiévreux and Martoja (1976) observed an initially solid finger-like outgrowth from the gonad. This meets the pallial anlage before metamorphosis, but fusion of the tissues occurs only afterwards. In the male, the organization comprises a third initially independent element, the extrapallial copulatory organs. Thiriot-Quiévreux (1967b, 1969) notes that in *Atlanta lesueuri* the penis and flagellum arise from a common anlage, which becomes manifest some time before metamorphosis. The same is true for *Oxygyrus keraudreni* (Thiriot-Quiévreux, 1971b). Although there are considerable differences between closely related species of Atlantidae (Thiriot-Quiévreux and Martoja,

1976), the mesodermal anlage is usually formed first (Thiriot-Quiévreux, 1969). Moritz (1939), however, records the presence of a penis in the hatching young of *Crepidula adunca,* whereas the gonadal anlage appears only after hatching. Nevertheless, it seems that in general a relatively long period elapses between the appearance of the anlagen and their eventual fusion in prosobranchs.

In pulmonates, some time after the PGC have appeared, a cellular proliferation grows from the rudimentary mantle cavity towards the PGC. According to Brisson and Régondaud (1971, 1977), a lumen is formed only secondarily, whereas in *Bradybaena* the outgrowth is tubular from its origin (Moor, 1977). The tip meets the PGC, which have meanwhile divided several times. The descendants of the PGC, together with the adjacent mesodermal cells become incorporated into the tip of the ectodermal proliferation (after which it is extremely difficult to recognize the mesodermal parts) (Fraser, 1946; Brisson and Régondaud, 1977). This may be the reason why Laviolette (1954) and Luchtel (1972) maintained the old theory of the entirely ectodermal origin of the pulmonate reproductive system

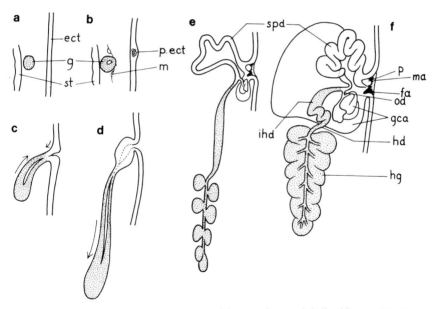

Fig. 17. Diagrams showing development of the gonad in *Aeolidiella alderi.* (a) Situation at the limapontioid stage, (b) at the early pseudovermis stage, slightly older than that represented in Fig. 12a, (c) the contact between the mesodermal and ectodermal anlagen is established, (d and e) stages of juvenile development, (f) adult condition. The parts that are mesodermal in origin are stippled. ect, ectoderm; fa, female atrium; g, gonadal anlage; gca, glandular complexes connected with the oviduct; hd, hermaphroditic duct; hg, hermaphroditic gland; ihd, inflated part of the hd (ampulla); m, mesenchyme; ma, male atrium; od, oviduct; p, penis; p.ect, ectodermal profiferation; spd, spermioduct; st, stomach. (After Tardy, 1970a.)

(for review, see Martoja, 1964). Most probably they have overlooked the essential early stages. Griffond and Bride (1981) have found that in *Helix aspersa,* the contribution of the mosodermal parts to the gonad formation only takes place shortly after hatching.

The results of the experiments in which the gonads have been removed in pulmonates (Brisson, 1971) and in opisthobranchs (Tardy, 1971a,b) throw some light on the question of the mesodermal contribution. Regeneration to a fertile state is only possible when the hermaphroditic duct is left in place.

The opisthobranchs studied so far (*Tritonia hombergi,* Thompson, 1962; *Trinchesia granosa,* Schönenberger, 1969; *Aeolidiella alderi* and *Amphorina doriae,* Tardy, 1967, 1970a,b 1971a,b), resemble pulmonates with respect to the fusion of separate anlagen in an early stage of genital organogenesis. The anlage of the gonad does not become manifest until metamorphosis or soon afterwards, that is, the limapontioid stage (Tardy, 1970b, 1971b) and then rests without any sign of differentiation until the early pseudovermis stage (Tardy, 1962). Thus, the entire morphogenesis of the opisthobranch genital tract (Fig. 17) is a matter of juvenile development. In contrast to the case in pulmonates and prosobranchs, pallial tissue does not contribute to the formation of the gonoduct in opisthobranchs. The entire ectodermal portion originates from outside the pallial area (Tardy, 1970a).

VI. Discussion and Concluding Remarks

How to deal with the striking variety of developmental phenomena is an outstanding problem in reviewing molluscan organogenesis. The choice of data was influenced by the aim to stress relevant research performed since Raven's synopsis (1966). Our present knowledge is based mainly on species that have been chosen without regard for their systematic position. Hence, extensive comparative morphological studies, which integrate systematic and ecological aspects are very promising for the evaluation of phylogenetic theories (Thiriot-Quiévreux, 1971a).

However, one must bear in mind that closely related species may differ fundamentally with respect to their development and that some features are only typical of a small systematic unit, for instance, the notch of the shell in *Crassostrea* and *Ostrea* (Carriker and Palmer, 1979; Waller, 1981), or the tooth and the socket of the Pholadidae (Boyle and Turner, 1976). As pointed out by Thiriot-Quiévreux (1974), the functional interpretation of these structures is very tricky.

What has not been adequately evaluated is that aspect of morphogenesis which is related to the dynamics of development, namely temporal relationships. The terms retardation and acceleration, and neoteny may characterize the problem

(Jägersten, 1972; but see Gould, 1977). A careful examination, however, will reveal that many of these concepts are arbitrary owing to the absence of comparative studies. It is worth considering the statement of Thompson (1962, p. 213) that acceleration is "a familiar but little-understood phenomenon," as our knowledge of the determinants involved in the dynamics of a particular ontogenesis is rather fragmentary. The interpretation of different patterns of radular development may illustrate the problem. In the neogastropod *Fusus*, a form which feeds on large amounts of albumen during embryonic development, radular development occurs only relatively late. In comparison with the pulmonates, Portmann and Sandmeier (1965) judge it as retarded. It appears that a relatively late beginning of radular development is rather common among Meso- and Neogastropoda (Fioroni, 1966b), whereas the Pulmonata are characterized by a relatively precocious organogenesis of the radula. In the species investigated by Kerth (1979) (see Section V,G) the structure of the initially formed radula differs from the parts formed later, and is thought to be recapitulating ancestral features. Jägersten (1972), however, has expressed the opinion that the early differentiation of definitive organs might be an acceleration (or adultation) representing a phylogenetic trend towards direct development. However, since important questions about the origin of the different classes are still obscure (Robertson, 1973), the phylogenetic significance of the early development of the radula in pulmonates cannot be evaluated.

The amazing varieties of developmental pathways should not distract the attention from the similarities between closely related species. Here, another aspect of temporal relationships may become important, namely the relative duration of different phases and the absolute duration of embryonic development. Kress (1975) has observed characteristic differences between morphologically similar ontogenies of three dotoid species. However, further comparative studies must reveal the significance of these phenomena. This is also true for the asymmetric development of originally bilaterally symmetrical organs. With respect to pallial organs in the Gastropoda, the phenomena of retarded formation, reduced differentiation, and final loss of the anlage at one side represent morphological aspects of the evolution of pallial asymmetry, namely torsion (Crofts, 1938, 1955). Nevertheless, temporal asymmetry is also evident in bilaterally symmetrical organs situated outside the area influenced by torsion. In *Buccinum*, the right statocyst forms slightly later than the left one (Giese, 1978). The initially unequal development of the tentacles is a well-known phenomenon (Thorson, 1940; Thiriot-Quiévreux, 1967a; Struhsaker and Costlow, 1968; D'Asaro, 1969). Heterochronism is also recorded in other classes, such as early gill development in *Ostrea*, where the right anlage is retarded with respect to the left one (Prytherch, 1934; Waller, 1981).

The above remarks on morphological research demonstrate the diversity in embryonic development and its significance for other disciplines. The motivation to go on with morphological analysis of developmental phenomena today may be expressed by Raven's closing words in his review in 1966 that "the exploration of the prodigy of animal development . . . is still in its initial phase [p. 268]."

Acknowledgments

I am indebted to Dr. C. A. M. van Dongen, Prof. P. Fioroni, Dr. E. Kniprath, Prof. E. Marcus, and Prof. L. Schmekel for having read certain parts of the manuscript. I thank Dr. S. M. Cragg for copies of a part of his unpublished thesis, and many colleagues for providing me the reprints I requested. I express my gratitude to Dr. van Dongen for giving me the original photomicrographs for examination in order to help the designing of the figures. The courtesy of the authors from whom the figures are quoted is greatly appreciated.

References

Allen, J. A. (1961). The development of *Pandora inaequivalvis* (L.). *J. Embryol. Exp. Morphol.* **9**, 252–268.

Allen, J. A., and Scheltema, R. S. (1972). The functional morphology and geographical distribution of *Planktomya henseni,* a supposed neotenous pelagic bivalve. *J. Mar. Biol. Assoc. U.K.* **52**, 19–31.

Andrews, J. D. (1979). Pelecypoda: Ostreidae. *In* "Reproduction of Marine Invertebrates" (A. C. Giese and J. S. Pearse, eds.), Vol. 5, pp. 293–341. Academic Press, New York.

Ansell, A. D. (1961). The development of the primary gonad in *Venus striatula* (da Costa). *Proc. Malacol. Soc. London* **34**, 243–247.

Ansell, A. D. (1962). The functional morphology of the larva, and the post-larval development of *Venus striatula* (da Costa). *J. Mar. Biol. Assoc. U.K.* **42**, 419–443.

Arni, P. (1973). Vergleichende Untersuchungen an Schlüpfstadien von neun Pulmonaten-Arten (Mollusca, Gastropoda). *Rev. Suisse Zool.* **80**, 323–402.

Arni, P. (1974). Licht- und elektronenmikroskopische Untersuchungen an Embryonen von *Lymnaea stagnalis* L. (Gastropoda, Pulmonata) mit besonderer Berücksichtigung der frühembryonalen Ernährung. *Z. Morphol. Tiere* **78**, 299–323.

Arni, P. (1975). Licht- und elektronenmikroskopische Untersuchungen zur Entwicklung und Degeneration transitorischer Speicherzellen der Mitteldarmregion von *Lymnaea stagnalis* L. (Gastropoda, Pulmonata). *Z. Morphol. Tiere* **81**, 221–240.

Atkinson, J. W. (1971). Organogenesis in normal and lobeless embryos of the marine prosobranch gastropod *Ilyanassa obsoleta. J. Morphol.* **133**, 339–352.

Baba, K. (1937). Contribution to the knowledge of a nudibranch, *Okadaia elegans* Baba. *Jpn. J. Zool.* **7**, 147–190.

Baba, K. (1938). The later development of a solenogastre, *Epimenia verrucosa* (Nierstrasz). *J. Dep. Agric., Kyushu Imp. Univ., Fukuoka, Jpn.* **6**, 21–40.

Baba, K. (1940). The early development of a solenogastre, *Epimenia verrucosa* (Nierstrasz). *Annot. Zool. Jpn.* **19**, 107–113.

Baba, K. (1951). General sketch of the development in a solenogastre, *Epimenia verrucosa* (Nierstrasz). *Misc. Rep. Res. Inst. Nat. Resour. (Tokyo)* **19–21**, 38–46 (in Japanese, with English summary).

Bandel, K. (1975). Entwicklung der Schale im Lebensablauf zweier Gastropodenarten; *Buccinum undatum* und *Xancus angulatus* (Prosobranchia, Neogastropoda). *Biomineralization* **8,** 67–91.

Bandel, K. (1982). In press.

Barber, V. C. (1968). The structure of mollusc statocysts, with particular reference to Cephalopods. *Symp. Zool. Soc. London* **23,** 37–62.

Bayne, B. L. (1964). The responses of the larvae of *Mytilus edulis* L. to light and to gravity. *Oikos* **15,** 162–174.

Bayne, B. L. (1965). Growth and the delay of metamorphosis of the larvae of *Mytilus edulis* L. *Ophelia* **2,** 1–47.

Beedham, G. E. (1958a). Observations on the mantle of the Lamellibranchia. *Q. J. Microsc. Sci.* **99,** 181–197.

Beedham, G. E. (1958b). Observations on the non-calcareous component of the shell of the Lamellibranchia. *Q. J. Microsc. Sci.* **99,** 341–357.

Belding, D. L. (1912). Report on the Quahaug and Oyster fisheries of Massachusetts. *Commonw. Mass., Mar. Fish. Ser.* **2,** 1–134.

Bernard, F. (1896). Troisième note sur le développement et la morphologie de la coquille chez les lamellibranches (Anisomyaires). *Bull. Soc. Geol. Fr.* Sér. [3] **24,** 412–449.

Bloch, S. (1938). Beitrag zur Kenntnis der Ontogenese von Süsswasserpulmonaten mit besonderer Berücksichtigung der Mitteldarmdrüse. *Rev. Suisse Zool.* **45,** 157–220.

Boer, H. H., and Joosse, J. (1975). Endocrinology. *In* "Pulmonates" (V. Fretter and J. Peake, eds.), Vol. 1, pp. 245–307. Academic Press, New York.

Boer, H. H., Slot, J. W., and Van Andel, J. (1968). Electron microscopical and histochemical observations on the relation between medio-dorsal bodies and neurosecretory cells in the basommatophoran snails *Lymnaea stagnalis, Ancylus fluviatilis, Australorbis glabratus* and *Planorbarius corneus. Z. Zellforsch. Mikrosk. Anat.* **87,** 435–450.

Bonar, D. B. (1976). Molluscan metamorphosis: A study in tissue transformation. *Am. Zool.* **16,** 573–591.

Bonar, D. B. (1978a). Fine structure of muscle insertions on the larval shell and operculum of the nudibranch *Phestilla sibogae* (Mollusca: Gastropoda) before and during metamorphosis. *Tissue Cell* **10,** 143–152.

Bonar, D. B. (1978b). Ultrastructure of a cephalic sensory organ in larvae of the gastropod *Phestilla sibogae* (Aeolidacea, Nudibranchia). *Tissue Cell* **10,** 153–165.

Bonar, D. B. (1978c). Morphogenesis at metamorphosis in opisthobranch molluscs. *In* "Settlement and Metamorphosis of Marine Invertebrate Larvae" (F.-S. Chia and M. E. Rice, eds.), pp. 177–196. Am. Elsevier, New York.

Bonar, D. B., and Hadfield, M. G. (1974). Metamorphosis of the marine gastropod *Phestilla sibogae* Bergh (Nudibranchia: Aeolidacea). I. Light and electron microscopic analysis of larval and metamorphic stages. *J. Exp. Mar. Biol. Ecol.* **16,** 227–255.

Boyle, P. J., and Turner, R. D. (1976). The larval development of the wood-boring piddock *Martesia striata* (L.) (Mollusca: Bivalvia: Pholadidae). *J. Exp. Mar. Biol. Ecol.* **22,** 55–68.

Boyle, P. R. (1969). Fine structure of the eyes of *Onithochiton neglectus* (Mollusca: Polyplacophora). *Z. Zellforsch. Mikrosk. Anat.* **102,** 313–332.

Brandenburg, J. (1966). Die Reusenformen der Cyrtocyten. *Zool. Beitr. N. F.* **12,** 345–417. (K. Herter, ed.). Duncker and Humblot, Berlin.

Brandenburg, J. (1975). The morphology of the protonephridia. *Fortschr. Zool.* **23** (2/3), 1–17.

Brisson, P. (1964a). Considérations sur la morphogénèse de la commissure viscérale chez deux espèces d'Ancyles (Mollusques Gastéropodes). *Bull. Soc Zool. Fr.* **89,** 166–173.

Brisson, P. (1964b). "Annexes embryonnaires" chez *Archachatina (Calachatina) marginata* (Swainson). *C.R. Hebd. Seances Acad. Sci., Ser. D* **259,** 3620–3623.

Brisson, P. (1968). Développement de l'embryon et de ses annexes et étude en culture in vitro chez les Achatines. *Arch. Anat. Microsc. Morphol. Exp.* **57,** 345–368.

Brisson, P. (1971). Castration chirurgicale et régénération gonadique chez quelques Planorbides (Gastéropodes Pulmonés). *Ann. Embryol. Morphog.* **4,** 189–210.

Brisson, P. (1973). Observation ultrastructurale des cellules germinales chez l'embryon d'*Acroloxus lacustris* (L.) (Gastéropode Pulmoné Basommatophore). *C.R. Hebd. Seances Acad. Sci., Ser. D* **277,** 2205–2208.

Brisson, P., and Besse, C. (1975). Etude ultrastructurale de l'ébauche gonadique chez l'embryon de *Lymnaea stagnalis* L. (Gastéropode Pulmoné Basommatophore). *Bull. Soc. Zool. Fr.* **100,** 345–349.

Brisson, P., and Régondaud, J. (1971). Observations relatives à l'origine dualiste de l'appareil génital chez quelques Gastéropodes Pulmonés Basommatophores. *C.R. Hebd. Seances Acad. Sci., Ser. D* **273,** 2339–2341.

Brisson, P., and Régondaud, J. (1977). Origine et structure de l'ébauche de la gonade chez les gastéropodes pulmonés basommatophores. *Malacologia* **16,** 457–466.

Caddy, J. F. (1969). Development of mantle organs, feeding and locomotion in postlarval *Macoma balthica* (L.) (Lamellibranchiata). *Can. J. Zool.* **47,** 609–617.

Carrick, R. (1938). The life-history and development of *Agriolimax agrestis* L., the Gray Field Slug. *Trans. R. Soc. Edinburgh* **59,** 563–597.

Carriker, M. R. (1961). Interrelation of functional morphology, behaviour, and autecology in early stages of the bivalve *Mercenaria mercenaria*. *J. Elisha Mitchell Sci. Soc.* **77,** 168–241.

Carriker, M. R., and Palmer, R. E. (1979). Ultrastructural morphogenesis of prodissoconch and early dissoconch valves of the Oyster *Crassostrea virginica*. *Proc. Natl. Shellfish. Assoc.* **69,** 103–128.

Carter, G. S. (1926). On the nervous control of the velar cilia of the nudibranch veliger. *J. Exp. Biol.* **4,** 1–26.

Carter, G. S. (1928). On the structure of the cells bearing the velar cilia of the nudibranch veliger. *J. Exp. Biol.* **6,** 97–109.

Cather, J., and Tompa, A. (1972). The podocyst in pulmonate evolution. *Malacol. Rev.* **5,** 1–3.

Chia, F.-S. (1971). Oviposition, fecundity and larval development of three sacoglossan opisthobranchs from the Northumberland Coast, England. *Veliger* **13,** 319–325.

Christiansen, M. E. (1954). The life history of *Lepidopleurus asellus* (Spengler). (Placophora). *Nytt Mag. Zool.* **2,** 52–72.

Coe, W. R., and Turner, H. J. (1938). Development of the gonads and gametes in the soft-shell clam (*Mya arenaria*). *J. Morphol.* **62,** 91–111.

Cole, H. A. (1938). The fate of the larval organs in the metamorphosis of *Ostrea edulis*. *J. Mar. Biol. Assoc. U.K.* **22,** 469–484.

Cragg, S. M. (1976). Some aspects of the behaviour and functional morphology of bivalve larvae. Ph. D. Thesis, University of Wales.

Cragg, S. M., and Nott, J. A. (1977). The ultrastructure of the statocysts in the pediveliger larvae of *Pecten maximus* (L.) (Bivalvia). *J. Exp. Mar. Biol. Ecol.* **27,** 23–36.

Cranfield, H. J. (1973a). A study of the morphology, ultrastructure, and histochemistry of the foot of the pediveliger of *Ostrea edulis*. *Mar. Biol.* **22,** 187–202.

Cranfield, H. J. (1973b). Observations on the behaviour of the pediveliger of *Ostrea edulis* during attachment and cementing. *Mar. Biol.* **22,** 203–209.

Cranfield, H. J. (1973c). Observations on the function of the glands of the foot of the pediveliger of *Ostrea edulis* during settlement. *Mar. Biol.* **22,** 211–223.

Cranfield, H. J. (1974). Observations on the morphology of the mantle folds of the pediveliger of *Ostrea edulis* L. and their function during settlement. *J. Mar. Biol. Assoc. U.K.* **54,** 1–12.

Cranfield, H. J. (1975). The ultrastructure and histochemistry of the larval cement of *Ostrea edulis* L. *J. Mar. Biol. Assoc. U.K.* **55**, 497–503.

Creek, G. A. (1960). The development of the lamellibranch *Cardium edule* L. *Proc. Zool. Soc. London* **135**, 243–260.

Crofts, D. R. (1938). The development of *Haliotis tuberculata*, with special reference to the organogenesis during torsion. *Philos. Trans. R. Soc. London, Ser. B* **228**, 219–268.

Crofts, D. R. (1955). Muscle morphogenesis in primitive gastropods and its relation to torsion. *Proc. Zool. Soc. London* **125**, 711–750.

Culliney, J. L., and Turner, R. D. (1976). Larval development of the deep-water wood boring bivalve, *Xylophaga atlantica* Richards (Mollusca, Bivalvia, Pholadidae). *Ophelia* **15**, 149–161.

Cumin, R. (1972). Normentafel zur Organogenese von *Limnaea stagnalis* L. (Gastropoda, Pulmonata) mit besonderer Berücksichtigung der Mitteldarmdrüse. *Rev. Suisse Zool.* **79**, 709–774.

D'Asaro, C. N. (1967). The morphology of larval and postlarval *Chione cancellata* Linné (Eulamellibranchia: Veneridae) reared in the laboratory. *Bull. Mar. Sci.* **17**, 949–972.

D'Asaro, C. N. (1969). The comparative embryogenesis and early organogenesis of *Bursa corrugata* and *Distorsio clathrata*. *Malacologia* **9**, 349–389.

Dautert, E. (1929). Die Bildung der Keimblätter von *Paludina*. *Zool. Jahrb., Abt. Anat. Ontog. Tiere* **50**, 433–496.

Dawydoff, C. (1940). Quelques véligères géantes de Prosobranches provenant de la Mer de Chine. *Bull. Biol. Fr. Belg.* **74**, 497–508.

de Lacaze-Duthiers, H. (1857). Histoire de l'organisation et du développement du Dentale. *Ann. Sci. Nat., Zool. Biol. Anim.* [4] **7**, 194–254.

Dell, R. K. (1962). Stages in the development of viviparity in the Amphineura. *Nature (London)* **195**, 512–513.

Demian, E. S., and Yousif, F. (1973a). Embryonic development and organogenesis in the snail *Marisa cornuarietis* (Mesogastropoda: Ampullariidae). I. General outlines of development. *Malacologia* **12**, 123–150.

Demian, E. S., and Yousif, F. (1973b). Embryonic development and organogenesis in the snail *Marisa cornuarietis*. II. Development of the alimentary system. *Malacologia* **12**, 151–174.

Demian, E. S., and Yousif, F. (1973c). Embryonic development and organogenesis in the snail *Marisa cornuarietis*. III. Development of the circulatory and renal systems. *Malacologia* **12**, 175–194.

Demian, E. S., and Yousif, F. (1973d). Embryonic development and organogenesis in the snail *Marisa cornuarietis*. IV. Development of the shell gland, mantle and respiratory organs. *Malacologia* **12**, 195–211.

Demian, E. S., and Yousif, F. (1975). Embryonic development and organogenesis in the snail *Marisa cornuarietis*. V. Development of the nervous system. *Malacologia* **15**, 29–42.

Drew, G. A. (1899a). The anatomy, habits, and embryology of *Yoldia limatula*, Say. *Mem. Biol. Lab. Johns Hopkins Univ.* **4**, 1–37.

Drew, G. A. (1899b). Some observations on the habits, anatomy, and embryology of members of the Protobranchia. *Anat. Anz.* **15**, 493–519.

Drew, G. A. (1901). The life history of *Nucula delphinodonta* (Mighels). *Q. J. Microsc. Sci.* **44**, 313–391.

Drew, G. A. (1906). The habits, anatomy and embryology of the giant scallop [*Pecten tenuicostatus* (Mighels)]. *Univ. Maine Stud.* **6**, 1–171.

Drummond, I. M. (1902). Notes on the development of *Paludina vivipara*, with special reference to the urinogenital organs and theories of gasteropod torsion. *Q. J. Microsc. Sci.* **46**, 97–143.

Eakin, R. M., and Brandenburger, J. L. (1967). Differentiation in the eye of a pulmonate snail *Helix aspersa*. *J. Ultrastruct. Res.* **18**, 391–421.

Erdmann, W. (1935). Ueber die Entwicklung und die Anatomie der 'ansatzreifen' Larve von *Ostrea edulis*, mit Bemerkungen über die Lebensgeschichte der Auster. *Wiss. Meeresunters., Abt. Helgol.* [N.S.] **19**(6), 1–25.

Fioroni, P. (1965a). Zur embryonalen Entwicklung und zum Schlüpfzustand von zwei mediterranen *Nassa*-Arten. *Rev. Suisse Zool.* **72**, 543–568.

Fioroni, P. (1965b). Zur embryonalen Entwicklung von *Philbertia* (Gastropoda, Prosobranchia, Conidae). *Verh. Naturforsch. Ges. Basel* **76**, 207–219.

Fioroni, P. (1966a). Un nouveau cas de rotation des oeufs nutritifs chez un gastéropode prosobranche marin. *Vie Milieu* **17**, 109–119.

Fioroni, P. (1966b). Zur Morphologie und Embryogenese des Darmtraktes und der transitorischen Organe bei Prosobranchiern (Mollusca, Gastropoda). *Rev. Suisse Zool.* **73**, 621–876.

Fioroni, P. (1971). Die Entwicklungstypen der Mollusken, eine vergleichend-embryologische Studie. *Z. Wiss. Zool.* **182**, 263–394.

Fioroni, P. (1980). Zur Signifikanz des Blastoporus-Verhaltens in evolutiver Hinsicht. *Rev. Suisse Zool.* **87**, 261–272.

Fioroni, P., and Meister, G. (1976). Zur embryonalen Entwicklung von *Lamellaria perspicua* L. (Gastropoda, Prosobranchia, Mesogastropoda, Lamellariacea). *Cah. Biol. Mar.* **17**, 323–336.

Fischer, F. P. (1978). Photoreceptor cells in chiton aesthetes. *Spixiana* **1**, 209–213.

Fischer, F. P. (1979). Die Aestheten von *Acanthochiton fascicularis* (Mollusca, Polyplacophora). *Zoomorphologie* **92**, 95–106.

Fischer, F. P. (1980). Fine structure of the larval eye of *Lepidochitona cinerea* L. (Mollusca, Polyplacophora). *Spixiana* **3**, 53–57.

Fischer, F. P., Maile, W., and Renner, M. (1980). Die Mantelpapillen und Stacheln von *Acanthochiton fascicularis* L. (Mollusca, Polyplacophora). *Zoomorphologie* **94**, 121–131.

Fol, H. (1880). Etudes sur le développement des Mollusques. IIIᵉ mémoire. Sur le développement des gastéropodes pulmonés. *Arch. Zool. Exp. Gen.* **8**, 103–232.

Fraser, L. A. (1946). The embryology of the reproductive tract of *Lymnaea stagnalis appressa* Say. *Trans. Am. Microsc. Soc.* **65**, 279–298.

Frenkiel, L., and Mouëza, M. (1979). Développement larvaire de deux Tellinacea, *Scrobicularia plana* (Semelidae) et *Donax vittatus* (Donacidae). *Mar. Biol.* **55**, 187–195.

Fretter, V. (1943). Studies in the functional morphology and embryology of *Onchidella celtica* (Forbes and Hanley) and their bearing on its relationships. *J. Mar. Biol. Assoc. U.K.* **25**, 685–720.

Fretter, V. (1967). The prosobranch veliger. *Proc. Malacol. Soc. London* **37**, 357–366.

Fretter, V. (1969). Aspects of metamorphosis in prosobranch gastropods. *Proc. Malacol. Soc. London* **38**, 375–386.

Fretter, V. (1972). Metamorphic changes in the velar musculature, head and shell of some prosobranch gastropods. *J. Mar. Biol. Assoc. U.K.* **52**, 161–177.

Fretter, V., and Graham, A. (1962). "British Prosobranch Molluscs. Their functional Anatomy and Ecology." Ray Society, London.

Fretter, V., and Montgomery, M. C. (1968). The treatment of food by prosobranch veligers. *J. Mar. Biol. Assoc. U.K.* **48**, 499–520.

Fretter, V., and Pilkington, M. C. (1971). The larval shell of some prosobranch gastropods. *J. Mar. Biol. Assoc. U.K.* **51**, 49–62.

Fullarton, J. H. (1890). On the development of the common scallop (*Pecten opercularis*, L.). *Fish. Board Scotland, 8th Annu. Rep., Part III., Sci. Invest.* pp. 290–299.

Geilenkirchen, W. L. M., Timmermans, L. P. M., van Dongen, C. A. M., and Arnolds, W. J. A.

(1971). Symbiosis of bacteria with eggs of *Dentalium* at the vegetal pole. *Exp. Cell Res.* **67,** 477–479.

Giese, K. (1978). Zur Embryonalentwicklung von *Buccinum undatum* L. (Gastropoda, Prosobranchia, Stenoglossa (Neogastropoda) Buccinacea). *Zool. Jahrb., Abt. Anat. Ontog. Tiere* **100,** 65–117.

Gould, S. J. (1977). "Ontogeny and Phylogeny." Belknap Press, Cambridge, Massachusetts.

Griffond, B. (1977). Individualisation et organogenèse de la gonade embryonnaire de *Viviparus viviparus* L. (Mollusque gastéropode prosobranche à sexes séparés). *Wilhelm Roux's Arch. Dev. Biol.* **183,** 131–147.

Griffond, B., and Bride, J. (1981). Etude histologique et ultrastructurale de la gonade d'*Helix aspersa* Müller à l'éclosion. *Reprod., Nutr., Dev.* **21,** 149–161.

Gruffydd, L. D., Lane, D. J. W., and Beaumont, A. R. (1975). The glands of the larval foot in *Pecten maximus* L. and possible homologues in other bivalves. *J. Mar. Biol. Assoc. U.K.* **55,** 463–476.

Guyomarc'h-Cousin, C. (1974). Etude descriptive de l'organogenèse du système nerveux chez *Littorina saxatilis* (Olivi): Gastéropode prosobranche. *Ann. Embryol. Morphog.* **7,** 349–364.

Guyomarc'h-Cousin, C. (1976). Organogénèse descriptive de l'appareil génital cehz *Littorina saxatilis* (Olivi) gastéropode prosobranche. *Bull. Soc. Zool. Fr.* **101,** 465–476.

Haas, W. (1976). Observations on the shell and mantle of the Placophora. *Belle W. Baruch Lib. Mar. Sci.* **5,** 389–402.

Haas, W. (1981). Evolution of calcareous hardparts in primitive molluscs. *Malacologia* **21,** 403–418.

Haas, W., and Kriesten, K. (1974). Studien über das Mantelepithel von *Lepidochitona cinerea* (L.) (Placophora). *Biomineralization* **7,** 100–109.

Haas, W., and Kriesten, K. (1975). Studien über das Perinotum–Epithel und die Bildung der Kalkstacheln von *Lepidochitona cinerea* (L.) (Placophora). *Biomineralization* **8,** 92–107.

Haas, W., and Kriesten, K. (1977). Studien über das Epithel und die kalkigen Hartgebilde des Perinotums bei *Acanthopleura granulata* (Gmelin) (Placophora). *Biomineralization* **9,** 11–27.

Haas, W., and Kriesten, K. (1978). Die Aestheten mit intrapigmentärem Schalenauge von *Chiton marmoratus* L. Zoomorphologie **90,** 253–268.

Haas, W., Kriesten, K., and Watabe, N. (1979). Notes on the shell formation in the larvae of the Placophora (Mollusca). *Biomineralization* **10,** 1–8.

Haas, W., Kriesten, K., and Watabe, N. (1980). Preliminary note on the calcification of the shell plates in *Chiton* larvae (Placophora). In "The Mechanisms of Biomineralization in Animals and Plants" (M. Omori and N. Watabe, eds.), pp. 67–72. Tokai University Press, Tokyo.

Hadfield, M. G. (1978). Metamorphosis in marine molluscan larvae: An analysis of stimulus and response. *In* "Settlement and Metamorphosis of Marine Invertebrate Larvae" (F.-S. Chia and M. E. Rice, eds.), pp. 165–175. Am. Elsevier, New York.

Hadfield, M. G. (1979). Aplacophora. *In* "Reproduction of Marine Invertebrates" (A. C. Giese and J. S. Pearse, eds.), Vol. 5, pp. 1–25. Academic Press, New York.

Hammarsten, O. D., and Runnström, J. (1925). Zur Embryologie von *Acanthochiton discrepans* Brown. *Zool. Jahrb., Abt. Anat. Ontog. Tiere* **47,** 261–318.

Harris, L. G. (1975). Studies on the life history of two coraleating nudibranchs of the genus *Phestilla. Biol. Bull. (Woods Hole, Mass.)* **149,** 539–550.

Hatschek, B. (1880). Entwicklungsgeschichte von *Teredo. Arb. Zool. Inst. Univ. Wien.* **3,** 1–44.

Heath, H. (1899). The development of *Ischnochiton. Zool. Jahrb., Abt. Anat. Ontog. Tiere* **12,** 567–656.

Heath, H. (1904). The larval eye of Chitons. *Proc. Acad. Nat. Sci. Philadelphia* **56,** 257–259.

Heath, H. (1918). Solenogastres from the eastern coast of North America. *Mem. Mus. Comp. Zool. Harv. Coll.* **45,** 185–263.

Heyder, P. (1909). Zur Entwicklung der Lungenhöhle bei *Arion*. Nebst Bemerkungen über die Entwicklung der Urniere und Niere, des Pericards und Herzens. *Z. Wiss. Zool.* **93**, 90–156.

Higley, R. M., and Heath, H. (1912). The development of the gonad and gonoducts in two species of chitons. *Biol. Bull. (Woods Hole, Mass.)* **22**, 95–97.

Hillman, R. E. (1964). The functional morphology of the fourth fold of the mantle of the Northern Quahog, *Mercenaria mercenaria* (L.). *J. Elisha Mitchell Sci. Soc.* **80**, 8–12.

Holmes, S. J. (1900). The early development of *Planorbis*. *J. Morphol.* **16**, 369–458.

Honegger, T. (1974). Die Embryogenese von *Ampullarius* (Gastropoda, Prosobranchia). *Zool. Jahrb., Abt. Anat. Ontog. Tiere* **93**, 1–76.

Horikoshi, M. (1967). Reproduction, larval features and life history of *Philine denticulata* (J. Adams) (Mollusca, Tectibranchia). *Ophelia* **4**, 43–84.

Jablonski, D., and Lutz, R. A. (1980). Molluscan larval shell morphology. Ecological and paleontological applications. *In* "Skeletal Growth of Aquatic Organisms" (D. C. Rhoads and R. A. Lutz, eds.), pp. 323–377. Plenum, New York.

Jackson, R. T. (1888). The development of the oyster with remarks on allied genera. *Proc. Boston Soc. Nat. Hist.* **23**, 531–557.

Jägersten, G. (1972). "Evolution of the Metazoan Life Cycle: A Comprehensive Theory." Academic Press, New York.

Jones, G. W., and Bowen, I. D. (1979). The fine structural localization of acid phosphatase in pore cells of embryonic and newly hatched *Deroceras reticulatum* (Pulmonata:Stylommatophora). *Cell Tissue Res.* **204**, 253–265.

Kändler, R. (1927). Muschellarven aus dem Helgoländer Plankton. Bestimmung ihrer Artzugehörigkeit durch Aufzucht. *Wiss. Meeresunter., Abt. Helgol. [N.S.]* **16**(5), 1–8.

Kerth, K. (1979). Phylogenetische Aspekte der Radulamorphogenese von Gastropoden. *Malacologia* **19**, 103–108.

Kerth, K., and Hänsch, D. (1977). Zellmuster und Wachstum des Odontoblastengürtels der Weinbergschnecke *Helix pomatia* L. *Zool. Jahrb., Abt. Anat. Ontog. Tiere* **98**, 14–28.

Kerth, K., Reder, J., and Zimmermann, R. (1981). Der Radulaabbau beim Embryo der Sumpfdeckelschnecke *Viviparus fasciatus Müll. (Gastropoda, Prosobranchia). Zool. Jahrb., Abt. Anat. Ontog. Tiere* **106**, 104–111.

Kniprath, E. (1970). Die Feinstruktur der Periostracumgrube von *Lymnaea stagnalis*. *Biomineralization* **2**, 23–37.

Kniprath, E. (1971). Die Feinstruktur des Drüsenpolsters von *Lymnaea stagnalis*. *Biomineralization* **3**, 1–11.

Kniprath, E. (1972). Formation and structure of the periostracum in *Lymnaea stagnalis*. *Calcif. Tissue Res.* **9**, 260–271.

Kniprath, E. (1975). Das Wachstum des Mantels von *Lymnaea stagnalis* (Gastropoda). *Cytobiologie* **10**, 260–267.

Kniprath, E. (1977). Zur Ontogenese des Schalenfeldes von *Lymnaea stagnalis*. *Wilhelm Roux's Arch. Dev. Biol.* **181**, 11–30.

Kniprath, E. (1978). Growth of the shell-field in *Mytilus* (Bivalvia). *Zool. Scr.* **7**, 119–120.

Kniprath, E. (1979a). The functional morphology of the embryonic shell-gland in the conchiferous molluscs. *Malacologia* **18**, 549–552.

Kniprath, E. (1979b). Ontogenèse de la région coquillière des Mollusques. Ph.D. Thesis, Univ. Pierre et Marie Curie, Paris.

Kniprath, E. (1980a). Functional and evolutionary aspects of the molluscan periostracum. *Haliotis* **10**, 79.

Kniprath, E. (1980b). Internal shell development in ectocochleate molluscs? *Haliotis* **10**, 80.

Kniprath, E. (1980c). Shell sac formation by cell delamination? *Haliotis* **10**, 81.

Kniprath, E. (1980d). Larval development of the shell and the shell gland in *Mytilus* Bivalvia). *Wilhelm Roux's Arch. Dev. Biol.* **188**, 201–204.

Kniprath, E. (1980e). Ontogenetic plate and plate field development in two Chitons, *Middendorffia* and *Ischnochiton*. *Wilhelm Roux's Arch. Dev. Biol.* **189**, 97–106.

Kniprath, E. (1980f). Sur la glande coquillière de *Helix aspersa* (Gastropoda). *Arch. Zool. Exp. Gen.* **121**, 207–212.

Kniprath, E. (1981). Ontogeny of the molluscan shell field: A review. *Zool. Scr.* **10**, 61–79.

Korschelt, E., and Heider, K. (1936). "Vergleichende Entwicklungsgeschichte der Tiere," Vol. 2, pp. 844–968. Fischer, Jena.

Kowalevsky, A. (1883a). Embryogénie du *Chiton polii* (Philippi) avec quelques remarques sur le développement des autres chitons. *Ann. Mus. Hist. Nat. Marseille, Zool.* **1**(5), 1–46.

Kowalevsky, A. (1883b). Embryologie du Dentale. *Ann. Mus. Hist. Nat. Marseille, Zool.* **1** (7), 1–54.

Kress, A. (1975). Observations during embryonic development in the genus *Doto* (Gastropoda, Opisthobranchia). *J. Mar. Biol. Assoc. U.K.* **55**, 691–701.

Kriegstein, A. R. (1977a). Development of the nervous system of *Aplysia californica*. *Proc. Natl. Acad. Sci. U.S.A.* **74**, 375–378.

Kriegstein, A. R. (1977b). Stages in the post-hatching development of *Aplysia californica*. *J. Exp. Zool.* **199**, 275–288.

Kümmel, G., and Brandenburg, J. (1961). Die Reusengeisselzellen (Cyrtocyten). *Z. Naturforsch., B: Anorg. Chem., Org. Chem., Biochem., Biophys., Biol.* **16B**, 692–697.

Kussakin, O. G. (1960). Not seen in original. Abstract quoted in Smith (1966).

LaBarbera, M. (1974). Calcification of the first larval shell of *Tridacna squamosa* (Tridacnidae: Bivalvia). *Mar. Biol.* **25**, 233–238.

Lane, D. J. W., and Nott, J. A. (1975). A study of the morphology, fine structure and histochemistry of the foot of the pediveliger of *Mytilus edulis* L.. *J. Mar. Biol. Assoc. U.K.* **55**, 477–495.

Laviolette, P. (1954). Etude cytologique et expérimentale de la régénération germinale après castration chez *Arion rufus* L. (Gastéropode Pulmoné). *Ann. Sci. Nat., Zool. Biol. Anim.* [11] **16**, 427–535.

Lebour, M. V. (1938). The life history of *Kellia suborbicularis*. *J. Mar. Biol. Assoc. U.K.* **22**, 447–451.

LePennec, M. (1974). Morphogenèse de la coquille de *Pecten maximus* (L.) élevé au laboratoire. *Cah. Biol. Mar.* **15**, 475–482.

LePennec, M., and Masson, M. (1976). Morphogénèse de la coquille de *Mytilus galloprovincialis* (Lamarck) élevé au laboratoire. *Cah. Biol. Mar.* **17**, 113–118.

Lever, J. (1958). On the occurrence of a paired follicle gland in the lateral lobes of the cerebral ganglia of some Ancylidae. *Proc. K. Ned. Akad. Wet., Ser. C* **61**, 235–242.

Lever, J., Boer, H. H., Duiven, R. J. T., Lammens, J. J., and Wattel, J. (1959). Some observations on follicle glands in pulmonates. *Proc. K. Ned. Akad. Wet. Ser. C* **62**, 139–144.

Lubet, P., Herlin-Houtteville, P., and Mathieu, M. (1976). La lignée germinale des mollusques pélécypodes. Origine et évolution. *Bull. Soc. Zool. Fr.* **101**, Suppl., 22–27.

Luchtel, D. (1972). Gonadal development and sex determination in pulmonate molluscs. I. *Arion circumscriptus*. *Z. Zellforsch. Mikrosk. Anat.* **130**, 279–301.

Martoja, M. (1964). Développement de l'appareil reproducteur chez les gastéropodes pulmonés. *Annee Biol.* [4] **3**, 199–232.

Meisenheimer, J. (1898). Entwicklungsgeschichte von *Limax maximus* L. II. Die Larvenperiode. *Z. Wiss. Zool.* **63**, 573–664.

Meisenheimer, J. (1899). Zur Morphologie der Urniere der Pulmonaten. *Z. Wiss. Zool.* **65**, 709–724.

Meisenheimer, J. (1901a). Entwicklungsgeschichte von *Dreissensia polymorpha* Pall. *Z. Wiss. Zool.* **69**, 1–137.

Meisenheimer, J. (1901b). Die Entwicklung von Herz, Pericard, Niere und Genitalzellen bei *Cyclas* im Verhältnis zu den übrigen Mollusken. *Z. Wiss. Zool.* **69**, 417–428.

Metcalf, M. M. (1893). Contributions to the embryology of *Chiton*. *Stud. Biol. Lab. Johns Hopkins Univ.* **5**, 249–267.

Minganti, A. (1950). Acidi nucleici e fosfatasi nello sviluppo della *Limnaea*. *Riv. Biol.* [N.S.] **42**, 295–319.

Moor, B. (1977). Zur Embryologie von *Bradybaena (Eulota) fruticum* Müller (Gastropoda, Pulmonata, Stylommatophora). *Zool. Jahrb., Abt. Anat. Ontog. Tiere* **97**, 323–399.

Moor, B. (1982). A transitory structure in early organogenesis of the nervous system in Stylommatophora (Gastropoda, Pulmonata). *Malacologia* **22**, 611–614.

Moritz, C. E. (1939). Organogenesis in the gastropod *Crepidula adunca* Sowerby. *Univ. Calif. Berkeley, Publ. Zool.* **43**, 217–248.

Morton, J. E. (1959). The habits and feeding organs of *Dentalium entalis*. *J. Mar. Biol. Assoc. U.K.* **38**, 225–238.

Nelson, T. C. (1918). On the origin, nature and function of the crystalline style of lamellibranchs. *J. Morphol.* **31**, 53–111.

Ockelmann, K. W. (1965). Developmental types in marine bivalves and their distribution along the Atlantic coast of Europe. *Proc. Eur. Malacol. Congr. 1st, 1962* pp. 25–35.

Okada, K. (1936). Some notes on *Sphaerium japonicum biwaense* Mori, a freshwater bivalve. IV. Gastrula and fetal larva. *Sci. Rep. Tohoku Imp. Univ., Ser. 4* **11**, 49–68.

Okada, K. (1939). The development of the primary mesoderm in *Sphaerium japonicum biwaense* Mori. *Sci. Rep. Tohoku Imp. Univ., Ser. 4* **14**, 25–47.

Oldfield, E. (1964). The reproduction and development of some members of the Erycinidae and Montacutidae (Mollusca, Eulamellibranchiata). *Proc. Malacol. Soc. London* **36**, 79–120.

Otto, H., and Tönniges, C. (1906). Untersuchungen über die Entwicklung von *Paludina vivipara*. *Z. Wiss. Zool.* **80**, 411–514.

Owen, G., Trueman, E. R., and Yonge, C. M. (1953). The ligament in the Lamellibranchia. *Nature (London)* **171**, 73–75.

Pelseneer, P. (1899). La condensation embryogénique chez un nudibranche. *Trav. Stn. Zool. Wimereux* **7**, 513–520.

Pelseneer, P. (1900). Les yeux céphaliques chez les lamellibranches. *Arch. Biol.* **16**, 97–103.

Pelseneer, P. (1908). Les yeux branchiaux des lamellibranches. *Bull. Cl. Sci., Acad. R. Belg.* pp. 773–779.

Pelseneer, P. (1911). Recherches sur l'embryologie des Gastropodes. *Mem. Acad. R. Belg., Cl. Sci.* [2] **3** (4), 1–167.

Perron, F. E., and Turner, R. D. (1977). Development, metamorphosis, and natural history of the nudibranch *Doridella obscura* Verrill (Corambidae: Opisthobranchia). *J. Exp. Mar. Biol. Ecol.* **27**, 171–185.

Portmann, A. (1930). Die Larvennieren von *Buccinum undatum*. *Z. Zellforsch. Mikrosk. Anat.* **10**, 401–410.

Portmann, A., and Sandmeier, E. (1965). Die Entwicklung von Vorderdarm, Macromeren und Enddarm unter dem Einfluss von Nähreiern bei *Buccinum, Murex* und *Nucella* (Gastropoda, Prosobranchia). *Rev. Suisse Zool.* **72**, 187–204.

Pruvot, G. (1890). Sur le développement d'un solénogastre. *C. R. Hebd. Seances. Acad. Sci.* **111**, 689–692.

Pruvot, G. (1892). Sur l'embryogénie d'une *Proneomenia*. *C. R. Hebd. Seances. Acad. Sci.* **114**, 1211–1214.

Prytherch, H. F. (1934). The role of copper in the settling, metamorphosis, and distribution of the American oyster, *Ostrea virginica. Ecol. Monogr.* **4**, 47–107.

Quattrini, D., and Sacchi, T. B. (1971). La podocisti di *Milax gagates* (Draparuaud). Mollusca Gastropoda Pulmonata. Ricerche al microscopio ottico ed elettronico. *Arch. Ital. Anat. Embriol.* **76**, 39–52.

Quayle, D. B. (1952). Structure and biology of the larva and spat of *Venerupis pullastra* (Montagu). *Trans. R. Soc. Edinburgh* **62**, 255–297.

Raven, C. P. (1952). Morphogenesis in *Limnaea stagnalis* L. and its disturbance by lithium. *J. Exp. Zool.* **121**, 1–77.

Raven, C. P. (1966). "Morphogenesis. The Anlaysis of Molluscan Development" (rev. and enlarged ed.). Pergamon, Oxford.

Raven, C. P. (1975). Development. *In* "Pulmonates" (V. Fretter and J. Peake, eds.), Vol. 1, pp. 367–400. Academic Press, New York.

Rees, C. B. (1950). The interpretation and classification of lamellibranch larvae. *Hull Bull. Mar. Ecol.* **3**, 73–104.

Régondaud, J. (1961a). Développement de la cavité pulmonaire et de la cavité palléale chez *Lymnaea stagnalis* L. C. R. Hebd. Seances. Acad. Sci. **252**, 179–181.

Régondaud, J. (1961b). Formation du système nerveux et torsion chez *Lymnaea stagnalis* L. *C. R. Hebd. Seances. Acad. Sci.* **252**, 1203–1205.

Régondaud, J. (1964). Origine embryonnaire de la cavité pulmonaire de *Lymnaea stagnalis* L. Considérations particulières sur la morphogénèse de la commissure viscérale. *Bull. Biol. Fr. Belg.* **98**, 433–471.

Régondaud, J. (1972). Observation ultrastructurale des cellules nucales de l'embryon de *Lymnaea stagnalis* L. (Gastéropode Pulmoné Basommatophore). *C. R. Hebd. Seances. Acad. Sci., Ser. D* **275**, 679–682.

Richardot, M. (1979). Calcium cells and groove cells in calcium metabolism in the freshwater limpet *Ferrissia wautieri* (Basommatophora: Ancylidae). *Malacol. Rev.* **12**, 67–78.

Richter, G. (1968). Heteropoden und Heteropodenlarven im Oberflächenplankton des Golfs von Neapel. *Pubbl. Stn. Zool. Napoli* **36**, 346–400.

Riedl, R. (1960). Beiträge zur Kenntnis der *Rhodope veranii*. II. Entwicklung. *Z. Wiss. Zool.* **163**, 237–316.

Rigby, J. E. (1965). *Succinea putris:* A terrestrial opisthobranch mollusc. *Proc. Zool. Soc. London* **144**, 445–486.

Robertson, R. (1971). Scanning electron microscopy of planctonic larval marine gastropod shells. *Veliger* **14**, 1–12.

Robertson, R. (1973). The biology of the Architectonicidae, gastropods combining prosobranch and opisthobranch traits. *Malacologia* **14**, 215–220.

Robertson, R. (1974). Marine prosobranch gastropods: Larval studies and systematics. *Thalassia Jugosl.* **10**, 213–238.

Rosen, M. D., Stasek, C. R., and Hermans, C. O. (1979). The ultrastructure and evolutionary significance of the ocelli in the larva of *Katharina tunicata* (Mollusca: Polyplacophora). *Veliger* **22**, 173–178.

Runham, N. W. (1975). Alimentary canal. *In* "Pulmonates" (V. Fretter and J. Peake, eds.), Vol. 1, pp. 53–104. Academic Press, New York.

Russell, L. (1929). The comparative morphology of the elysioid and aeolidoid types of the molluscan nervous system and its bearing on the relationships of the ascoglossan nudibranchs. *Proc. Zool. Soc. London* **14**, 197–233.

Sastry, A. N. (1965). The development and external morphology of pelagic larval and post-larval stages of the Bay scallop *Aequipecten irradians concentricus* Say, reared in the laboratory. *Bull. Mar. Sci.* **15**, 417–435.

Sastry, A. N. (1979). Pelecypoda (excluding Ostreidae). *In* "Reproduction of Marine Invertebrates" (A. C. Giese and J. S. Pearse, eds.), Vol. 5, pp. 113–292. Academic Press, New York.

Saunders, A. M. C., and Poole, M. (1910). The development of *Aplysia punctata*. *Q. J. Microsc. Sci.* **55**, 497–539.

Scheltema, R. S. (1971). Larval dispersal as a means of genetic exchange between geographically separated populations of shallow-water benthic marine gastropods. *Biol. Bull. (Woods Hole, Mass.)* **140**, 284–322.

Schönenberger, N. (1969). Beiträge zur Entwicklung und Morphologie von *Trinchesia granosa* Schmekel (Gastropoda, Opisthobranchia). *Pubbl. Stn. Zool. Napoli* **37**, 236–292.

Sigerfoos, C. P. (1908). Natural history, organization and late development of the Teredinidae or ship-worms. *Bull. U.S. Bur. Fish.* **27**, 191–231.

Smith, A. G. (1966). The larval development of chitons (Amphineura). *Proc. Calif. Acad. Sci.* **32**, 433–446.

Smith, F. G. W. (1935). The development of *Patella vulgata*. *Philos. Trans. R. Soc. London, Ser. B* **225**, 95–125.

Smith, S. T. (1967). The development of *Retusa obtusa* (Montagu) (Gastropoda, Opisthobranchia). *Can. J. Zool.* **45**, 737–764.

Struhsaker, J. W., and Costlow, J. D. (1968). Larval development of *Littorina picta* (Prosobranchia, Mesogastropoda), reared in the laboratory. *Proc. Malacol. Soc. London*, **38**, 153–160.

Switzer-Dunlap, M. (1978). Larval biology and metamorphosis of Aplysiid gastropods. *In* "Settlement and Metamorphosis of Marine Invertebrate Larvae" (F.-S. Chia and M. E. Rice, eds.), pp. 197–206. Am. Elsevier, New York.

Switzer-Dunlap, M., and Hadfield, M. G. (1977). Observations on development, larval growth and metamorphosis of four species of Aplysiidae (Gastropoda:Opisthobranchia) in laboratory culture. *J. Exp. Mar. Biol. Ecol.* **29**, 245–261.

Tardy, J. (1962). Cycle biologique et métamorphose d'*Eolidina alderi* (Gastéropode, Nudibranche). *C. R. Hebd. Seances. Acad. Sci.* **255**, 3250–3252.

Tardy, J. (1967). Organogenèse de l'appareil génital du Mollusque Nudibranche *Aeolidiella alderi* (Cocks). *C. R. Hebd. Seances Acad. Sci., Ser. D* **265**, 2013–2014.

Tardy, J. (1970a). Organogenèse de l'appareil génital chez les mollusques. *Bull. Soc. Zool. Fr.* **95**, 407–428.

Tardy, J. (1970b). Contribution à l'étude des métamorphoses chez les nudibranches. *Ann. Sci. Nat., Zool. Biol. Anim.* [12] **12**, 299–370.

Tardy, J. (1971a). Etude expérimentale de la régénération germinale après castration chez les Aeolidiidae. *Ann. Sci. Nat., Zool. Biol. Anim.* [12] **13**, 91–147.

Tardy, J. (1971b). Embryologie et organogenèse sexuelle. *Haliotis* **1**, 151–166.

Tardy, J. (1974). Morphogenèse du système nerveux chez les Mollusques Nudibranches. *Haliotis* **4**, 61–75.

Thiriot-Quiévreux, C. (1967a). Observations sur le développement larvaire et post-larvaire de *Simnia spelta* Linné (Gastéropode Cypraeidae). *Vie Milieu, Ser. A* **18**, 143–151.

Thiriot-Quiévreux, C. (1967b). Apparition précoce de l'ébauche de l'appareil copulateur chez *Atlanta lesueuri* Souleyet (Mollusques Hétéropodes). *C. R. Hebd. Seances Acad. Sci., Ser. D* **265**, 130–132.

Thiriot-Quiévreux, C. (1969). Organogenèse larvaire du genre *Atlanta* (Mollusque Hétéropode). *Vie Milieu, Ser. A* **20**, 347–395.

Thiriot-Quiévreux, C. (1970). Transformations histologiques lors de la métamorphose chez *Cymbulia peroni* de Blainville (Mollusca, Opisthobranchia). *Z. Morphol. Tiere* **67**, 106–117.

Thiriot-Quiévreux, C. (1971a). Les véligères planctoniques de prosobranches de la région de Banyuls-sur-Mer (Méditerranée Occidentale): Phylogénie et métamorphose. *Eur. Mar. Biol. Symp.* [*Proc.*], *4th, 1969* pp. 221–225.

Thiriot-Quiévreux, C. (1971b). Contribution à l'étude de l'organogenèse des hétéropodes (Mollusca, Prosobranchia). *Z. Morphol. Tiere* **69**, 363–384.

Thiriot-Quiévreux, C. (1972). Microstructures de coquilles larvaires de prosobranches au microscope électronique à balayage. *Arch. Zool. Exp. Gen.* **113**, 553–564.

Thiriot-Quiévreux, C. (1974). Anatomie interne de véligères planctoniques de prosobranches mésogastropodes au stage proche de la métamorphose. *Thalassia Jugosl.* **10**, 379–399.

Thiriot-Quiévreux, C. (1975). Observations sur les larves et les adultes de Carinariidae (Mollusca: Heteropoda) de l'Océan Atlantique Nord. *Mar. Biol.* **32**, 379–388.

Thiriot-Quiévreux, C. (1977). Véligère planctotrophe du Doridien *Aegires punctilucens* (D'Orbigny) (Mollusca:Opisthobranchia): Description et métamorphose. *J. Exp. Mar. Biol. Ecol.* **26**, 177–190.

Thiriot-Quiévreux, C., and Martoja, M. (1976). Appareil génital femelle des Atlantidae (Mollusca, Heteropoda). II. Etude histologique des structures larvaires, juvéniles et adultes. Données sur la fécondation et la ponte. *Vie Milieu, Ser. A* **26**, 201–233.

Thompson, T. E. (1958). The natural history, embryology, larval biology and postlarval development of *Adalaria proxima* (Alder and Hancock) Gastropoda, Opisthobranchia. *Philos. Trans. R. Soc. London, Ser. B* **242**, 1–58.

Thompson, T. E. (1959). Feeding in nudibranch larvae. *J. Mar. Biol. Assoc. U.K.* **38**, 239–248.

Thompson, T. E. (1961a). The development of *Neomenia carinata* Tullberg (Mollusca, Aplacophora). *Proc. R. Soc. London, Ser. B* **153**, 263–278.

Thompson, T. E. (1961b). The importance of the larval shell in the classification of the Sacoglossa and the Acoela (Gastropoda, Opisthobranchia). *Proc. Malacol. Soc. London* **34**, 233–238.

Thompson, T. E. (1962). Studies on the ontogeny of *Tritonia hombergi* Cuvier (Gastropoda, Opisthobranchia). *Philos. Trans. R. Soc. London, Ser. B* **245**, 171–218.

Thompson, T. E. (1967). Direct development in a Nudibranch, *Cadlina laevis*, with a discussion of developmental processes in Opisthobranchia. *J. Mar. Biol. Assoc. U.K.* **47**, 1–22.

Thompson, T. E. (1972). Eastern Australian Dendronotoidea (Gastropoda, Opisthobranchia). *Zool. J. Linn. Soc.* **51**, 63–77.

Thompson, T. E. (1976). "Biology of Opisthobranch Molluscs," Vol. I. Ray Society, London.

Thorson, G. (1940). Studies on the egg masses and larval development of gastropoda from the Iranian Gulf. *Dan. Sci. Invest. Iran* **2**, 159–238.

Thorson, G. (1946). Reproduction and larval development of Danish marine bottom invertebrates with special reference to the planctonic larvae in the Sound (Oresund). *Medd. Komm. Dan. Fisk.- Havunders., Ser. Plankton* **4**, 1–523.

Timmermans, L. P. M., Geilenkirchen, W. L. M., and Verdonk, N. H. (1970). Local accumulation of feulgen-positive granules in the egg cortex of *Dentalium dentale* L. *J. Embryol. Exp. Morphol.* **23**, 245–252.

Underwood, A. J. (1972). Spawning, larval development and settlement behaviour of *Gibbula cineraria* (Gastropoda: Prosobranchia) with a reappraisal of torsion in gastropods. *Mar. Biol.* **17**, 341–349.

van Dongen, C. A. M. (1976a). The development of *Dentalium* with special reference to the significance of the polar lobe. V. Differentiation of the cell pattern in lobeless embryos of *Dentalium vulgare* (da Costa) during late larval development. *Proc. K. Ned. Akad. Wet. Ser. C* **79**, 245–255.

van Dongen, C. A. M. (1976b). The development of *Dentalium* with special reference to the significance of the polar lobe. VI. Differentiation of the cell pattern in lobeless embryos of *Dentalium vulgare* (da Costa) during late larval development. *Proc. K. Ned. Akad. Wet., Ser. C* **79**, 256–266.

van Dongen, C. A. M. (1976c). The development of *Dentalium* with speical reference to the significance of the polar lobe. VII. Organogenesis and histogenesis in lobeless embryos of

Dentalium vulgare (da Costa) as compared to normal development. *Proc. K. Ned. Akad. Wet., Ser. C* **79**, 454–465.

van Dongen, C. A. M. (1977). Mesoderm formation during normal development of *Dentalium dentale. Proc. K. Ned. Akad. Wet., Ser. C* **80**, 372–376.

van Dongen, C. A. M., and Geilenkirchen, W. L. M. (1974a). The development of *Dentalium* with special reference to the significance of the polar lobe. I. Division chronology and development of the cell pattern in *Dentalium dentale* (Scaphopoda). *Proc. K. Ned. Akad. Wet., Ser. C* **77**, 57–70.

van Dongen, C. A. M., and Geilenkirchen, W. L. M. (1974b). The development of *Dentalium* with special reference to the significance of the polar lobe. II. Division chronology and development of the cell pattern in *Dentalium dentale* (Scaphopoda). *Proc. K. Ned. Akad. Wet., Ser. C* **77**, 71–84.

van Dongen, C. A. M., and Geilenkirchen, W. L. M. (1974c). The development of *Dentalium* with special reference to the significance of the polar lobe. III. Division chronology and development of the cell pattern in *Dentalium dentale* (Scaphopoda). *Proc. K. Ned. Akad. Wet., Ser. C* **77**, 85–100.

van Dongen, C. A. M., and Geilenkirchen, W. L. M. (1975). The development of *Dentalium* with special reference to the significance of the polar lobe. IV. Division chronology and development of the cell pattern in *Dentalium dentale* after removal of the polar lobe at first cleavage. *Proc. K. Ned. Akad. Wet., Ser. C* **78**, 358–375.

van Mol, J.-J. (1967). Etude morphologique et phylogénétique du ganglion cérébroïde des gastéropodes pulmonés (Mollusques). *Mem. Acad. R. Belg., Cl. Sci., Collect. 8°* **37** (5), 1–168.

Verdonk, N. H. (1965). Morphogenesis of the head region in *Limnaea stagnalis* L.. Ph.D. Thesis, pp. 1–135. University of Utrecht.

von Salvini-Plawen, L. (1969). Solenogastres und Caudofoveata (Mollusca:Aculifera): Organisation und phylogenetische Bedeutung. *Malacologia* **9**, 191–216.

von Salvini-Plawen, L. (1972). Zur Morphologie und Phylogenie der Mollusken: Die Beziehungen der Caudofoveata und der Solenogastres als Aculifera, als Mollusca und als Spiralia (nebst einem Beitrag zur Phylogenie der coelomatischen Räume). *Z. Wiss. Zool.* **184**, 205–394.

von Salvini-Plawen, L. (1973). Zur Klärung des ''Trochophora''-Begriffes. *Experientia* **29**, 1434–1435.

von Salvini-Plawen, L. (1980). Was ist eine Trochophora? Eine Analyse der Larventypen mariner Protostomier. *Zool. Jahrb. Abt. Anat. Ontog. Tiere* **103**, 389–423.

Wada, S. K. (1968). Mollusca. 1. Amphineura, Gastropoda, Scaphopoda, Pelecypoda. *In* ''Invertebrate Embryology'' (K. Kume and K. Dan, eds.), pp. 485–525. Nolit Publ. House, Belgrade.

Waller, T. R. (1981). Functional morphology and development of veliger larvae of the European Oyster, *Ostrea edulis* L. *Smithson. Contribu. Zool.* **328.**

Weiss, M. (1968). Zur embryonalen und postembryonalen Entwicklung des Mitteldarmes bei Limaciden und Arioniden (Gastropoda, Pulmonata). *Rev. Suisse Zool.* **75**, 157–226.

Werner, B. (1939). Ueber die Entwicklung und Artunterscheidung von Muschellarven des Nordseeplanktons, unter besonderer Berücksichtigung der Schalenentwicklung. *Zool. Jahrb., Abt. Anat. Ontog. Tiere* **66**, 1–54.

Werner, B. (1955). Ueber die Anatomie, die Entwicklung und Biologie des Veligers und der Veliconcha von *Crepidula fornicata* L. (Gastropoda, Prosobranchia). *Helgol. Wiss. Meeresunters.* **5**, 169–217.

Wierzejski, A. (1905). Embryologie von *Physa fontinalis* L. *Z. Wiss. Zool.* **83**, 502–706.

Williams, L. G. (1971). Veliger development in *Dendronotus frondosus* (Ascanius, 1774) (Gastropoda: Nudibranchia). *Veliger* **14**, 166–171.

Wilson, E. B. (1904). Experimental studies on germinal localization. I. The germ regions in the egg of *Dentalium. J. Exp. Zool.* **1**, 1–72.

Yonge, C. M. (1926). Structure and physiology of the organs of feeding and digestion in *Ostrea edulis*. *J. Mar. Biol. Assoc. U.K.* **14**, 295–386.

Yonge, C. M. (1957). Mantle fusion in the Lamellibranchia. *Pubbl. Stn. Zool. Napoli* **29**, 151–171.

Yonge, C. M. (1962). On the primitive significance of the byssus in the Bivalvia and its effects in evolution. *J. Mar. Biol. Assoc. U.K.* **42**, 113–125.

Ziegler, H. E. (1885). Die Entwicklung von *Cyclas cornea* Lam. (Sphaerium corneum L.). *Z. Wiss. Zool.* **41**, 525–569.

Zunke, U. (1978). Bau und Entwicklung des Auges von *Succinea putris* (Linné, 1758) (Mollusca, Stylommatophora). Lichtmikroskopische Ergebnisse. *Zool. Anz.* **201**, 220–244.

Zunke, U. (1979). Bau und Entwicklung des Auges von *Succinea putris* (Linné, 1758) (Mollusca, Stylommatophora). Elektronenmikroskopische Ergebnisse. *Zool. Jahrb., Abt. Anat. Ontog. Tiere* **101**, 27–103.

5

Origin of Spatial Organization

J. A. M. VAN DEN BIGGELAAR

Zoological Laboratory
State University of Utrecht
Utrecht, The Netherlands

P. GUERRIER

Station Biologique
Roscoff, France

I. Introduction

In most organisms, the fertilized egg has some degree of cytoplasmic localization (Davidson, 1976). By the cleavage process, this leads to the formation of qualitatively different lines of cells. Most, and probably all, molluscan eggs only show an animal–vegetal differentiation in the distribution of the various ooplasmic constituents and in the architecture of the egg membrane. This animal–vegetal polarity has a definite relationship to the anteroposterior axis of the developing embryo. The animal pole corresponds with the cephalic or anterior pole, and the vegetal pole corresponds with the caudal or posterior pole, that is, the place of origin of the blastopore. Thus, the mechanisms that determine the cytoplasmic localization finally determine the embryonic axis.

THE MOLLUSCA, VOL. 3
Development

In the absence of a bilateral symmetrical structure in the undivided egg, an additional mechanism must be involved in the determination of the plane of bilateral symmetry, thus in the determination of the dorsoventral axis. In many molluscs, the first division is unequal. It appears that the first cleavage plane directly influences the position of the plane of bilateral symmetry (Morgan and Tyler, 1930; Guerrier, 1970a). Apparently, in unequally dividing species, the mechanisms that determine division inequality also determine the dorsoventral axis.

In other species, the first two cleavages are equal and produce qualitatively equal quadrants. Here, cellular interactions seem to be involved in the differentiation of the quadrants.

Finally, gastropods are only bilaterally symmetrical during the early part of their life cycle. The adult form is asymmetric. This asymmetry is related to the chirality of the spiral cleavage.

In this chapter we will discuss (1) the development of the animal–vegetal polarity; (2) the mechanisms that determine division inequality at the first two cleavages; (3) the mechanisms that, in equally dividing species, determine the differentiation of the initially equal quadrants, and hence the formation of the dorsoventral axis; and (4) the development of asymmetry in gastropods.

II. General Considerations on Polarity

Polarity is a general feature of cell morphology. Somatic cells usually have a special form which is related to their function. With respect to that function, they are often polarized. Good examples are epithelial cells, neurons, muscle cells, and fibroblasts. The great variety of their specific morphologies is maintained by a cytoplasmic architecture, the cytoskeleton. Cells can transmit the instructions for a special form to their offspring. For instance, the detailed neurite morphologies of a neuroblastoma cell appear to be heritable (Solomon, 1980). After division of a neuronal cell, the shapes of both sister cells are mirror images. If these cells are treated with a microtubule-depolymerizing agent, the cells lose their typical form and become spherical. The treatment is reversible; after being washed, the majority of the cells exactly reassume their original morphology. This suggests that morphological determinants persist in an unexpressed form. The character of the morphological determinants and also its morphogenesis in somatic cells remain enigmatic phenomena.

The special form of a cell, as well as its polarity, is accompanied by specific localizations of cytoplasmic components. A number of investigations have shown that not only the form, but also cytoplasmic segregation, depends upon the presence of microtubules (Hyams and Stebbings, 1979). A representative example is the regularly changing distribution pattern of pigment granules in chromatophores (Schliwa, 1981). Intracellular transport of pigment granules appears to be sensitive to cytochalasin B and mitotic inhibitors, which suggests a possible involvement of microfilaments and microtubules. Chromatophores also nicely demonstrate the selectivity of transport. Although pigment granules are transported, other organelles such as mitochondria do not move (Murphy, 1975). The chromatophores of the glass shrimp are polychromatic and probably contain blue, white, yellow, and red pigments (Robinson and Charlton, 1973). On a dark background, all granules except the white ones are displaced from the center of the cell. On a white environment, blue, red, and yellow move back to the center again, whereas white pigment is transported in the opposite direction.

Another mechanism involved in cell polarization and cytoplasmic localization might be related to the cell membrane in which a segregation of ionic leaks and ion pumps occurs, which produces an electric current through the cell and back through the conductive environment (Jaffe, 1981; Jaffe and Nuccitelli, 1974, 1977). A voltage difference of one or a few millivolts appears to be sufficient for the self-electrophoresis of a number of charged molecules or organelles. In peculiar, Woodruff and Telfer (1980) provided strong evidence for a unidirectional transport of charged molecules from nurse cells to oocytes as a result of such a voltage difference.

It may be expected that as soon as the mechanisms for intracellular transport in somatic cells are fully understood, the phenomenon of ooplasmic segregation will no longer remain an enigmatic event. Since polarization of the egg cell obviously represents a special case of cell polarity in general, the understanding of the phenomenon of ooplasmic segregation will necessarily imply consideration of experimental evidence for the polarizing agents at work in a variety of cell types. A few selected examples will be sufficient to demonstrate the influence of external factors on such a polarization process.

In thyroid cells, the polarity appears to depend on the presence and localization of extracellular collagen, which is normally apposed to the basal poles of the cells (Mauchamp et al., 1980; Chambard et al., 1981). Indeed, after an exposure of the apical poles to collagen, the original apical–basal polarity is reversed. This reversion does not require dissociation of the epithelial cell layer. This example shows that an extracellularly applied agent can easily orient the polarity of a cell.

Another example of cell polarization can be derived from the early development of the mouse embryo. At the beginning of the eight-cell stage, the individual blastomeres are apolar, whereas they become bipolar during the interval between third and fourth cleavage (Ziomek and Johnson, 1980; Johnson and Ziomek, 1981). Cell polarization is accompanied by a variety of surface and cytoplasmic differentiations (Mulnard and Huygens, 1978; Lehtonen and Badley, 1980; Ziomek and Johnson, 1980). This arising polarization is not an autonomous process, but actually depends on cellular interactions, that is, again on extracellular signals. The apparently required intercellular contacts do not seem to be specific, because contact with debris or with the bottom of the culture dish is as effective as contact with a neighboring cell. Whatever object is touched, the point of contact always functions as a reference point for the polarity to arise. Prior to compaction of the eight-cell mouse embryo, the mitochondria are more or less randomly distributed in the cytoplasm of the blastomeres. During compaction they become localized in those parts of the cell cortex that are associated with the newly established common cell borders (Ducibella et al., 1977). This polarity appears to be labile during a certain initial phase but becomes progressively irreversible.

Another illustration of cell polarization resulting from external influence is the formation of a rhizoid in the fucoid egg (Jaffe, 1966, 1968, 1969, 1977, 1979; Jaffe and Nuccitelli, 1974). Environmental gradients of pH and temperature, electric fields, and light can induce gradients or a polarity in the unicellular fucoid zygote. Just as in the mouse embryo, this polarity is not immediately fixed, but is labile, at least during a short period following fertilization.

After this short, and necessarily far from complete, general introduction to cell polarity and polarizing mechanisms, as derived from observations performed on nonmolluscan material, we hope the reader is on the scent of an explanation of the general mechanisms involved in the polarization of the molluscan egg cell.

III. Animal–Vegetal Polarity

A. Morphological and Physiological Expression

After ovulation, the eggs of most molluscan species start maturation up to the first meiotic metaphase independent of fertilization (Chapter 2, this volume). The freshly ovulated egg usually has a more or less homogenous distribution of the ooplasmic substances. Immediately after ovulation and the beginning of the maturation divisions, a redistribution occurs, and a remarkably heterogenous distribution of the cytoplasmic components is attained. This redistribution process correlated with maturation and fertilization has been called ooplasmic segregation (Costello, 1945) and bipolar differentiation (Spek, 1930). Spek assumed that cytoplasmic localization is an electric phenomenon accompanied by the appearance of an intracellular pH gradient. By means of metachromatic vital dyes with the properties of pH indicators, he demonstrated an intracellular pH gradient in the fertilized egg of the prosobranch *Columbella* (Spek, 1934). Starting from an initial uniform distribution, the more alkaline substances concentrate in the animal hemisphere, whereas the more acidic components accumulate in the vegetal half.

Segregation of ooplasmic substances, resulting in the formation of an intracellular pH gradient, has also been found in the egg of the sea hare *Aplysia* (Ries,

1939; Ries and Gersch, 1936; Attardo, 1957). During the maturation divisions of the *Aplysia* egg, the pH in the animal hemisphere apparently becomes higher (~pH 8) than in the vegetal region of the egg (~pH 6). Besides pH gradients, a gradient in the oxidoreduction potential could also appear some time after oviposition and shortly before the first maturation division. After treatment with the vital dye leucomethylene blue prior to maturation, the egg is homogenously stained. As soon as ooplasmic segregation starts, the cytoplasm of the animal pole becomes blue due to an oxidation of leucomethylene blue. This localized oxidation would be the result of an accumulation of oxidases in the animal region. After staining of the egg with Janus green, reduction of this basic vital dye to a red color appears to start at the vegetal pole. Ries (1937) also demonstrated a homogenous distribution of vitamin C granules prior to maturation, and rearrangement of these granules into a slightly supraequatorial zone during ripening of the egg. A similar distribution process of cytochrome oxidases was observed. According to Attardo (1957), this zonation must be due to a concentration of most of the mitochondria in the same supraequatorial region.

A number of additional examples can be given in which the mature egg gradually obtains a bipolar organization. For the egg of *Lymnaea*, a differential distribution of the yolk, mitochondria, Golgi complexes, glycogen, and different enzymes has been described (Raven, 1945). In general, the proteid yolk is concentrated in the vegetal hemisphere (Fig. 1a,b) as in the opistobranchs *Cymbulia* (Fol, 1875), *Aplysia* (Ries, 1939; Ries and Gersch, 1936), and *Navanax* (Worley and Worley, 1943), in the prosobranchs *Neritina* (Blochmann, 1882), *Trochus* (Robert, 1902), *Haliotis* and *Patella* (J. A. M. van den Biggelaar, personal observations), *Ilyanassa* (Morgan, 1933), *Busycon* (formerly called *Fulgur*, Conklin, 1907) and in the pulmonates *Planorbis* (Rabl, 1879), *Lymnaea* (Raven, 1945), and *Limax* (Guerrier, 1968). In the scaphopod *Dentalium*, the

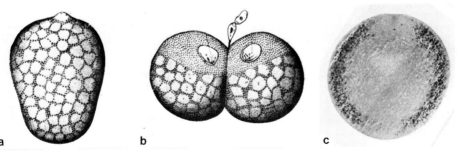

a   b   c

Fig. 1. (a) Homogeneous distribution of the yolk of the egg of *Cymbulia* prior to the first maturation division. (b) Heterogeneous distribution of the yolk in an egg of *Cymbulia* at the two-cell stage. (c) Concentration of the yolk in a broad circumferential zone between the animal and vegetal pole of the egg of *Dentalium*. (From Fol, 1875 (**a** and **b**) and by courtesy of N. H. Verdonk (**c**).)

animal and vegetal pole are more or less free from yolk, which is concentrated in a broad ring between the animal and vegetal pole (Wilson, 1904) (Fig. 1c). In small eggs, the animal–vegetal gradient in the distribution of yolk is often less clear as can be seen in the usually little-yolk-containing eggs of lamellibranchs and in eggs of the opistobranch *Fiona* (Casteel, 1904).

A striking localization is apparent in the position of the maturation spindles in the animal hemisphere. In the egg of *Crepidula* (Conklin, 1938), the first maturation spindle is formed at some distance from the animal pole, where it may have any position relative to the egg axis. A stream of hyaloplasm and nucleoplasm in the direction of the animal pole places the spindle in a position parallel to the egg axis. Finally, the layer of hyaloplasm in the animal region flows down immediately beneath the surface of the egg. In lamellibranchs, both asters first appear in contact with the dissolving germinal vesicle at equal distance from the future animal pole. Then, they rotate while the spindle is built up giving rise to a superficial and a central aster (Guerrier, 1969; Chapter 2, this volume).

In *Crepidula, Buccinum,* and *Littorina,* a special example of cytoplasmic localization has been found at the vegetal pole (Dohmen and Verdonk, 1979a; Chapter 1, this volume). Dohmen and Lok (1975) have observed aggregates of vesicles that are set apart in the polar lobe during first cleavage. The egg of *Bithynia* is endowed with a special cup-shaped cytoplasmic inclusion that is rich in RNA (Verdonk, 1973). Observations with the electron microscope showed that this so-called vegetal body consists of a mass of small vesicles between which ribosomes, mitochondria, and dense bodies can be found (Dohmen and Verdonk, 1974, 1979a,b; Chapter 1, this volume).

It was not our attempt to give an exhaustive review of ooplasmic segregation. The examples have only been chosen as representative demonstrations of cytoplasmic localizations as related to bipolarity of the molluscan egg. The origin of bipolarity in the molluscan egg will be discussed further in light of recent experimental results concerning cell polarity in general.

B. Origin of Bipolarity in the Molluscan Egg

Theoretically, there are two possible explanations for the origin of the ordered distribution of egg components and the bipolarity in the molluscan ovum. They may be inherited from the parent and become more pronounced during maturation and cleavage, or both may arise *de novo,* independently of an inherited internal and bipolar organization. Elements of both possibilities may normally cooperate and need not exclude one another.

Especially in lamellibranchs, a number of observations point to an inherited polarity. During oogenesis, the oocyte is surrounded by a membrane. After ovulation, a small opening, the micropyle, often persists in this membrane,

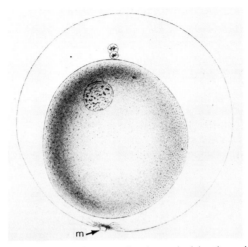

Fig. 2. A mature egg of *Unio*. The animal pole marked by the polar bodies is always opposite the micropyle in the egg capsule. m, micropyle. (From Lillie, 1895.)

which marks the point of attachment to the ovarian wall, for example, *Unio* (Fig. 2) (Lillie, 1895), *Sphaerium* (Woods, 1932), and *Anodonta* (Herbers, 1914). In *Pholas*, the oocyte is surrounded by a chorion. In the free egg, the point of attachment of this chorion to the ovarian wall is marked by the presence of a peduncle (Fig. 3) (Guerrier, 1968). The germinal vesicle is situated opposite the micropyle or peduncle. Thus, the ovulated egg has an animal–vegetal polarity that corresponds with the apical–basal polarity of the ovarian egg. This original apical–basal polarity normally becomes the definite animal–vegetal polarity as

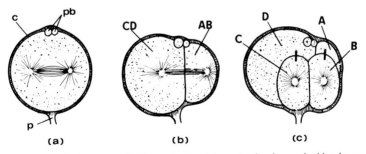

Fig. 3. Eggs of *Pholas* surrounded by a chorion. The animal pole, marked by the presence of the polar bodies, is always opposite the peduncle. (a) Symmetrical position of the first mitotic spindle. (b) Unequal two-cell stage resulting from a unilateral displacement of the first mitotic spindle. (c) Four-cell stage, reached by an equal division of the smaller AB cell into the two equally small cells A and B, and an unequal division of the CD cell into a small C and a big D cell. c, chorion; p, peduncle; pb, polar bodies. (From Guerrier, 1968.)

the polar bodies are always extruded diametrically opposite the micropyle or peduncle.

An inherited polarity related to the position of the oocyte in the ovary has also been assumed for the egg of *Lymnaea*. In this gastropod, Raven (1963, 1967) reported the presence of a regular pattern of six patches of a special cortical ooplasm around the equator of the freshly laid egg. These subcortical accumulations (SCA) of a special cytoplasm are arranged to an almost bilaterally symmetrical pattern. Four patches are relatively small and lie close together, occupying about half of the equator of the egg. The other two SCA are larger and more separated from each other and opposite the other four SCA. A plane of symmetry can thus be drawn between the two larger patches and between the middle of the smaller patches. This pattern arises during the passage of the ovum through the genital tract and shows a great similarity with the configuration of the nuclei of the follicle cells that surround the oocyte during oogenesis. From the similarity between the pattern of the follicle cells and the pattern of the SCA, Raven (1967) concluded that the animal–vegetal polarity and also dorsoventral symmetry were determined during oogenesis. In the following paragraphs, we will see that this assumption cannot be maintained in its original form in order to come into harmony with more recent experimental observations.

Special differentiations on the vegetal surface of the eggs of *Nassarius, Buccinum*, and *Crepidula* have been observed by Dohmen and van der Mey (1977); the vegetal pole surface of the *Dentalium* egg specifically binds extracellular bacteria (Geilenkirchen et al., 1971). At the vegetal pole of eggs of *Bithynia, Buccinum*, and *Crepidula*, aggregates of peculiar vesicles have been found (Dohmen and Verdonk, 1974, 1979a). In *Bithynia*, these vesicles are organized in the vegetal body that can already be observed in the ovarian oocyte. Again, this observation points to a correspondance between the apical–basal polarity of the ovarian oocyte and the embryonic animal–vegetal polarity (Chapter 1, this volume).

C. Experimental Evidence on the Irreversibility versus Reversibility of the Original Apical–Basal Polarity

It is legitimate to ask whether the inherited apical–basal polarity is necessarily correlated with the animal–vegetal polarity, or whether the egg might succeed in developing an animal–vegetal polarity independent of the original apical–basal polarity, without any consequence for normal development. In our general introduction to cell polarity, we have already referred to the epigenetic determinations of bipolarity in the fucoid egg, the blastomeres of the eight-cell mouse embryo and the thyroid cell. These examples clearly demonstrate indeed that a variety of agents or stimuli may polarize a supposed apolar cell or may overrule an original polarity and establish a new one.

1. Centrifugation Experiments

Experimental evidence for the introduction of a secondary animal–vegetal polarity in molluscs is scarce. Many authors have tried to disturb the normal distribution pattern of the ooplasmic constituents by centrifugal force. Centrifugation experiments have mostly been performed to obtain specific answers to three main questions. (1) Is maintenance of a normal distribution of the constituents essential for the determination of the developmental capacities of the blastomeres? (2) Is it possible to induce an animal–vegetal polarity that is independent of the apical–basal polarity of the oocyte? (3) What forces are involved in the redistribution of the egg substances from a stratified pattern into the normal one?

The first question is not of special interest for this chapter, and can be answered quickly. From a number of stratification experiments on eggs of organisms such as *Hydatina* (Morgan, 1910), *Archidoris, Diaulula, Triopha* (Costello, 1939), *Cumingia* (Pease, 1940), *Lymnaea* (Raven and Bretschneider, 1942), *Pholas* (Guerrier, 1968), *Ilyanassa* (Clement, 1968), and *Dentalium* (Verdonk, 1968) it can be concluded that a highly abnormal distribution of the ooplasmic components does not interfere with normal development. In all these experiments, it was observed that after centrifugation, the stratification gradually disappeared and at least a number of substances reattained a more or less normal distribution. Costello (1939) is the only author who reports that the shorter the time between centrifugation and first cleavage, hence the more abnormal the distribution of egg components prior to cleavage, the higher the percentage of abnormally developing embryos. As Costello's observation has no direct bearing on the problem of polarity we will not discuss it any further.

More relevant for our discussion on the determination of animal–vegetal polarity are those experiments that might give information on the eventual relationship between the direction of the centrifugal force and the direction of the resulting stratification with the definite embryonic polarity. Generally, it is assumed that during normal development the place where the polar bodies are extruded corresponds with the apical pole of the ovarian egg. In order to investigate whether under the influence of the centrifugal force the polar bodies are extruded at some point of the egg surface other than the animal pole, it is essential to have a reliable landmark of the primary egg axis. This condition is fulfilled in the eggs of *Pholas* (Guerrier, 1968). In these eggs, the original basal, that is, vegetal, pole is characterized by the presence of a peduncle. It appeared that, during maturation, most eggs orient perpendicular to the centrifugation axis (Fig. 4a,d). At the centrifugal pole, the heavy yolk is accumulated, whereas the lighter oil and fat is concentrated at the centripetal pole. The clear cytoplasm is located in an intermediary central layer. As in a number of other centrifugation experiments, the polar bodies were extruded in or in the near vicinity of the hyaloplasmic zone. In the case of *Pholas,* however, it appeared that the polar bodies were not only extruded in the central layer of cytoplasm, but also diametrically opposite the peduncle (Fig. 4a,d), thus opposite the original basal or vegetal pole. During further development, the first two cleavage planes crossed near both points, as in normal development. Evidently, the primary egg polarity has been unaffected by the centrifugal force. This was further affirmed by the low percentage of eggs in which the primary egg axis and the axis of centrifugation were not parallel (Fig. 4b,e). In these eggs, the spindles initially had a position parallel to the stratification, but then rotated and the bodies were formed relative to the original egg axis, independent of the direction of the centrifugal force (Fig. 4c,f), and the position of the hyaloplasm.

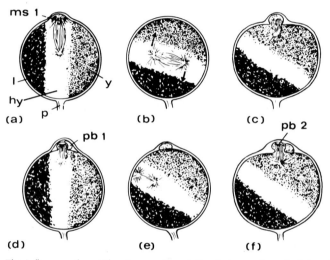

Fig. 4. The influence of centrifugation on the relation between the apical–basal and animal–vegetal polarity in eggs of *Pholas dactylus.* (a) Egg centrifuged during the first maturation division. The first polar body will be extruded in the hyaloplasmic equatorial zone. The spindle is placed at right angles to the axis of centrifugation, and opposite the peduncle. (b) The axis of stratification makes an angle with the original apical–basal polarity. After centrifugation the first meiotic spindle is reoriented into the original apical–basal direction. (c) Formation of the first polar body irrespective of the direction of stratification, but definitely related to the apical–basal polarity by reorientation. (d) Egg centrifuged after formation of the first polar body. The egg has been oriented during centrifugation with the apical–basal polarity at right angles to the direction of the centrifugal force. The second polar body will be extruded at its normal position, under the first polar body and opposite the peduncle. (e) The axis of stratification makes an angle with the apical–basal polarity, and the second meiotic spindle is forced into an oblique position. (f) After centrifugation the second meiotic spindle leaves its position at right angles to the axis of stratification in the hyaloplasmic layer and returns to its normal position opposite the peduncle underneath the first polar body. hy, hyaloplasm; 1, lipid; ms1, first meiotic spindle; p, peduncle; pb1, first polar body; pb2, second polar body; y, yolk. (From Guerrier, 1968.)

These results indicate that the original bipolarity of the oocyte is independent of the localization of those egg components that can be stratified with the applied centrifugal force. The inherited polarity must depend on a still more or less normal distribution of lighter components that were not (or not sufficiently) stratified, or on the presence of a cytoskeleton that, possibly in cooperation with the assumed unaffected egg cortex would maintain the original polarity and cause a normal redistribution of the egg substances.

Of special interest are the centrifugation experiments on eggs with a polar lobe. From deletion experiments, it is known that the polar lobe is of crucial importance for the formation of normal embryos (Chapter 6, this volume). Irrespective of a seriously disturbed composition of the polar lobe contents, the lobe maintained its developmental significance and in none of the centrifugation experiments did it appear possible to change the primary egg axis. Often it was concluded that the morphogenetic importance of the lobe depended on cortical properties. The assumption of regional cortical differences is strengthened with observed differences in electrical properties between the polar lobe and the rest of the egg as observed in *Dentalium* (Jaffe and Guerrier, 1981), but not in *Ilyanassa* (Moreau and Guerrier, 1981).

This restriction shows that one must be very careful in attributing the importance of the polar lobe to its membrane properties. As mentioned before, the polar lobe of *Bithynia* is characterized by the presence of the so-called vegetal body. After centrifugation with a force of about 1400 g it appeared that in 30 out of 70 eggs the vegetal body was not located in the polar lobe, but somewhere in one of the two blastomeres. After removal of the first polar lobe of such centrifuged eggs, completely normal embryos were obtained, whereas control lobeless embryos always developed abnormally (Cather and Verdonk, 1974; Chapter 6, this volume). This result indicates that, at least in *Bithynia,* the loss of a special cytoplasmic component can deprive the polar lobe of its usual developmental significance. No general conclusion can be drawn, however, from this specific observation since different mechanisms may well account for the activation of polar lobe components in unrelated species.

In conclusion, after centrifugation, the original egg polarity seems to be maintained. The velocity with which the various cytoplasmic components are redistributed after stratification shows that the egg must have a mechanism for the selective and bipolar distribution of ooplasmic substances. If this mechanism is associated with a cytoskeleton, it is clear why it is almost impossible to change the primary polarity of the undivided egg. It may be expected that a combination of the influence of the centrifugal force and a treatment with substances that destroy the cytoskeleton will give additional information about the constituents upon which the egg polarity will depend. If the cytoskeleton is anchored to special sites of the egg membrane, this membrane may also influence ooplasmic segregation.

2. Pressure Experiments

Pressure experiments have been performed in an attempt to analyze the polarizing capacity of the maturation divisions. The central question is whether the egg cytoplasm becomes bipolar because of the position of the meiotic spindles, or whether the egg has an irreversibly fixed and inherited bipolarity that determines the position of the maturation spindles. One may ask if after displacement of the meiotic spindles, the embryonic axes will be in correlation with the primary apical–basal polarity or in correlation with the abnormal place where the polar bodies are extruded?

Eggs of *Crepidula* (Conklin, 1912) and of *Cumingia* (Browne, 1910) have been compressed in the direction of the egg axis during the two maturation divisions. The polar bodies always appeared at the original animal pole, and, as a consequence, no alteration of the original egg polarity was obtained.

The eggs of *Limax* seem to be more suitable for pressure experiments (Guerrier, 1968, 1971), because they can be compressed maximally, changing the optical diameter of the egg from 180 to 290 μm, which corresponds to a shortening of the egg axis from 180 to 50 μm and an increase in surface of about 70%. The experiments were performed after the formation of the first polar body in order to maintain a reliable landmark for the original animal pole. Pressure was applied perpendicular to the egg axis, and the second polar body was most often extruded beneath the first. However, in 10% of the compressed eggs, the second maturation spindle was found in the center of the egg and was placed at right angles to it and to the direction of the first maturation spindle. This abnormal position led to the formation of a giant second polar body at a place without any compulsory relationship to the original animal pole. In normal eggs, ooplasmic segregation becomes prominent simultaneously with the extrusion of the second polar body. The large granular inclusions and fat droplets then move away from the place where the first polar body has been formed (Fig. 5a). Besides, a distinct cortical layer develops, which is most thick at the animal pole. The same segregation phenomena have been observed in compressed eggs in which the second meiotic spindle was oriented perpendicular to the original animal–vegetal polarity (Fig. 5b). In those compressed eggs, ooplasmic segregation was related to the position of the second maturation spindle (Fig. 5b). Thus, the axis of segregation corresponded with a secondary animal–vegetal polarity. These eggs, which developed a giant second polar body, thus were divided into two cells of nearly equal size. It should be stressed that, in both cells, the normal ooplasmic segregation took place, but in diametrically opposite directions. It looks as if each of the two asters acts as a reference point for the direction of the ooplasmic segregation (Fig. 5b). As only one of both cells contains the male pronucleus, only one of these two cells finally gives rise to a normal embryo. The polarity of that embryo actually corresponds to the secondary egg axis. The first two cleavages

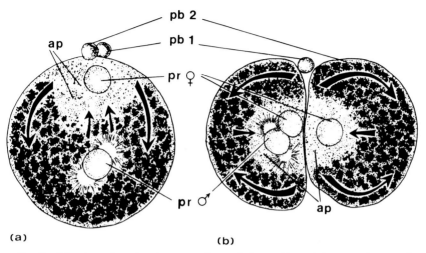

Fig. 5. Direction of ooplasmic segregation and the position of the second maturation spindle in *Limax maximus*. (**a**) Control egg with female pronucleus in a yolk-free area of animal pole plasm beneath the two polar bodies. (**b**) Formation of a second polar body of the same size as the zygote after a displacement of the second maturation spindle at right angles to the original polarity of the egg. Note (arrows) the diametrically opposed directions of ooplasmic segregation in both halves, which is not related to the original polarity, but with the position of the second maturation spindle. ap, animal pole plasm; pb1, 2, first and second polar body; pr♀, female pronucleus; pr♂, male pronucleus. (From Guerrier, 1968.)

pass through the point of extrusion of the second polar body, without any relationship to the place of extrusion of the first. At third cleavage, the first quartet of micromeres is also formed toward the new animal pole.

On the contrary, after achievement of the second maturation division and using the same pressure conditions, it was no longer possible to change the animal–vegetal polarity of the egg. Any displacement of the first mitotic spindle from a position perpendicular to the egg axis invariably led to the development of abnormal embryos.

These pressure experiments permit the conclusion that, at least in the egg of *Limax*, cytoplasmic localization is regulated epigenetically relative to the second maturation division. If one realizes that the second polar body may be formed at any point on the egg surface, but normally always appears beneath the first, it may be concluded that the egg polarity, as determined by the first maturation spindle, is normally directional for the position of the second meiotic spindle, by which the egg bipolarity is then irreversibly fixed.

If the position of the first spindle is directional for the second meiotic spindle, it may be asked which factors are directional for the first. One may ask if this in turn normally depends upon the previous apical–basal polarity of the oocyte in the ovary. It is merely a point of subjective appreciation to conclude whether the egg polarity is predetermined during oogenesis or arises epigenetically after ovulation and fertilization. In some species, the original apical–basal polarity may be more directive than in others. In any case, it is not allowed to attribute a highly organized architecture to the ovarian oocyte.

Whether the ooplasmic segregation is the result of a polar organization of the egg membrane, or the organization of the membrane is the result of an interaction with differentially distributed underlying ooplasmic components is open to discussion. Nevertheless, the latter may be more likely than the former. The pressure experiments performed on eggs of *Limax* have shown indeed that the localization of the second meiotic spindle determines the ooplasmic polarity, irrespective of any eventual polarity present in the egg membrane. Only secondarily, the egg membrane may become polarized definitely. A similar influence on the properties of the cell membrane can be derived from experiments on sea urchin eggs in which the first mitotic spindle was displaced at different moments during mitosis (Rappaport, 1971). Prior to anaphase, the cleavage plane is determined by the position of the spindle. After anaphase, the influence of the spindle upon the cell membrane has become irreversible, and cleavage has become independent of the actual position of the spindle.

D. The Possible Role of Cytoskeletal Elements in Ooplasmic Segregation

The degree of cytoplasmic localization in the ripe unfertilized molluscan egg may differ significantly according to the species (Huebner and Anderson, 1976). Irrespective of the initial segregations, as soon as fertilization has taken place and the egg prepares for maturation and division, the segregation of egg components is strengthened (Raven, 1966). For a number of different types of somatic cells it has been assumed that the presence of microtubules, microfilaments, and intermediate fibers is essential for the intracellular transport (Hyams and Stebbings, 1979). In the fertilized and maturing egg, the development of asters and spindles may be essential for the specific distribution of the egg substances. The significance of fertilization and the following meiotic processes for a directed movement of egg components can be derived from the observations of Rebhun (1959, 1960) on eggs of *Spisula*. Time-lapse cinematography and direct observation of the unfertilized egg showed that metachromatic granules are uniformly distributed. Prior to fertilization, these particles move in a saltatory way, whereas other inclusions like yolk, lipid, and mitochondria do not show such saltations. After fertilization, the homogenous distribution of the metachromatic granules completely disappears. Whereas prior to fertilization the saltations were at random and without any apparent direction, after fertilization the saltatory movements are oriented. A segregation of the particles from other inclusions occurs. Most of the granules move into the central aster at maturation, whereas the polar bodies are deprived from those particles. The segregation of the metachromatic granules concomitantly with the presence of asters and spindles may very well indicate a causal relationship between these processes. A possible role of cytoskeletal elements in the process of ooplasmic segregation can be analyzed by the application of mitotic inhibitors such as colchicine, cytochalasin B, and nocodazole. Zalokar (1974) could not block ooplasmic segregation in the egg of the ascidian *Phallusia* with colchicine, whereas cytochalasin B inhibited segregation at concentrations that were at least five times higher than those necessary for prevention of mitosis. Thus, the assembly of microtubules into the mitotic apparatus cannot be involved in segregation in this type of egg, as it still goes on when cleavage is arrested.

In the annelid *Chaetopterus,* however, ooplasmic segregation appears to be sensitive for colchicine and not for cytochalasin B. Here it is concluded that microtubule, and not microfilament organization, is required for ooplasmic segregation (Eckberg, 1981).

In an attempt to analyze the possible role of microtubules and microfilaments in the orientation and stabilization of polarity in the zygote of *Fucus,* Quatrano (1973) also treated eggs with colchicine and cytochalasin B. Although colchicine inhibited cell division, thus seriously affecting the microtubules, rhiziod formation and photopolarization were not prevented. With cytochalasin, rhizoid formation was inhibited, although cell division was not affected. As it is still difficult to develop a synthetic explanation of the results of these first experiments on the possible influence of microfilament and microtubule inhibitors on ooplasmic segregation, it will be necessary to perform a series of experiments specifically designed to evaluate the possible role of the cytoskeletal elements in the normal distribution or redistribution of the egg components.

IV. Determination of Dorsoventral Polarity

A. Species with Unequal Quadrants

Dorsoventral polarity, just as animal–vegetal polarity, might be regarded as a special case of cytoplasmic localization. First, we will describe a few examples indicating that cytoplasmic localization determines dorsoventral symmetry. They all derive from species with an unequal division in which the four quadrants can definitely be distinguished after the second cleavage. Usually the D cell is bigger than cells A, B, and C, for example, in most lamellibranchs, scaphopods, and in a number of gastropods (Chapter 3, this volume).

The size difference between the four quadrants is a reliable landmark for the denomination of the four quadrants. As the usually larger D cell will develop the dorsal quadrant of the embryo, it immediately follows that in unequally dividing

eggs the position of the first two cleavage planes is essential for the determination of the dorsoventral axis. The position of these planes depends upon the
position of the first two cleavage spindles. Consequently, the mechanisms that
determine the positions of the first spindles, necessarily determine the embryonic
axes.

In *Cumingia* (Morgan and Tyler, 1930), the sperm entrance point has a definite relationship with the position of the first two cleavage planes. The first
spindle is always located perpendicular to the pathway of the sperm to the center
of the egg (Fig. 6a). Immediately preceding cleavage, the spindle is shifted into
an excentrical position, to the right (Fig. 6e) or to the left (Fig. 6b) of the plane
through the sperm entrance point and the egg axis. As a result, two symmetrical
two-cell stages occur (Fig. 6c,f). In both types, the smaller AB cell divides
equally, whereas the bigger CD cell always divides unequally. Both types of the
two-cell stage develop a four-cell stage in which the smaller C cell is directed
toward the sperm entrance point, whereas the D cell, thus the furture dorsal
quadrant, is found in a diametrically opposite position (Fig. 6d,g). Morgan and
Tyler assumed that the position of the mesentoblast 4d would determine the plane

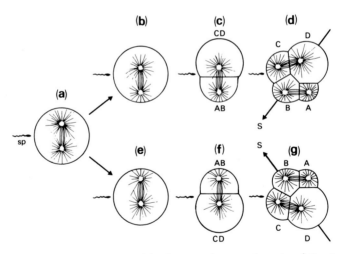

Fig. 6. The alternative positions of the first two cleavage planes in relation to the sperm
entrance point in *Pholas dactylus*. Animal views. (**a**) Initial position of the first mitotic spindle at
right angles to the penetration path of the sperm. At later mitotic stages the spindle will be
displaced to the right (**e**) or to the left (**b**) from the plane through sperm entrance point and the
egg axis. (**b**) Displacement of the spindle to the left. (**c**) Two-cell stage after displacement of the
spindle towards the left. (**d**) Four-cell stage with a clockwise arrangement of the quadrants. (**e**)
Displacement of the spindle to the right. (**f**) Two-cell stage after displacement of the spindle
toward the right. (**g**) Four-cell stage with a counterclockwise direction of the quadrants. sp,
indication of sperm entrance point. S and arrow in (**d**) and (**g**) indicate the future plane of
bilateral symmetry. The arrow points ventrally. (Modified after Guerrier, 1970b.)

of bilateral symmetry. In both types of eggs, 4d is formed by a sinistral division. As according to the displacement of the first cleavage spindle to the right or to the left of the plane through the sperm entrance point and the egg axis, the succession of the quadrants is counterclockwise or clockwise, respectively, 4d will be formed adjacent to the A or the C quadrant. As a consequence, the quoted authors conclude that the plane of bilateral symmetry would either correspond with the first or the second cleavage plane.

In our opinion, it is more likely that the plane of bilateral symmetry passes through the centers of the B and D quadrants as shown for *Pholas* and *Spisula*. But even then, two types of eggs can be distinguished, in which the plane of bilateral symmetry deviates about 45° to the left or to the right from the first cleavage plane (Fig. 6d,g). In either of the resulting four-cell stages, the D quadrant is opposite the sperm entrance point.

If the sperm can enter at any particular point of the egg surface, the constancy of the relationship between the sperm entrance point and the position of the D quadrant necessarily indicates that each part of the egg may become the dorsal quadrant. If the entrance of the sperm is limited to a particular point of the egg surface, it would be predetermined which part of the unfertilized egg would become the dorsal quadrant. Guerrier's (1970b) experiments on eggs of *Pholas* and *Spisula* exclude the possibility of a predetermined dorsoventrality and are in favor of an epigenetic origin. By pressure experiments, he obtained an equal first cleavage (Fig. 7a,b,e). The division plane corresponded with a meridional plane passing through the egg axis and through the sperm entrance point. This affirms the relationship between sperm entrance and first cleavage. The following second cleavage may occur in three different ways. Second cleavage may be equal, leading to an equal four-cell stage (Fig. 7f). Second cleavage may be unequal, and then two types of the four-cell stage occur. In one type, both spindles have moved toward the sperm entrance point. In the other type, the spindle of one cell moves toward the sperm entrance point (Fig. 7d) and in the other it moves away from it (Fig. 7c). In the majority of the eggs with an equal first cleavage, the two small cells are formed toward the point of sperm penetration (79% in *Pholas;* 98% in *Spisula*) (Fig. 7d).

Whatever its position, during further division each big cell behaves as a D cell in normal development. This result permits two conclusions. First, if the sperm entrance point is predetermined, it cannot determine at the same time which part of the unfertilized egg will become dorsal and which ventral, as large D cells may be formed toward and away from that point and in both cases behave as perfectly normal dorsal quadrants. Second, it shows that although during normal development the sperm entrance point determines the position of the dorsal quadrant, this determination can only have a preliminary character. It can be assumed that the undivided egg is radially symmetrical and that the differentiation of the quadrants is only the result of division inequality. The bigger cell

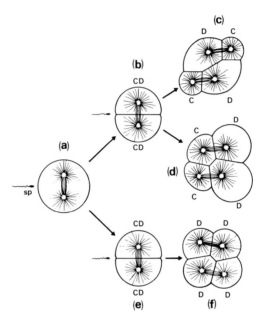

Fig. 7. Variation in cleavage of compressed eggs of *Pholas dactylus*. Animal view. (a) Normal central position of the cleavage spindle. (b) Equal two-cell stage. (c) Unequal four-cell stage reached by migration of the second cleavage spindle towards sperm entrance point in one cell (left) and away from it (right) in the other. (d) Unequal four-cell stage reached by migration of both cleavage spindles in the direction of the sperm entrance point. (e) Equal two-cell stage. (f) Equal four-cell stage after the absence of spindle displacement. sp marks the sperm entrance point. (Modified after Guerrier, 1970b.)

becomes the dorsal quadrant because most of the structures and substances of the vegetal pole that are essential for the development of dorsal characters are segregated into it. Inheritance of part of this vegetal material then determines the developmental capacity of a cell, whether formed opposite or toward the entrance point of the sperm.

The preliminary role of the point of penetration of the sperm also follows from centrifugation experiments performed on undivided eggs of *Pholas*, the development of which was followed by time-lapse cinematography (Guerrier, 1970b). As discussed before, the eggs orient with the animal–vegetal axis at right angles to the axis of centrifugation. Polarity of cleavage and further development are always related to the original animal–vegetal polarity. Dorsoventral polarity, however, can no longer be relative to the sperm entrance point—the position of which could always be recognized even in these conditions—as the position of the first cleavage plane was always parallel to the axis of centrifugation and since second cleavage was such that the big D cell usually appeared at the centripetal

end. This result strengthens the hypothesis that the sperm entrance point determines dorsoventral polarity in a preliminary way, and that dorsoventral polarity
can actually be determined *de novo,* merely by fixation of the position of the
cleavage planes.

The same conclusion probably holds for the species in which division inequality is preceded by the formation of a polar lobe at first and second cleavages. Blastomere inequality then results from the fusion of the lobe with only one
of the cells. In eggs of *Dentalium,* the formation of the first polar lobe has been
suppressed with cytochalasin B (Guerrier et al., 1978). At second cleavage, both
cells behave as a CD cell and both form a polar lobe, which then fuses with one
of the daughter cells. At the four-cell stage, two cells contain half of the polar
lobe material, which is normally segregated in only one blastomere. These cells
then divide as a normal D quadrant and also develop typical dorsal structures
even if they received only one-eighth of the original lobe volume (unpublished
observation of the authors). Most embryos with such an experimentally dupli-
cated dorsal quadrant show an alternate succession of the D quadrant and have
been described as CDCD embryos. Only rarely CCDD embryos appeared. This
result again favors the hypothesis that prior to division the egg is radially symmetrical, the determination of the dorsal quadrant depending merely upon the
fusion of the polar lobe with one of the originally equal quadrants.

B. Species with Equal Quadrants

A completely different mechanism than division inequality, accompanied by
the segregation of most of the egg components of the vegetal pole into one out of
four quadrants, must determine dorsoventral polarity in equally dividing eggs. It
may be assumed that, also in these types of eggs, the undivided zygote has a
radially symmetrical, or axially symmetrical segregation of the ooplasm.

It is logical to presume that as long as corresponding cells in the four equal
quadrants divide synchronously and follow the same division pattern, they will
also share identical developmental capacities. In accordance with this presump-
tion, it can be expected that only at the stage when the behavior of corresponding
cells in the four quadrants becomes different, does the determinant of that difference come into existence. As far as we know in equally dividing gastropod
eggs, the development of the quadrants remains essentially equal up to the
beginning of the interval between fifth and sixth cleavage (Fig. 8a,b), whereupon
the 3D macromere divides unequally and asynchronously with the other macromeres. During this mitotic interval the cleavage cavity disappears and the cells
of the animal and vegetal pole approach each other. Finally, a situation is
attained in which the macromere of one of the four quadrants exclusively makes
contact with the micromeres at the opposite animal pole (Fig. 10). This arrangement is the first asymmetric event in the development of the quadrants (van den

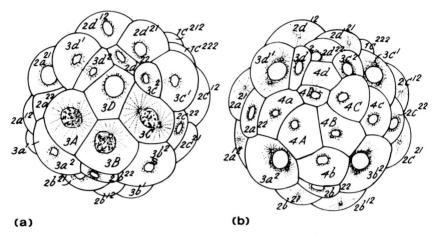

(a)   **(b)**

Fig. 8. Ventral views of *Pattella* embryos. (**a**) 60-cell stage, showing division asynchrony of the macromeres 3A–3D. The cells 3A, 3B, and 3C are entering mitosis, whereas 3D is still in interphase. A second aspect of bilateral symmetry is shown by the cells of the third quartet of micromeres 3a–3d. The micromeres 3a and 3b at the future ventral side of the embryo divide unequally with the bigger cell towards the vegetal pole, whereas the micromeres 3c and 3d divide unequally with the smaller cell towards the vegetal pole. (**b**) 64-cell stage. (From van den Biggelaar, 1977.)

Biggelaar, 1976a, 1977). The quadrant to which the central macromere belongs becomes the dorsal side, the opposite quadrant becomes ventral; the other two quadrants contribute to the left and right side of the embryo.

We have shown that in *Patella* the differential behavior of one of the four macromeres cannot be attributed to special properties of that macromere, and thus finally to qualitative differences between the four quadrants, but depends upon an influence that the animal micromeres exert on the central macromere (van den Biggelaar and Guerrier, 1979). Normally, one of the two macromeres at the vegetal cross-furrow attains a central position during the 32-cell stage and develops into the stem cell of the mesoderm. If some kind of qualitative difference between this macromere and the others determined this peculiar behavior, then after elimination of one of the two cross-furrow macromeres, 50% of the embryos would be able to form the mesoderm stem cell, whereas the other 50% would not. We found that after elimination of one of the two cross-furrow macromeres in 24 embryos immediately after fifth cleavage, in 19 of these embryos the other attained the central position and developed as a normal dorsal macromere, whereas in three embryos one of the two non-cross-furrow (i.e., presumptive lateral) macromeres obtained the central position and behaved as the macromere of a normal dorsal quadrant. This result indicates that the central position determines a macromere to develop as the dorsal macromere and to become the mother cell of the mesentoblast. Due to its central position only the central macromere comes into touch with a great number of the previously formed micromeres. It appeared that the contacts with the micromeres of the first quartet occupying the most animal part of the embryo in particular are essential for the induction of the mesodermal qualities in the central macromere. After deletion of all four first quartet cells at the eight-cell stage, no mesentoblast was formed and the macromeres maintained their originally equal cleavage pattern. After deletion of three, two or one micromere(s), progressively more embryos showed a central macromere, which then showed a more or less normal cleavage delay and the normal deviating cleavage pattern.

Another demonstration of initial equivalence of the quadrants comes from deletion experiments on embryos of *Patella* and *Lymnaea*. In the eight-cell embryo one can distinguish two polar cross-furrows. These cross-furrows are borderlines where two opposite quadrants touch each other. The micromeres of the A and C quadrant form the animal cross-furrow (Fig. 9), and the macromeres of the B and D quadrant form the vegetal cross-furrow (Fig. 9). From cell lineage studies, we know that the quadrants at the vegetal cross-furrow become the median quadrants, whereas the quadrants at the animal cross-furrow become the lateral quadrants. In the deletion experiments, one cell of the first quartet of micromeres was deleted at the eight-cell stage and the effect on the organization of the embryo studied. It was expected that when the quadrants were qualitatively different from the beginning, removal of one of the micromeres at the animal cross-furrow (Fig. 9a) (i.e., the micromere of one of the two presumptive lateral quadrants) would lead to the development of embryos either with an

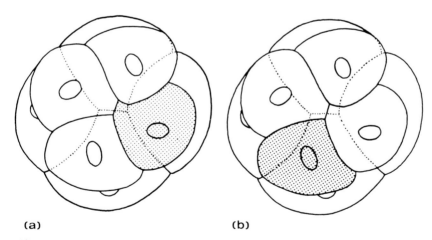

(a)   (b)

Fig. 9. Diagrams of eight-cell stages showing the position of the micromeres at the animal cross-furrow and the macromeres at the vegetal cross-furrow. (**a**) An animal cross-furrow micromere (stippled). (**b**) A non-cross-furrow micromere (stippled).

affected right or an affected left quadrant. It appeared that in the great majority, the left quadrant showed deficiencies, whereas the right quadrant was completely normal. Apparently, deletion of a micromere of a presumptive lateral quadrant forces the median quadrant counterclockwise to it to become the dorsal quadrant. This means that deletion of a micromere of a presumptive lateral quadrant favors the macromere of the quadrant counterclockwise to it to become the central macromere (3D) and to differentiate into the stem cell of the mesoderm.

In a corresponding series of deletion experiments the micromeres of one of the vegetal cross-furrow macromeres was deleted at the eight-cell stage (Fig. 9b). Again, if one of these quadrants were predetermined to produce the stem cell of the mesoderm, in 50% of the cases the dorsal quadrant should be affected and in 50% the ventral quadrant should have a deficiency. In contrast to this possibility, almost without any exception, the embryos appeared to have a defect in the ventral quadrant. Apparently, by deletion of a first quartet cell, one can direct with certainty the developmental diversification between the quadrants (Arnolds et al., 1983).

The above result completely differs from the results of comparable experiments of Clement (1967), in which he deleted each of the first quartet micromeres of *Ilyanassa obsoleta*. In *Ilyanassa*, the determination of the quadrants results from the fusion of the polar lobe with one of the quadrants at the four-cell stage. In this egg it is possible to delete either 1a or 1c at the eight-cell stage, followed by the formation of embryos in which either the left or the right eye is missing, respectively.

The above results do not necessarily exclude that, during normal development, one of the quadrants is favored to develop the central cell. At the four-cell stage, two cells make a cross-furrow at the vegetal pole and usually the other two make a cross-furrow at the animal pole. Almost without exception, the central macromere derives from a quadrant that is involved in the formation of the vegetal cross-furrow. This implies that the position of the first two cleavage planes strongly influences the plane of bilateral symmetry of the future embryo. In this aspect, equally and unequally dividing eggs do not differ.

The position of the first cleavage plane may be influenced by some kind of asymmetry in the undivided egg. According to Raven (1970, 1974), the pattern of the follicle cells around the ovarian oocyte determines a mirror pattern of subcortical accumulations (SCA) which in turn predisposes the first cleavage plane to occupy a certain position. But again, exactly like for the directive influence of the sperm penetration point in bivalves, the eventual directive influence of the follicle cells can easily be overruled by any aspecific agent that changes the position of the first cleavage plane. In correspondance with this explanation are the experiments of Guerrier (1970a, 1971) on the equally dividing egg of *Limax*. It appeared that the first cleavage spindle can be rotated in any direction perpendicular to the egg axis without any consequences for normal development. This result convincingly shows that any meridional plane can become the first cleavage plane, and any combination of two perpendicular meridional planes can divide the egg into the future four quadrants of the embryo. Finally, it has been shown that in *Lymnaea* a single blastomere of a two-cell embryo can develop into a normal embryo, provided it did not cleave as a half but as a whole embryo, by which normal cell contact relationships are reestablished (Morrill et al., 1973; Verdonk, 1979).

We can only speculate about the mechanism underlying centralization of only one of the four macromeres after fifth cleavage. It probably represents the only possible equilibrium configuration (van den Biggelaar, 1976a, 1977). The way in which the diversification of the cells in the quadrants starts makes it unlikely that dorsoventral symmetry depends on some kind of field property. It must be the result of a progressive increase of asymmetric cellular interactions. In *Lymnaea* and *Physa*, aggregates of pyroninophilic granules appear in the vegetal region of the four blastomeres at the four-cell stage (Wierzejski, 1905; Minganti, 1950). These aggregates have been called ectosomes. During further division, these granules accumulate along the vegetal cross-furrow (Fig. 10a,b). After fifth cleavage, the ectosomes are displaced toward the central tips of the macromeres along their mutual cell borders. In the noncentral macromeres, these ectosomes become more and more aggregated and are finally extruded in the cleavage cavity, whereas in the central macromere they redisperse (van den Biggelaar, 1976b). In *Lymnaea*, *Physa*, and *Patella*, the central macromere divides asynchronously with the other macromeres. Its division pattern also

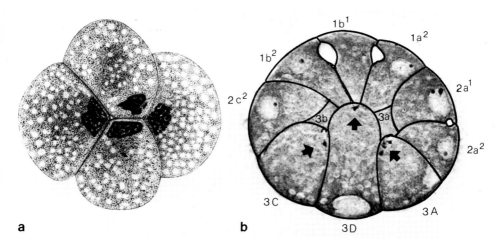

Fig. 10. (a) Vegetal view of an eight-cell stage of *Physa*. The ectosomal material is concentrated at the vegetal cross-furrow in each of the four macromeres. (From Wierzejski, 1905.) (b) Meridional section through a 24-cell *Lymnaea* embryo showing the ectosomal material in the central tips of the macromeres. (Arrows indicate the position of the ectosomes.)

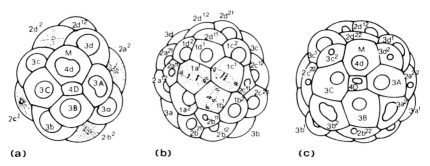

Fig. 11. Three successive stages during the sixth cleavage round of a *Physa* embryo. (**a**) Vegetal view of a 29-cell stage, showing the division asynchrony between the macromeres. In the D quadrant the macromere has divided, whereas the macromeres 3A, 3B, and 3C are still undivided. Note the division inequality of the D macromere: 4D is a small and 4d or M is a big cell. (**b**) Animal view of a 33-cell stage, showing the division asynchrony in the four animal micromeres $1a^1-1d^1$; the dorsal micromere $1d^1$ has a significant delay. (**c**) Vegetal view of a 41-cell embryo. In addition to the division asynchrony between the macromeres, a second indication of dorsoventrality can be observed in the division pattern of the dorsal micromeres 3c and 3d in comparison with the ventral micromeres 3a and 3b. The cells 3c and 3d divide unequally with the smallest cell toward the vegetal pole, whereas 3a and 3b divide unequally with the biggest cell towards the vegetal pole. (After Wierzejski, 1905.)

differs. Whereas the other macromeres (3A, 3B, and 3C) divide more or less equally, the central macromere (3D) gives rise to a small macromere (4D) and a bigger mesentoblast (4d or M) (Fig. 11a,c). In connection with this typical division pattern of 3D, the third quartet cells at the future dorsal and ventral side divide differently (Fig. 8a). At the ventral side, the vegetal cells ($3a^2$ and $3b^2$) are big and the animal sister cells ($3a^1$ and $3b^1$) are small. The opposite division pattern is observed dorsally; there, the vegetal cells ($3d^2$ and $3c^2$) are small and the animal sister cells ($3d^1$ and $3c^1$) are larger (van den Biggelaar, 1976a; van den Biggelaar and Guerrier, 1979). This pattern is causally related to the determination of the central macromere, and is by no means an independent property of the third quartet cells. A similar division pattern can be inferred from Wierzejski's figure of a 40-cell stage of *Physa*, in which he also stresses the discordancy of the division pattern of the ventral cells 3a and 3b and the dorsal cells 3c and 3d (Wierzejski, 1905) (Fig. 11c).

The differentiation of the quadrants in the vegetal hemisphere starts with the appearance of a difference in position and a division asynchrony between the macromeres. These differences are causally related to the contacts between the macromeres and the animal micromeres. They fail to develop after deletion of the first quartet of micromeres, and the quadrant in which they will appear can be predetermined by a specific deletion of one of the four micromeres of the first quartet. In *Lymnaea* and *Physa*, the embryo counts 24-cells after fifth cleavage.

The first quartet of micromeres then consists of the animal tier, $1a^1–1d^1$, and the more vegetal tier of primary trochoblasts, $1a^2–1d^2$ (Fig. 11b). The micromeres $1a^1–1d^1$ have a radial position, $1d^1$ is dorsal, $1b^1$ is ventral, $1a^1$ is right and $1c^1$ is left. The primary trochoblasts have an interradial position; $1c^2$ and $1d^2$ are placed interradially right and left from the dorsal cell $1d^1$: $1a^2$ and $1b^2$ are placed interradially, right and left from the ventral $1b^1$ cell. Both tiers of micromeres divide after the determination of the central 3D macromere, thus after the first event in the determination of dorsoventral symmetry, which is immediately followed by the determination of dorsoventral symmetry in the animal region. There it becomes apparent by the division delay of the dorsal cells $1d^1$, $1c^2$ and $1d^2$ (Fig. 11b) (Wierzejski, 1905; van den Biggelaar, 1971a,b, 1976a; van den Biggelaar and Boon-Niermeijer, 1973). This division asynchrony is the first indication of different developmental pathways for these cells. The primary trochoblasts $1c^2$ and $1d^2$ never become true trochoblasts, but contribute to the head vesicle and do not develop cilia. The corresponding cells $1a^2$ and $1b^2$ produce ciliary cells of the prototroch. The derivatives of $1d^1$ mainly form the head vesicle and apical plate and a minor amount of the left cephalic plate. The derivatives of the lateral micromeres $1a^1$ and $1c^1$ are the main sources for the formation of the cephalic plates from which the eyes and tentacles will develop. The derivatives of $1b^1$ give a minor contribution to both cephalic plates and never form head vesicle cells (Verdonk, 1965; Chapter 3, this volume).

The developmental differences of the first quartet cells are no autonomic properties, but depend upon the interaction with cells of the underlying vegetal region, primarily with 3D.

The experimental analysis of the development of dorsoventrality in equally dividing molluscan eggs reveals the mutual dependence of the developmental fate of the different blastomeres. In this aspect, the molluscan embryo does not differ from the other developmental systems in which cellular interactions are essential for the induction of special properties.

The mechanism of cellular interactions has been examined in embryos of *Lymnaea* and *Patella*. The previously described morphological observations and deletion experiments on the development of dorsoventrality indicate that the mechanism for cellular interaction we are looking for requires mutual cell contact between the blastomeres. From electron microscopical investigations on the occurrence of different types of cellular junctions, it appeared that septate junctions and intermediate junctions are present as early as the two-cell stage in *Lymnaea* (Berendsen, 1971; Dorresteijn et al., 1981). In *Patella*, intermediate junctions are present from the two-cell stage, septate junctions appear shortly after fifth cleavage (van den Biggelaar and Dorresteijn, 1982; van den Biggelaar et al., 1981; Dorresteijn et al., 1982). Although it cannot be excluded that septate or tight junctions play a role in the direct exchange of molecules between cells (Sheridan and Larson, 1982), the best candidate to function as a channel for direct intercellular transport are the gap junctions (Pitts and Finbow, 1982). Gap junctions have been found in *Lymnaea* (Dorresteijn et al., 1981) and *Patella* (Dorresteijn et al., 1982) as early as the four-cell stage, but experimental evidence for an actual cell communication prior to the fifth cleavage is still lacking. Analyzing the transport of iontophoretically applied lucifer yellow, it appeared indeed that this fluorescent probe was not transferred from an impaled macromere to any other neighboring cell prior to fifth cleavage. After fifth cleavage, however, the dye easily passed from a labelled macromere to surrounding cells (de Laat et al., 1980) (Fig. 12a,b). Apparently, the gap junctions only start to become physiologically active intercellular communication channels a few divisions after their appearance. The transition from closed into open channels cannot be taken as the cause of the induction of one of the four macromeres into the mesentoblast mother cell, as it appeared that the dye was not transported from the central macromere to the inducing micromeres, or vice versa (van den Biggelaar and Dorresteijn, 1981; Dorresteijn et al., 1983). The opening of the gap junctions between the macromeres and adjacent cells may be a time-dependent phenomenon which automatically occurs after fifth cleavage.

At the moment, no experimental data are available for the explanation of the primary mechanism by which the animal micromeres induce the central macromere to differentiate as the mesentoblast mother cell.

With respect to the development of dorsoventrality and thus to the determination of the developmental pathways of the quadrants, it may be concluded that

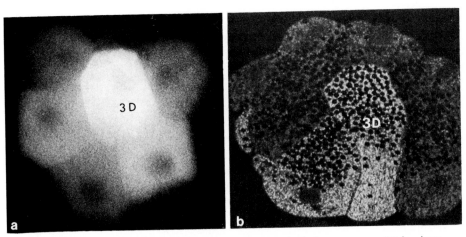

Fig. 12. Micrographs of 32-cell *Patella* embryos in which the macromere 3D has been injected iontophoretically with lucifer yellow. (a) Vegetal view of a living embryo showing dye transport from the impaled macromere 3D toward neighboring cells. (b) Median section showing the fluorescent central 3D macromere and secondarily fluorescent cells at the vegetal pole. The animal micromeres overlying the 3D cell are not labeled.

any quartet of the undivided egg has the capacity to develop any quadrant of the embryo. In unequally dividing eggs, division inequality causes a differential segregation of the vegetal substances into one particular quadrant. In equally dividing eggs, differentiation of the quadrants appears as the epigenetic result of cellular interactions. In any case, the experimental evidence indicates that the fertilized egg is able to develop a plane of bilateral symmetry, in the complete absence of any initial dorsoventral pattern.

V. The Development of Left–Right Asymmetry

A remarkable characteristic of gastropod development is the transition from an initially bilateral symmetry into a left–right asymmetry. This phenomenon has been discussed by Verdonk (1979). His final conclusion was that this asymmetry is the result of an unequal contribution of the quadrants to the development of the posttrochal region. In molluscs other than gastropods, the posttrochal region is formed nearly exclusively by the D quadrant. One might therefore suppose that clockwise or counterclockwise succession of the quadrants is not important for the organization of the embryo. This is indeed the case in lamellibranchs, where both arrangements occur (Chapter 3, this volume). In gastropods, an additional contribution is derived from the C quadrant. Species with a clockwise succession of the quadrants develop into dextral snails, whereas species with a counterclockwise succession of the quadrants develop into sinistral snails (Fig. 13). In dextral snails the C quadrant is to the left of (i.e., counterclockwise from) D and, in sinistral snails, it is to the right (i.e., clockwise) of D, the additional somatoblast formed by the C quadrant would give rise to an asymmetric outgrowth of the embryo. This assumption is very attractive as it directly relates left–right asymmetry to the clockwise-counterclockwise arrangement of the quadrants.

The assumption of an additional somatoblast (2c) apart from the first somatoblast (2d) is derived from deletion experiments on eggs of gastropods which form a polar lobe. In *Bithynia,* the C quadrant inherits the properties essential for the development of an additional somatoblast, by means of ooplasmic segregation (Cather and Verdonk, 1974, Cather et al., 1976; Verdonk and Cather, 1973). In *Ilyanassa,* the somatoblast properties of 2c are the result of an inductive influence originating from the D quadrant (Clement, 1962). In species with an equal division, the special potencies of the 2c cell cannot depend on autonomous properties, but again they must result from some kind of cellular interactions. As the central 3D macromere is the center from which dorsoventrality starts, it may be expected that the different positions of 2c and 2a relative to 3D must be the intimate cause for the discordancy in their contribution to the posttrochal region.

Presently, we do not have any information about the positional relationship between 2c and 3D. In any case, such a differential position can only result from

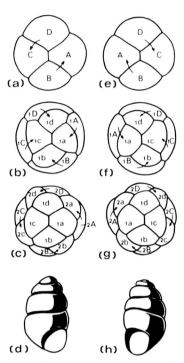

Fig. 13. The relation between chirality of cleavage and coiling of the shell in gastropods. (a–c) Cleavage stages of dextral species. (d) Shell of a dextral species. (e–g) Corresponding cleavage stages in sinistral species. (h) Shell of a sinistral species. (From Verdonk, 1979.)

previous alternation in left and right handed cleavages, which therefore, finally refers back to the first two cleavages. Chirality of the later embryo actually depends upon the respective positions of the first two mitotic spindles.

Guerrier (1968, 1970a) assumed that the overall egg cortex might be provided with an ensemble of repetitive units with some kind of circular vector, since in experimental conditions, in which the orientation of the spindles of dextral or sinistral species has been changed with respect to the polar axis, the specific direction of spiral cleavage was preserved. The direction of this vector thus needs to be different according to the chirality of the cleavages and of the adult animal. However, recent experiments by Freeman and Lundelius (1982) have shown that the direction of cleavage in eggs of sinsistral individuals of *Lymnaea peregra* can be reversed by the injection of ooplasm from the dextral form. This inversion can be induced in the stages preceding the second maturation division.

If the cortex is the site of determination of the direction of spiral cleavage, it appears that this direction can be easily overruled by some cytoplasmic, genetically determined component and is only fixed irreversibly after second meiotic division.

Another interesting hypothesis on the origin of chirality is the assumption of a spiral structure in the contractile ring of microfilaments that are involved in the formation of the cleavage furrow. In dextral species this spiral ring would cause a dextrally oriented rotation, whereas in sinistral forms it would give rise to a sinistral rotation of the blastomeres (Mescheryakov, 1978a,b; Mescheryakov and Belusov, 1975; Mescheryakov and Veryasova, 1979). The results obtained by Freeman and Lundelius suggest that the cytoplasmic components of a dextral species definitely influence chirality of the cleavages, if injected prior to second-polar-body formation. This moment corresponds with the moment at which animal–vegetal polarity in the egg of *Limax,* and presumably of a number of other species, becomes irreversibly fixed. It is not illogical to assume that the flexibility of the polar structure of the egg membrane is associated with its flexibility relative to the determination of chirality, as both can be influenced up to the formation of the second polar body.

We will probably have to wait for further details on the determination of helical structures in general, in chromatin, microtubules, microfilaments, and intermediate filaments, which will all be determined genetically. Moreover, from a study on the organization of actin filaments in the stellocilium of the cochlear hair cell of the cochlea (De Rosier et al., 1980), it was concluded that the conditions found at the time of assembly of the subunits are decisive for the determination of the resulting pattern. Similarly, injection of cytoplasm of a dextral donor egg into a recipient sinistral egg might alter the conditions required at the time of assembly of the eventually responsible actin subunits. If these subunits become fixed to the membrane, this might become irreversibly endowed with some kind of spiral character, independent of the polar structure of the egg, as supposed by Guerrier (1968, 1970a). The observation of Freeman and Lundelius (1982) that, with succession of the cleavage, the cytoplasm of the donor cells became less and less effective in reversing sinistrality into dextrality, is in agreement with a gradual fixation of a cytoplasmic component to the egg membrane.

A difficult but not insuperable point in this attractive hypothesis, however, is the explanation of sinistrality, which does not appear as the most frequent feature among spiralian embryos. Despite that, the work of Freeman and Lundelius (1982) seems to indicate that the cytoplasm of a sinistral form would lack the cytoplasmic product that is coded only by a dextral gene. In the absence of this product, cleavage would automatically become sinistral.

VI. Conclusions

The origin of the spatial organization in the molluscan embryo can be traced back to the uncleaved egg where the only clearly apparent differentiation resides in its apical–basal polarity. This polarity is related to the position taken by the oocyte in the ovary during oogenesis. In some species, this inherited polarity is maintained after ovulation and appears decisive for the orientation of the maturation spindles by which an exact correspondence is insured between the apical–basal and the animal–vegetal polarity which first expresses itself in the process of ooplasmic segregation. In other species, the apical–basal polarity of the oocyte architecture appears less stable and can be modified. Experimental evidences have been obtained in this case, which indicate that the orientation of the second maturation spindle can function as a mechanism by which the primary polarity is definitively determined in an irreversible fashion. A possible role for the membrane and its associated cytoskeletal elements in these early processes of polarization is suggested, which appeals for the design of further ingenious and specifically directed experiments.

The possibility of a direct relationship between the pattern of structures that surround the oocyte during oogenesis and the emergence of dorsoventrality during embryogenesis has not been confirmed by experiments. Instead, it was found that dorsoventral polarity can easily be determined by external or internal influences completely independent of the position of the egg in the ovary or of the existence of any eventual differentiation in the oocyte which would appear apart from the apical–basal polarity.

Finally, another important structural feature, necessary for the orientation of cleavage and further development of the asymmetry characteristic of the gastropods, seems to arise quite early under the influence of a gene product. This factor, which does not rely upon spatially defined regions in the oocyte, might eventually influence spindle orientation, either by acting at the membrane and the cortical level or at the level of those structural kinetic elements that are responsible for cytokinesis. Direct relevant experiments are needed in order to clarify this question.

References

Arnolds, W. J. A., van den Biggelaar, J. A. M., and Verdonk, N. H. (1983). *Wilhelm Roux's Arch. Dev. Biol.* In press.

Attardo, C. (1957). I mitocondri e la citochromo-ossidasi nello sviluppo di *Aplysia. Acta Embryol. Morphol. Exp.* **1,** 65–70.

Berendsen, W. (1971). Morphologische analyse van de eerste celdeling van het ei van *Lymnaea stagnalis* L. Ph. D. Thesis, University of Utrecht, Utrecht, The Netherlands.

Blochmann, F. (1882). Über die Entwicklung der *Neritina fluviatilis* Müll. *Z. Wiss. Zool.* **36,** 125–174.

Browne, E. N. (1910). Effects of pressure on *Cumingia* eggs. *Arch. Entwicklungs Mech.* **29,** 243–254.

Casteel, D. B. (1904). The cell-lineage and early larval development of *Fiona marina*, a nudibranch mollusk. *Proc. Acad. Nat. Sci. Philadelphia* **56,** 325–405.

Cather, J. N., and Verdonk, N. H. (1974). The development of *Bithynia tentaculata* (Prosobranchia, Gastropoda) after removal of the polar lobe. *J. Embryol. Exp. Morphol.* **31,** 415–422.

Cather, J. N., Verdonk, N. H., and Dohman, M. R. (1976). Role of the vegetal body in the regulation of development in *Bithynia tentaculata* (Prosobranchia, Gastropoda). *Am. Zool.* **16**, 455–468.

Chambard, M., Gabrion, J., and Mauchamp, J. (1981). Influence of collagen on the orientation of epithelial cell polarity: Follicle formation from isolated thyroid cells and from preformed monolayers. *J. Cell Biol.* **91**, 157–166.

Clement, A. C. (1962). Development of *Ilyanassa* following removal of the D macromere at successive cleavage stages. *J. Exp. Zool.* **149**, 193–216.

Clement, A. C. (1967). The embryonic value of the micromeres in *Ilyanassa obsoleta,* as determined by deletion experiments. I. The first quartet cells. *J. Exp. Zool.* **166**, 77–88.

Clement, A. C. (1968). Development of the vegetal half of the *Ilyanassa* egg after removal of most of the yolk by centrifugal force, compared with the development of animal halves of similar visible composition. *Dev. Biol.* **17**, 165–186.

Conklin, E. G. (1907). The embryology of *Fulgur:* A study of the influence of yolk on development. *Proc. Acad. Nat. Sci. Philadelphia* **59**, 320–359.

Conklin, E. G. (1912). Experimental studies on nuclear and cell division in the eggs of *Crepidula. J. Acad. Nat. Sci. Philadelphia* **15**, 503–591.

Conklin, E. G. (1938). Disorientations of development in *Crepidula plana* produced by low temperature. *Proc. Am. Philos. Soc.* **79**, 179–211.

Costello, D. P. (1939). Some effects of centrifuging eggs of nudibranchs. *J. Exp. Zool.* **80**, 473–499.

Costello, D. P. (1945). Segregation of ooplasmic constituents. *J. Elisha Mitchell Sci. Soc.* **61**, 277–289.

Davidson, E. H. (1976). "Gene Activity in Early Development," 2nd ed. Academic Press, New York.

de Laat, S. W., Tertoolen, L. G. J., Dorresteijn, A. W. C., and van den Biggelaar, J. A. M. (1980). Intercellular communication patterns are involved in cell determination in early molluscan development. *Nature (London)* **287**, 546–548.

De Rosier, D. J., Tilney, L. G., and Egelman, E. (1980). Actin in the inner ear: The remarkable structure of the stereocilium. *Nature (London)* **287**, 291–296.

Dohmen, M. R., and Lok, D. (1975). The ultrastructure of the polar lobe of *Crepidula fornicata* (Gastropoda, Prosobranchia). *J. Embryol. Exp. Morphol.* **34**, 419–438.

Dohmen, M. R., and van der Mey, J. C. A. (1977). Local surface differentiations at the vegetal pole of the eggs of *Nassarius reticulatus, Buccinum undatum,* and *Crepidula fornicata* (Gastropoda, Prosobranchia). *Dev. Biol.* **61**, 104–113.

Dohmen, M. R., and Verdonk, N. H. (1974). The structure of a morphogenetic cytoplasm, present in the polar lobe of *Bithynia tentaculata* (Gastropoda, Prosobranchia). *J. Embryol. Exp. Morphol.* **31**, 423–433.

Dohmen, M. R., and Verdonk, N. H. (1979a). The ultrastructure and role of the polar lobe in development of molluscs. *In* "Determinants of Spatial Organization" (S. Subtelny and I. R. Konigsberg, eds.), pp. 3–27. Academic Press, New York.

Dohmen, M. R., and Verdonk, N. H. (1979b). Cytoplasmic localizations in mosaic eggs. *In* "Maternal Effects in Development" (D. R. Newth and M. Balls, eds.), pp. 127–145. Cambridge Univ. Press, London and New York.

Dorresteijn, A. W. C., van den Biggelaar, J. A. M., Bluemink, J. G., and Hage, W. (1981). Electronmicroscopical investigations of the intercellular contacts during the early cleavage stages of *Lymnaea stagnalis* (Mollusca, Gastropoda). *Wilhelm Roux's Arch. Dev. Biol.* **190**, 215–220.

Dorresteijn, A. W. C., Bilinski, S. M., van den Biggelaar, J. A. M., and Bluemink, J. G. (1982). The presence of gap junctions during early *Patella* embryogenesis. An electronmicroscopical study. *Dev. Biol.* **91**, 397–401.

Dorresteijn, A. W. C., Wagemaker, H. A., de Laat, S. W. and van den Biggelaar, J. A. M. (1983). In preparation.

Ducibella, T., Ukena, T., Karnovsky, M., and Anderson, E. (1977). Changes in cell surface and cortical cytoplasmic organization during early embryogenesis in the preimplantation mouse embryo. *J. Cell Biol.* **74**, 153–167.

Eckberg, W. R. (1981). The effect of cytoskeleton inhibitors on cytoplasmic localization in *Chaetopterus pergamentaceus*. *Differentiation (Berlin)* **19**, 55–58.

Fol, H. (1875). Etudes sur le développement des mollusques. I. Sur le développement des Ptéropodes. *Arch. Zool. Exp. Gen.* **4**, 1–214.

Freeman, G., and Lundelius, J. W. (1982). The developmental genetics of dextrality and sinistrality in the gastropod *Lymnaea peregra*. *Wilhelm Roux's Arch. Dev. Biol.* **191**, 69–83.

Geilenkirchen, W. L. M., Timmermans, L. P. M., van Dongen, C. A. M., and Arnolds, W. J. A. (1971). Symbiosis of bacteria with eggs of *Dentalium* at the vegetal pole. *Exp. Cell Res.* **67**, 477–479.

Guerrier, P. (1968). Origine et stabilité de la polarité animale-végétative chez quelques Spiralia. *Ann. Embryol. Morphog* **1**, 119–139.

Guerrier, P. (1969). ''L'orientation dorsoventrale et la formation d'embryos doubles chez deus Spiralia.'' Film distributed by the S.F.R.S., 96 Bᵈ Raspail, 75272 Paris, Cédex 06.

Guerrier, P. (1970a). Les caractères de la segmentation et la détermination de la polarité dorsoventrale dans le développement de quelques Spiralia. I. Les formes à clivage égal. *J. Embryol. Exp. Morphol.* **23**, 611–637.

Guerrier, P. (1970b). Les caractères de la segmentation et la détermination de la polarité dorsoventrale dans le développement de quelques Spiralia. III. *Pholas dactylus* et *Spisula subtruncata* (Mollusques, Lamellibranchs). *J. Embryol. Exp. Morphol.* **23**, 667–692.

Guerrier, P. (1971). La polarisation cellulaire et la caractère de la segmentation au cours de la morphogenèse spirale. *Annee Biol.* **10**, 151–192.

Guerrier, P., van den Biggelaar, J. A. M., van Dongen, C. A. M. and Verdonk, N. H. (1978). Significance of the polar lobe for the determination of dorsoventral polarity in *Dentalium vulgare* (da Costa). *Dev. Biol.* **63**, 233–242.

Herbers, K. (1914). Entwicklungsgeschichte von *Anodonta cellensis* Schröt. *Z. Wiss. Zool.* **108**, 1–35.

Huebner, E., and Anderson, E. (1976). Comparative spiralian oogenesis—structural aspects: Overview. *Am. Zool.* **16**, 315–343.

Hyams, J. S., and Stebbings, H. (1979). Microtubule associated cytoplasmic transport. *In* ''Microtubules'' (K. Roberts and J. S. Hyams, eds.), pp. 487–530. Academic Press, New York.

Jaffe, L. A., and Guerrier, P. (1981). Localization of electrical excitability in the early embryo of *Dentalium*. *Dev. Biol.* **83**, 370–373.

Jaffe, L. F. (1966). Electrical currents through the developing *Fucus* egg. *Proc. Natl. Acad. Sci. U.S.A.* **56**, 1102–1109.

Jaffe, L. F. (1968). Localization in the developing *Fucus* egg and the general role of localizing currents. *Adv. Morphog.* **7**, 295–328.

Jaffe, L. F. (1969). On the centripetal course of development, the *Fucus* egg, and self-electrophoresis. *Dev. Biol., Suppl.* **3**, 83–111.

Jaffe, L. F. (1977). Electrophoresis along cell membranes. *Nature (London)* **265**, 600–602.

Jaffe, L. F. (1979). Control of development by ionic currents. *In* ''Membrane Transduction Mechanisms'' (R. A. Cone and J. Dowling, eds.), pp. 199–231. Raven Press, New York.

Jaffe, L. F. (1981). The role of ion currents in establishing developmental gradients. *In* ''International Cell Biology 1980–1981'' (H. G. Schweiger, ed.), pp. 505–511. Springer Verlag, Berlin and New York.

Jaffe, L. F., and Nuccitelli, R. (1974). An ultrasensitive vibrating probe for measuring steady extracellular currents. *J. Cell Biol.* **63**, 614–628.

Jaffe, L. F., and Nuccitelli, R. (1977). Electrical controls of development. *Annu. Rev. Biophys. Bioeng.* **6**, 445–476.

Johnson, M. H., and Ziomek, C. A. (1981). The foundation of two distinct cell lineages within the mouse embryo. *Cell* **24**, 71–80.

Lehtonen, E., and Badley, R. A. (1980). Localization of cytoskeletal proteins in preimplantation mouse embryos. *J. Embryol. Exp. Morphol.* **55**, 211–225.

Lillie, F. R. (1895). The embryology of the Unionidae. A study in cell lineage. *J. Morphol.* **10**, 1–100.

Mauchamp, J., Chambard, M., Gabrion, J., and Pelassy, C. (1980). Polarisation morphologique et foncitonnelle d'un épithélium simple en culture: Le modèle thyroidiën. *C. R. Seances Soc. Biol. Ses Fil.* **174**, 241–256.

Mescheryakov, V. N. (1978a). Orientation of cleavage spindles in pulmonate molluscs. I. Role of blastomere form in orientation of the second cleavage spindles. *Ontogenez* **9**, 558–566.

Mescheryakov, V. N. (1978b). Orientation of cleavage spindles in pulmonate molluscs. II. Role of architecture of intercellular contacts in orientation of the third and fourth cleavage spindle. *Ontogenez* **9**, 567–575.

Mescheryakov, V. N., and Belusov, L. V. (1975). Asymmetrical rotations of blastomeres in early cleavage of Gastropoda. *Wilhelm Roux's Arch. Dev. Biol.* **177**, 193–203.

Mescheryakov, V. N., and Veryasova, G. V. (1979). Orientation of cleavage spindles in pulmonate molluscs. III. Form and localization of mitotic apparatus in binucleate zygotes and blastomeres. *Ontogenez* **9**, 24–35.

Minganti, A. (1950). Acidi nucleici e fosfatasi nello sviluppo della *Limnaea*. *Riv. Biol.* **42**, 295–319.

Moreau, M., and Guerrier, P. (1981). Absence of regional differences in the membrane properties from the embryo of the mud snail *Ilyanassa obsoleta*. *Biol. Bull. (Woods Hole, Mass.)* **161**, 335–336.

Morgan, T. H. (1910). Cytological studies of centrifuged eggs. *J. Exp. Zool.* **9**, 593–655.

Morgan, T. H. (1933). The formation of the antipolar lobe in *Ilyanassa*. *J. Exp. Zool.* **64**, 433–467.

Morgan, T. H., and Tyler, A. (1930). The point of entrance of the spermatozoön in relation to the orientation of the embryo in eggs with spiral cleavage. *Biol. Bull. (Woods Hole, Mass.)* **58**, 59–73.

Morrill, J. B., Blair, C. A., and Larsen, W. (1973). Regulative development in the pulmonate gastropod, *Lymnaea palustris*, as determined by blastomere deletion experiments. *J. Exp. Zool.* **183**, 47–55.

Mulnard, J., and Huygens, R. (1978). Ultrastructural localization of nonspecific alkaline phosphatase during cleavage and blastocyst formation in the mouse. *J. Embryol. Exp. Morphol.* **44**, 121–131.

Murphy, D. B. (1975). The mechanism of microtubule-dependent movement of pigment granules in teleost chromatophores. *Ann. N.Y. Acad. Sci.* **253**, 692–701.

Pease, D. C. (1940). The influence of centrifugal force on the bilateral determination and the polar axis of *Cumingia* and *Chaetopterus* eggs. *J. Exp. Zool.* **84**, 387–415.

Pitts, J. D., and Finbow, M. E., eds. (1982). "Functional Integration of Cells in Animal Tissues." Cambridge Univ. Press, London and New York.

Quatrano, R. S. (1973). Separation of processes associated with differentiation of two-celled *Fucus* embryos. *Dev. Biol.* **30**, 209–213.

Rabl, C. (1879). Über die Entwicklung der Tellerschnecke. *Morphol. Jahrb., Abt. Anat. Ontog. Tiere* **5**, 562–660.

Rappaport, R. (1971). Cytokinesis in animal cells. *Int. Rev. Cytol.* **31**, 169–213.

Raven, C. P. (1945). The development of the egg of *Limnaea stagnalis* L. from oviposition till first cleavage. *Arch. Neerl. Zool.* **7**, 91–121.

212   J. A. M. van den Biggelaar and P. Guerrier

Raven, C. P. (1963). The nature and origin of the cortical morphogenetic field in *Limnaea*. *Dev. Biol.* **7**, 130–143.

Raven, C. P. (1966). "Morphogenesis: The Analysis of Molluscan Development," 2nd ed. Pergamon, Oxford.

Raven, C. P. (1967). The distribution of special cytoplasmic differentiations of the egg during early cleavage in *Limnaea stagnalis*. *Dev. Biol.* **16**, 407–437.

Raven, C. P. (1970). The cortical and subcortical cytoplasm of the *Lymnaea* egg. *Int. Rev. Cytol.* **28**, 1–44.

Raven, C. P. (1974). Further observations on the distribution of cytoplasmic substances among the cleavage cells in *Lymnaea stagnalis*. *J. Embryol. Exp. Morphol.* **31**, 37–59.

Raven, C. P., and Bretschneider, L. H. (1942). The effect of centrifugal force upon the eggs of *Limnaea stagnalis* L. *Arch. Neerl. Zool.* **6**, 255–278.

Rebhun, L. I. (1959). Studies of early cleavage in the surf clam, *Spisula solidissima*, using the methylene blue and toluidine blue as vital stains. *Biol. Bull. (Woods Hole, Mass.)* **117**, 518–545.

Rebhun, L. I. (1960). Aster-associated particles in the cleavage of marine invertebrate eggs. *Ann. N.Y. Acad. Sci.* **90**, 357–380.

Ries, E. (1937). Die Verteilung von Vitamin C, Glutathion, Benzidin-Peroxydase, Phenolase (Indophenolblauoxydase) und Leukomethylenblau-Oxyredukase während der frühen Embryonalentwicklung verschiedener wirbelloser Tiere. *Pubbl. Stn. Zool. Napoli* **16**, 363–401.

Ries, E. (1939). Histochemische Sonderungsprozesse während der frühen Embryonalentwicklung verschiedener wirbelloser Tiere. *Arch. Exp. Zellforsch.* **22**, 569–586.

Ries, E., and Gersch, M. (1936). Die Zelldifferenzierung und Zellspezialisierung während der Embryonalentwicklung von *Aplysia limacina* L. Zugleich ein Beitrag zu Problemen der vitalen Färbung. *Pubbl. Stn. Zool. Napoli* **15**, 223–273.

Robert, A. (1902). Recherches sur le développement des troques. *Arch. Zool. Exp. Gen.* **10**, 269–359.

Robinson, W. G., and Charlton, J. S. (1973). Microtubules, microfilaments, and pigment movement in the chromatophores of *Palaemonetes vulgaris* (Crustacea). *J. Exp. Zool.* **186**, 279–304.

Schliwa, M. (1981). Microtubule dependent intracellular transport in chromatophores. *In* "International Cell Biology 1980–1981" (H. G. Schweiger, ed.), pp. 275–285. Springer-Verlag, Berlin and New York.

Sheridan, J. D., and Larson, D. M. (1982). Junctional communication in the peripheral vasculature. *In* "Functional Integration of Cells in Animal Tissues" (J. D. Pitts and M. E. Finbow, eds.), pp. 263–283. Cambridge Univ. Press, London and New York.

Solomon, F. (1980). Neuroblastoma cells recapitulate their detailed neurite morphologies after reversible microtubule disassembly. *Cell* **21**, 333–338.

Spek, J. (1930). Zustandsänderungen der Plasmakolloide bei Befruchtung und Entwicklung des Nereis-Eies. *Protoplasma* **9**, 370–427.

Spek, J. (1934). Die bipolare Differenzierung des Cephalopoden- und des Prosobranchiereies. *Wilhelm Roux's Arch. Dev. Biol.* **131**, 362–373.

van den Biggelaar, J. A. M. (1971a). Timing of the phases of the cell cycle during the period of asynchronous division up to the 49-cell stage in *Lymnaea*. *J. Embryol. Exp. Morphol.* **26**, 367–391.

van den Biggelaar, J. A. M. (1971b). Development of division asynchrony and bilateral symmetry in the first quartet of micromeres in eggs of *Lymnaea*. *J. Embryol. Exp. Morphol.* **26**, 393–399.

van den Biggelaar, J. A. M. (1976a). Development of dorsoventral polarity preceding the formation of the mesentoblast in *Lymnaea stagnalis*. *Proc. K. Ned. Akad. Wet., Ser. C* **79**, 112–126.

van den Biggelaar, J. A. M. (1976b). The fate of maternal RNA containing ectosomes in relation to the appearance of dorsoventrality in the pond snail, *Lymnaea stagnalis*. *Proc. K. Ned. Akad. Wet. Ser. C* **79**, 421–426.

van den Biggelaar, J. A. M. (1977). Development of dorsoventral polarity and mesentoblast determination in *Patella vulgata*. *J. Morphol.* **154**, 157–186.

van den Biggelaar, J. A. M., and Boon-Niermeijer, E. K. (1973). Origin and prospective significance of division asynchrony during early molluscan development. *In* "The Cell Cycle in Development and Differentiation" (M. Balls and F. S. Billett, eds.), pp. 215–228. Cambridge Univ. Press, London and New York.

van den Biggelaar, J. A. M., and Dorresteijn, A. W. C. (1982). Cellular organisation in the early molluscan embryo. *In* "Functional Integration of Cells in Animal Tissues" (J. D. Pitts and M. E. Finbow, eds.), pp. 181–193. Cambridge Univ. Press, London and New York.

van den Biggelaar, J. A. M., and Guerrier, P. (1979). Dorsoventral polarity and mesentoblast determination as concomitant results of cellular interactions in the mollusk *Patella vulgata*. *Dev. Biol.* **68**, 462–471.

van den Biggelaar, J. A. M., Dorresteijn, A. W. C., de Laat, S. W., and Bluemink, J. G. (1981). The role of topographical factors in cell interaction and determination of cell lines in molluscan development. *In* "International Cell Biology 1980–1981" (H. G. Schweiger, ed.), pp. 526–538. Springer-Verlag, Berlin and New York.

Verdonk, N. H. (1965). Morphogenesis of the head region in *Limnaea stagnalis*. Ph.D. Thesis, University of Utrecht, Utrecht, The Netherlands.

Verdonk, N. H. (1968). The effect of removing the polar lobe in centrifuged eggs of *Dentalium*. *J. Embryol. Exp. Morphol.* **19**, 33–42.

Verdonk, N. H. (1973). Cytoplasmic localization in *Bithynia tentaculata* and its influence on development. *Malacol. Rev.* **6**, 57.

Verdonk, N. H. (1979). Symmetry and asymmetry in the embryonic development of molluscs. *In* "Pathways in Malacology" (S. van der Spoel, A. C. van Bruggen, and J. Lever, eds.), pp. 25–45. Bohn, Scheltema & Holkema, Utrecht.

Verdonk, N. H., and Cather, J. N. (1973). The development of isolated blastomeres in *Bithynia tentaculata* (Prosobranchia, Gastropoda). *J. Exp. Zool.* **186**, 47–62.

Wierzejski, A. (1905). Embryologie von *Physa fontinalis* L. *Z. Wiss. Zool.* **83**, 502–706.

Wilson, E. B. (1904). Experimental studies on germinal localization. *J. Exp. Zool.* **1**, 1–72.

Woodruff, R. I., and Telfer, W. H. (1980). Electrophoresis of proteins in intercellular bridges. *Nature (London)* **286**, 84–86.

Woods, F. H. (1932). Keimbahn determinants and continuity of the germ cells in *Sphaerium striatinum*. *J. Morphol.* **53**, 345–365.

Worley, L. G., and Worley, E. K. (1943). Studies on the supravitally stained Golgi-apparatus. I. Its cycle in the tectibranch mollusc, *Navanax inermis* (Cooper). *J. Morphol.* **73**, 365–390.

Zalokar, M. (1974). Effect of colchicine and cytochalasin B on ooplasmic segregation of ascidian eggs. *Wilhelm Roux's Arch. Dev. Biol.* **175**, 243–248.

Ziomek, C. A., and Johnson, M. H. (1980). Cell surface interaction induces polarization of mouse 8-cell blastomeres at compaction. *Cell* **21**, 935–942.

6

Morphogenetic Determination and Differentiation

N. H. VERDONK

Zoological Laboratory
State University of Utrecht
Utrecht, The Netherlands

J. N. CATHER

Department of Zoology
The University of Michigan
Ann Arbor, Michigan

I. Introduction

The constant character of cleavage in spiralian molluscs creates the impression that in this group of animals the blastomeres are determined to distinct pathways even at the time of their origin. By this constant course of cleavage, morphogenetic determinants localized in the cytoplasm of the uncleaved egg could be parceled out to distinct blastomeres and hence determine the fate of these

THE MOLLUSCA, VOL. 3
Development

blastomeres. Cleavage would then lead directly to a mosaic of cells with different developmental capacities. This view received strong support from cell-lineage studies, which showed that a fixed relationship exists between individual blastomeres and the formation of definite structures in the embryo. Cell-lineage studies alone are not proof of the capacity of the cells for self differentiation, even though it was these studies that led to the concept of a pure mosaicism of the molluscan embryo. Cell-lineage studies have not been made possible by an exact analysis of the segregation of special cytoplasmic substances into the stem cells of the different cell lines, but primarily by the constancy of the geometrical relationship between the successive generations of cleavage cells (van den Biggelaar et al., 1981). However, it is probable that the same cytoplasmic region of the egg is incorporated into the same blastomere in every embryo. The only conclusive argument for a mosaic development is the capacity of an isolated blastomere to develop *in vitro* the same structures that it would have formed in normal development in the embryo. As we shall see, this condition is fulfilled only in exceptional cases. Therefore, it must be true that correlative differentiations, which depend on interactions between cells, must also play an important role in mollusc development. These interactions must be based, however, on a prior diversification of the interacting cells (Davidson and Britten, 1971). The occurrence of special cleavage mechanisms such as polar-lobe formation and unequal cleavage, by which different cytoplasmic regions of the uncleaved egg are shunted to distinct blastomeres suggests that cytoplasmic morphogenetic determinants play a major role in the primary diversification of the molluscan embryo.The crucial questions are Do morphogenetic determinants exist? Are they differentially distributed? Do cell interactions occur to modify the developmental fate of a cell? Do morphogenetic determinants play a role in cell interactions? What is the character and chemical nature of morphogenetic determinants? How is a nucleus that is associated with a particular set of morphogenetic determinants directed to activate or regulate particular genes or gene products which determine the differentiated state of the cell? Evidence related to the first four of these questions will be evaluated in this chapter; the others will be considered by Collier in Chapter 7 of this volume.

II. Experimental Evidence for the Existence of Morphogenetic Determinants

Information on the influence of morphogenetic determinants on the development of molluscs is largely based on eggs that form polar lobes during the first cleavages. This is because polar lobes can be easily removed without damaging the blastomeres, which continue their cleavage and development. Development of lobeless larvae is not chaotic but follows a characteristic and regular course. Studying the effects of polar-lobe removal on cleavage and differentiation has contributed greatly to our understanding of the influence of morphogenetic determinants on normal development.

A. Role of the Polar Lobe in Cleavage and Differentiation

In normal development of eggs with a polar lobe, this structure is segregated at first cleavage to one of the blastomeres, which is then denominated the CD blastomere. The other blastomere, AB, is devoid of lobe substance. At second cleavage CD forms a second polar lobe, which subsequently fuses only with the D blastomere (see Chapter 3, pp. 95–96, this volume). In this way the D quadrant, which will become the dorsal quadrant of the embryo, can be distinguished from quadrants A, B, and C from its inception. In the D quadrant both the pattern of cleavage and the chronology of divisions are different from A, B, and C. In normal development of the snail *Ilyanassa* some cells of the D quadrant (e.g., 1d, $1d^1$ and $1d^{12}$) are smaller than the corresponding cells in the other quadrants (Clement, 1952). Similar phenomena were observed in *Dentalium* by Wilson (1904a) and van Dongen and Geilenkirchen (1974, 1975). In normal development, cells 2d, 3d, and 4d are larger than the corresponding micromeres in the other quadrants. The cell line 2d is characterized by a succession of very unequal divisions, so that the cells $2d^{11}$, $2d^{111}$, and $2d^{112}$ are very large. If the polar lobe is removed at first cleavage, the second cleavage is equal; no second polar lobe forms and the cleavage pattern of all quadrants is exactly the same.

The polar lobe not only influences the cleavage pattern but also the chronology of divisions in the D quadrant. In *Ilyanassa,* Clement (1952) showed that in normal development 4d appears 3 h ahead of 4a–4c and that the division of 1d and its descendants $1d^1$, $1d^{11}$, and $1d^{12}$ lags behind the corresponding cells in the other quadrants. In *Dentalium,* the divisions in the D quadrant succeed each other at a faster rate than in the A, B, or C quadrant, according to van Dongen and Geilenkirchen (1974, 1975). After removal of the polar lobe at first cleavage in both *Ilyanassa* and *Dentalium,* divisions in the four quadrants are synchronized. These observations suggest that the polar lobe contains factors that control the initiation of cell division and the position and direction of cleavages in cells of the D quadrant, to which the lobe is segregated. Under its influence the D quadrant becomes different from the other quadrants.

The influence of the lobe is not restricted, however, to the D quadrant. Painstaking cell-lineage studies of normal and lobeless embryos by van Dongen and Geilenkirchen (1974, 1975) have revealed that in *Dentalium* the influence of the lobe also reaches the A and C quadrants. During later phases of cleavage both the time schedule and the pattern of cleavage in particular cells of the A and C quadrants start to deviate from corresponding cells in the B quadrant. After removal of the lobe at first cleavage these differences also disappear; the embryo remains radially symmetrical and all quadrants behave as the ventral B quadrant does in normal development.

The influence of the polar lobe on cleavage is not only apparent from excision experiments but also from experiments in which its formation is suppressed during first cleavage. Guerrier et al. (1978) treated eggs of *Dentalium* with cytochalasin B prior to first cleavage. In certain concentrations the formation of a polar lobe was inhibited, but the egg cleaved and two equal blastomeres were formed. When the eggs were transferred to seawater, each blastomere behaved as a CD cell and formed a polar lobe at second cleavage. This resulted in a four-cell stage consisting of two larger D and two smaller C blastomeres. In most cases the two D blastomeres were situated diagonally opposite each other (CDCD embryos); in a few cases they were adjacent cells (CCDD embryos). The two large D cells, each of which had now received a polar lobe, showed all characteristic features of the normal D quadrant both with respect to the time schedule and the pattern of cleavage (Fig. 1).

The role of the polar lobe in morphogenesis has been studied extensively since Wilson (1904a) carried out his classic experiments at the Naples Zoological Station with eggs of the scaphopod *Dentalium*. After removal of the polar lobe at first cleavage actively swimming larvae were obtained, but they lacked the apical tuft, the mesoderm bands, and the posttrochal region from which the mantle fold, shell, and foot develop during normal development. Removal of the lobe at second cleavage had nearly the same effects except that as a rule the apical tuft was formed in these lobeless embryos. Similar effects were obtained by Ratten-

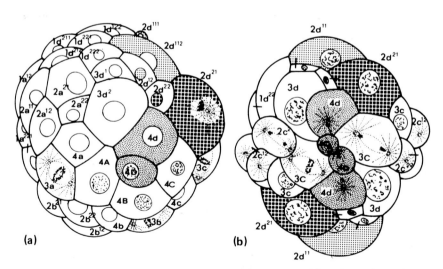

Fig. 1. Comparison of a normal (**a**) and a cytochalasin B-treated (**b**) embryo of *Dentalium* at about sixth cleavage. Note the two pairs of large second-quartet cells ($2d^{11}$ and $2d^{21}$) and the two second somatoblasts (4d) in the two D quadrants of the cytochalasin B-treated embryo. In the normal embryo a single D quadrant is present. (From Guerrier et al., 1978.)

bury and Berg (1954) after removal of the first polar lobe in the lamellibranch *Mytilus*. In lobeless embryos the apical tuft, mesoderm bands, and shell were missing, whereas stomodaeum and velum were present. The missing structures in lobeless embryos of *Dentalium* and *Mytilus* belong to the D quadrant of the embryo, to which polar lobe material is segregated. One might infer, therefore, that there are cytoplasmic determinants prelocalized in the uncleaved egg, and that these are partitioned into cells of the D quadrant through the polar lobe and thereby directly determine the fate of the cells in this quadrant.

Studies of gastropod development have provided a more complete picture of the influence of the lobe on morphogenesis. This is mainly due to the detailed investigations of Clement (1952, 1956, 1962, 1967, 1968, 1976) on the embryogenesis of the prosobranch snail *Ilyanassa*. Crampton (1896) was the first to remove the polar lobe in this species, but he was unable to rear the embryos far enough to study the influence of the lobe on morphogenesis. He only observed that the mesoderm bands failed to form. Clement succeeded in rearing embryos far enough to study the influence of polar-lobe removal on organogenesis. He found that in lobeless embryos adult organs such as shell, foot, operculum, otocysts, eyes, tentacles, and a beating heart were missing (Fig. 6b). Histological studies of Clement (1952) and Atkinson (1971) revealed that velar cells, ganglia, muscle cells, internal shell masses, stomodaeum, esophagus, stomach, and digestive gland were present. Intestine and heart were never observed. Essentially the same results were observed by Cather and Verdonk (1974) after removal of the very small polar lobe in the snail *Bithynia tentaculata*. Lobeless embryos of this species lack ectodermal structures such as eyes, tentacles, shell, and foot but may develop ganglia, muscle cells, liver, mid and hindgut structures (Fig. 2). A recognizable stomodaeum was never observed, but in some cases stomodaeal entrance cells were present.

Although cell-lineage studies of *Ilyanassa* and *Bithynia* have never been continued beyond the formation of the mesentoblasts, information available from other gastropods indicates that eyes and tentacles originate from 1a and 1c with a contribution from 1b. As eyes and tentacles are always missing in lobeless embryos, one may conclude that the influence of the lobe is not restricted to the D quadrant, but also reaches the lateral quadrants A and C. This is corroborated by studies of van Dongen and Geilenkirchen (1974, 1975) and van Dongen (1976a,b) on the influence of the polar lobe on morphogenesis in *Dentalium*. Following the cell lineage of both normal and lobeless embryos up to the trochophore stage, these authors not only confirmed the data of Wilson (1904a) that in lobeless embryos the dorsal structures derived from the D quadrant (shell, foot, mesoderm bands) are missing, but in addition they showed that the development of the lateral quadrants (A and C) is affected after removal of the polar lobe. In lobeless embryos the radially symmetrical pattern of cleavage leads to a radially symmetrical embryo. In each quadrant cleavage and organogenesis fol-

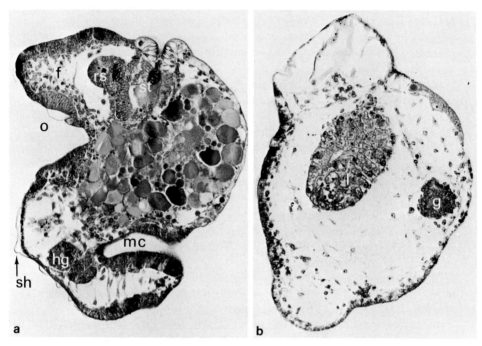

Fig. 2. Sections through a normal 7-day-old embryo (**a**) and a lobeless embryo (**b**) aged 14 days of *Bithynia*. In the normal embryo several adult structures are visible such as foot (f), operculum (o), mantle cavity (mc), shell (sh), stomodaeum (st), radular sac (rs) and hindgut (hd). In the balloon-like lobeless embryo only interspersed mesenchyme cells, larval liver (l) and a ganglion(g) can be observed.

low the pattern that in normal development is observed in the B quadrant of the embryo. In the normal embryo the stomodaeum is formed by the invagination of the stomatoblasts, derived exclusively from the micromere 2b^{22}. In lobeless embryos the micromeres 2^{22} in each quadrant form stomatoblasts, which unite into a ring-shaped area and invaginate to form a radialized stomodaeum. Similarly, in lobeless embryos derivatives of each of the four arms of the cross constitute four interradial placodes from which four cerebral ganglia originate. In a normal embryo, only two cell plates are formed in an interradial position on the ventral side of the pretrochal region. These cell plates originate from the a, b, and c arm of the cross, the b arm contributing to the formation of both the right and the left cephalic plate, from which the right and left cerebral ganglion originate by invagination. In lobeless embryos the archenteron is formed by the endodermal components from each quadrant. Thus, in the lobeless embryo only those structures that in normal development originate in part (ganglia, liver) or exclusively (stomodaeum) from the B quadrant are formed. In normal development morphogenetic determinants localized at the vegetal pole of the egg are enclosed in the polar lobe and at the four-cell stage are segregated to one quadrant, D, which in this way gets special developmental properties. This view is corroborated by experiments of Guerrier et al. (1978) on *Dentalium*. After equalization of first cleavage, the vegetal region of the egg was divided approximately equally between both blastomeres. At second cleavage this region was segregated to two blastomeres of the four-cell stage (see p. 218). Both blastomeres that had received a part of the vegetal region formed dorsal structures, resulting in development of a double embryo with two posttrochal regions each with a well-developed shell (Fig. 3).

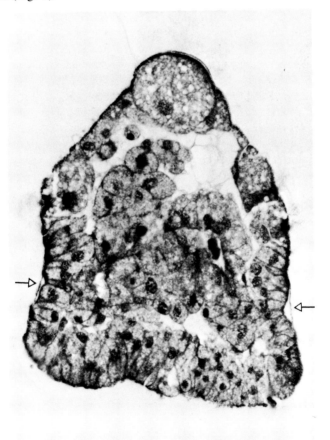

Fig. 3. Longitudinal section of a cytochalasin B–treated embryo of *Dentalium* with the apical organ at the top and two posttrochal regions each with a shell gland and a shell anlage indicated by arrows. (From Guerrier et al., 1978.)

B. Unequal Cleavage and Segregation of Morphogenetic Determinants

In many bivalves first cleavage is unequal. The plane of cleavage divides the egg into a smaller blastomere AB and a larger blastomere CD. Second cleavage is unequal in CD so that at the four-cell stage D is the largest blastomere, whereas A, B, and C are the same size. In this way the vegetal region of the egg is incorporated into the D quadrant. First cleavage may be made equal by various treatments, such as compression, centrifugation, ultraviolet irradiation, and chemical treatment. After an equal first cleavage the second cleavage is usually unequal in both blastomeres, the four-cell stage consists of two large cells situated either side by side or in alternating positions. As a consequence, duplication of the dorsal structures may occur, as was first shown by Tyler (1930) in eggs of *Cumingia*. He showed that after equalization of first cleavage, embryos might develop two shells. More detailed information was obtained by Guerrier (1970b) who closely followed the development of eggs of *Pholas* and *Spisula,* in which first cleavage was equalized by compression or centrifugation. He observed that each of the two large blastomeres formed at the four-cell stage behaved as a D quadrant in normal development. Two first somatoblasts (2d) and two second somatoblasts (4d) were formed, resulting in a twin embryo with two shells. It thus appears that morphogenetic determinants situated in the vegetal region of the egg are shunted by unequal cleavage to the D quadrant, which is then provided with special developmental capacities.

C. Morphogenetic Localizations in Equally Cleaving Eggs

In many species of molluscs the cleavage planes of the first two divisions separate the egg into four blastomeres of the same size. The question, therefore, arises whether or not morphogenetic determinants are prelocalized and differentially distributed to the blastomeres during the first two cleavages in equally cleaving eggs, as they are in unequally cleaving eggs. Guerrier's (1970a) experiments on the determination of dorsoventral polarity in the pulmonate *Limax* have a bearing on this question. By compressing eggs between two planes parallel to the egg axis just prior to the first cleavage, he was able to rotate the spindle, thus changing the positions of the first and second cleavages. Even though both the sequence of partitioning of the egg cytoplasm and the direction of the cleavage planes to form the blastomeres of the four-cell stage were changed, when the first two cleavage planes intersected along the egg axis, morphogenesis was normal. This indicates that, if morphogenetic determinants are present in the egg, they are not differentially distributed in a significant way by the first two cleavage planes. One might even infer that in equally cleaving eggs the blastomeres of the four-cell stage have equal developmental capacities and that although each of the blastomeres will form a quadrant of the future embryo, the quadrants are not yet specified. As discussed by van den Biggelaar and Guerrier (Chapter 5, this volume), this situation persists until after the fifth cleavage when the dorsal quadrant is determined by an interaction between the animal micromeres and one of the vegetal macromeres. Even if one quadrant is predetermined to become dorsal, under experimental conditions each quadrant may become the dorsal quadrant of the embryo in equally cleaving eggs. In unequally cleaving eggs, as shown earlier, morphogenetic factors present at the vegetal pole are shunted exclusively to one of the blastomeres at the four-cell stage, and in this way determine the dorsal quadrant of the embryo. This does not mean, however, that morphogenetic determinants are absent in equally cleaving eggs of molluscs or even that they are equally distributed. It indicates that all of the primary macromeres have the determinants necessary to assume the important organizing role of the D quadrant. The situation in these molluscs may be similar to the situation in echinoderms, where each of the blastomeres of the two- and four-cell stage has equal potential and may form a normal larva. However, only the vegetal halves of the egg gastrulate and develop into normal embryos (Hörstadius, 1937). Direct evidence for the presence of morphogenetic determinants at the vegetal pole of equally cleaving eggs of molluscs is not available because experiments in which the uncleaved eggs were dissected in various directions have never been performed. Nevertheless Morrill's experiments (1964) point in this direction. After centrifugation of *Lymnaea* eggs, prior to first cleavage, egg fragments that lacked the vegetal region of the egg always developed abnormally. His interpretation at the time, that the morphogenetic factors present at the vegetal pole become localized in the vegetal region of the D quadrant, is possibly too restrictive because the quadrants in *Lymnaea* seem (from the evidence of van den Biggelaar et al., 1981) to have equal potential until after the fifth cleavage.

III. Prelocalization of Morphogenetic Determinants

A. Time and Site of Localization

Experiments on the influence of polar lobes on morphogenesis have convincingly shown that important morphogenetic determinants are indeed segregated through the polar lobe, and thus that these determinants are already localized in the vegetal region of the egg at the beginning of first cleavage when this region is incorporated into the polar lobe. Evidence is available that localization of determinants takes place at a much earlier stage. When freshly laid eggs of *Dentalium* are cut into two pieces, both fragments may develop after fertilization. Animal fragments form polar bodies, but no polar lobe or lobe-dependent structures, whereas fragments that contain a major part of the vegetal region do form a polar lobe and lobe-dependent structures (Wilson, 1904a; Verdonk et al., 1971). Indirectly this indicates that in *Dentalium* morphogenetic determinants become localized in the oocyte during oogenesis.

More direct evidence for a prelocalization of morphogenetic determinants during oogenesis comes from observations of Dohmen (Chapter 1, this volume) on the origin of the vegetal body in *Bithynia* (Fig. 4). This structure consists of a large cup-shaped aggregate of electron-dense vesicles. The vegetal body originates during oogenesis when it appears at the basal side of the oocyte. In freshly laid eggs it is situated at the vegetal pole, which is incorporated during cleavage in the polar lobe. Similar aggregates of vesicles have been observed in the small polar lobes of the snails *Crepidula* (Dohmen and Lok, 1975), *Buccinum,* and *Littorina* (Dohmen and Verdonk, 1979). From these observations alone one cannot conclude that these structures in small polar lobes contain the morphogenetic determinants, as they are absent in the large polar lobes of *Mytilus, Dentalium,* and *Nassarius.* There is experimental evidence, however, which seems to support the conclusion that the special structures in small polar lobes do carry morphogenetic determinants.

B. Morphogenetic Significance of the Special Structures in Small Polar Lobes

The presence of special cytoplasmic structures in eggs with small polar lobes offers an opportunity to investigate whether the morphogenetic determinants are indeed localized in these special cytoplasms. Eggs of *Bithynia* seem to be an especially favorable material because they contain one large aggregate of vesicles, the vegetal body, which can be observed easily in histological sections. Dohmen and Verdonk (1979) centrifuged uncleaved eggs of *Bithynia* in an attempt to disperse or displace the constituents of the vegetal body. The body appeared to be bound rather strongly to the cortex, as it could only be removed by a relatively strong centrifugal force (about 1400 *g*), which separates part of the eggs into two fragments. When the vegetal body is removed from the vegetal pole of the egg it is never dispersed but always displaced as a whole. Seventy eggs, centrifuged about 1 h before first cleavage, were fixed at the moment of first cleavage and studied in sections. In 30 eggs the vegetal body was no longer found in the polar lobe, but in the cytoplasm of one of the blastomeres (Fig. 4).

The vegetal body cannot be observed in the living egg. We know, however, from experiments of Cather and Verdonk (1974) that after removal of the polar lobe in normal uncentrifuged eggs, adult structures such as shell, foot, eyes, and tentacles are never found (Fig. 5a). After removal of the polar lobe in centrifuged eggs not only were abnormal embryos of the characteristic lobeless type obtained

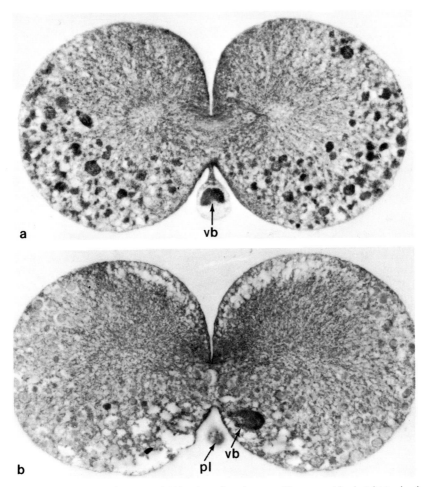

Fig. 4. (a) Section of an egg of *Bithynia* at first cleavage. The vegetal body (vb) is clearly seen to be situated in the polar lobe. In the egg centrifuged 1 h before first cleavage and fixed at first cleavage (**b**), the vegetal body is absent from the polar lobe and located in one of the blastomeres. pl, Tangential section through the polar lobe. (Fig. 4b is from Dohmen and Verdonk, 1979.)

but also completely normal embryos (Fig. 5b). These data indicate that in *Bithynia* the morphogenetic determinants are localized in the vegetal body, which in normal development is segregated through the polar lobe but which in centrifuged eggs may be displaced so that the polar lobe no longer contains it. Devoid of the vegetal body, the polar lobe loses its influence on development, thus it can be removed without any effect on development.

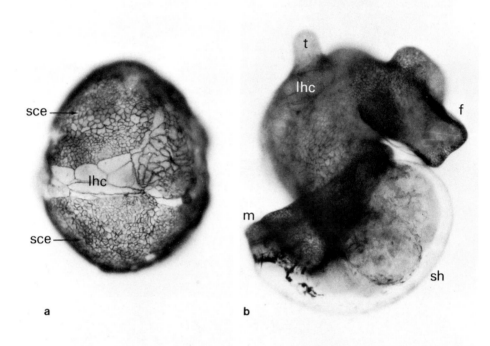

Fig. 5. (a) Embryo of *Bithynia* showing very defective development after removal of the polar lobe without prior centrifugation. In (b) a normal embryo is shown obtained after removal of the polar lobe of a centrifuged egg. Centrifugation (1400 *g*) 1 h before first cleavage. lhc, Larval head cells; t, tentacle; f, foot; m, mantle; sce, small-celled ectoderm; sh, shell. (From Dohmen and Verdonk, 1979.)

C. Cortical Localization in Large Polar Lobes

Evidence for a cortical localization of morphogenetic determinants in eggs with large polar lobes comes from centrifugation experiments. Verdonk (1968a) centrifuged eggs of *Dentalium* just prior to first cleavage. Although the constituents of the egg were stratified in various configurations and consequently the polar lobe received different cytoplasms, centrifugation did not change the development of intact or lobeless embryos. In *Ilyanassa,* the yolk is concentrated at the vegetal pole of the egg. Taking advantage of the distribution of the yolk as an indicator of polarity, Clement (1968) centrifuged the eggs in animal–vegetal direction or in reverse so that the yolk either remained concentrated at the vegetal pole or was displaced toward the animal pole. Upon continued centrifugation eggs could be separated into halves so that nucleated animal or vegetal halves without yolk could be obtained. The animal halves never formed polar lobes and always developed as lobeless embryos without eyes, shell, foot, etc. The vegetal halves, from which the yolk was removed, formed polar lobes of approximately the correct relative size. About half of these vegetal fragments developed the usual lobe-dependent structures. It appears then that in *Ilyanassa* the morphogenetic determinants are not bound to displaceable components of the polar lobe.

An electron-microscopic study of normal and centrifuged eggs of *Dentalium* and *Nassarius* did not reveal any nondisplaceable component to which the morphogenetic role of the polar lobe can be ascribed. As discussed by Dohmen (Chapter 1, this volume), all discernible organelles present in the polar lobe are displaced by the centrifugal forces applied by Clement and Verdonk in their experiments.

The question therefore arises whether in eggs with large polar lobes the morphogenetic determinants are localized in the interior cytoplasm or, as suggested earlier by Morgan (1933, 1935), in the cortex of the vegetal region of the egg. Cortical localization has received support by observations of J. A. M. van den Biggelaar (unpublished) who, with a small pipette, sucked the cytoplasm out of the polar lobe at the trefoil stage of *Dentalium*. The cytoplasm of the lobe is then gradually replaced by cytoplasm flowing in from the rest of the egg. Although nearly the whole content of the lobe was removed, the eggs continued cleaving and developed into normal embryos. Van den Biggelaar also transplanted cytoplasm of the second polar lobe of *Dentalium* to the B blastomere of another egg at the four-cell stage. This attempt to give the B blastomere, which in normal development does not receive a part of the polar lobe, the quality of a D blastomere by injecting lobe cytoplasm failed. No duplication of dorsal structures was found, whereas 79% of the injected embryos showed normal development. The above data strongly support the view that the morphogenetic determinants present in large polar lobes as found in *Dentalium* and *Ilyanassa* are not situated in the general endoplasm. They are either bound to or situated in the cortical cytoplasm or plasma membrane of the egg or are linked to endoplasmic fibrillar structures which are not discernible by currently used methods.

Scanning and transmission electron-microscopic studies have shown that eggs forming polar lobes show a special surface morphology at the vegetal region even before protrusion of the lobe starts, consisting of villi, folds or ridges (see Chapter 1, this volume). During formation of the polar lobe these structures cover the lobe. They do not disappear with the polar lobe, but persist on the D macromere. In *Crepidula* they remain visible on the macromere 5D until this cell disappears into the interior of the embryo (see Fig. 6 of Chapter 3, this volume). It is obvious that a correlation exists between this special surface architecture at the vegetal pole and the presence of morphogenetic determinants in that region, but the nature of the relationship between both phenomena is still obscure. A direct relationship between the local surface differentiation and the activity of the morphogenetic determinants is unlikely as these structures eventually end up on one of the entomeres, which are not essential for normal development. Another argument against a direct relationship comes from experiments with eggs of *Bithynia,* discussed earlier, showing that after centrifugation the polar lobe with its surface structures can be removed without influencing development.

Dohmen and Verdonk (1979) supposed that surface differentiations charac- terized by an increased surface area may indicate an increased transport of ions through the membrane of the vegetal pole resulting in an electrical current. As in *Fucus* (Jaffe, 1966) and *Pelvetia* (Nuccitelli, 1978), these currents may control the localization of charged molecules, vesicles, or particles. Jaffe and Guerrier (1981) have shown that in *Dentalium* at the trefoil stage, the membrane of the polar lobe and that of the blastomeres respond differently to applied current. The membrane of an isolated lobe appears to be excitable, whereas the membrane of an isolated blastomere is not. Their results indicate that the membrane of the blastomeres is more permeable to ions than the membrane of the polar lobe. How these observations are related to the morphogenetic influence exerted by the lobe is still unclear.

D. Special Structures Observed in Equally Cleaving Eggs

Observations on the presence of putative morphogenetic localizations in equal- ly cleaving eggs are only available from the pulmonates. In the egg of *Lymnaea stagnalis,* Raven (1963, 1967, 1970, 1974, 1976) described the presence of subcortical patches of cytoplasm staining differently from surrounding cyto- plasm. These so-called subcortical accumulations (SCA), always six in number, are situated in the equatorial zone of the egg, arranged in a regular pattern that is nearly bilaterally symmetrical. Just before first cleavage the SCA extend in latitudinal direction and start to fuse. During the first two cleavages the SCA are divided over the four blastomeres, where they gradually shift toward the vegetal pole. At the eight-cell stage the SCA substance is situated at the most vegetal part of each macromere. At the same time, granules rich in RNA become visible in this cytoplasm, and condense into special bodies, called *ectosomes.* Similar structures have been described by Wierzejski (1905) in the egg of *Physa.* In *Lymnaea* they consist of clusters of ribosomes, endoplasmic reticulum, and small electron-dense vesicles (Dohmen and van der Mast, 1978). After the fifth cleav- age at the onset of the 24-cell stage in *Lymnaea* these ectosomes start moving up to the tips of the macromeres (Minganti, 1950; Raven, 1974; van den Biggelaar, 1976a,b). After the determination of one of the macromeres to 3D, which takes place during the 24-cell stage in *Lymnaea,* the ectosomes disperse in 3D, and gradually disappear, whereas in 3A, 3B, and 3C they remain compact bodies, which are segregated to 4a, 4b, and 4c at the next cleavage and then are probably extruded into the cleavage cavity. Although it has been repeatedly suggested that the ectosomes may play a role in the determinative events, which occur during the 24-cell stage in *Lymnaea,* direct evidence for a morphogenetic role is still lacking.

IV. Segregation of Morphogenetic Determinants during Cleavage

Cell-lineage studies of several species of molluscs have provided exact information on the fate of the blastomeres in normal development. These studies do not show, however, to what extent a particular cell is determined by internal cellular factors inherited from the egg, for they are not based on a direct analysis of the segregation of morphogenetic determinants. Rather they are based on the constancy of geometric relations between the blastomeres at successive cleavages (van den Biggelaar et al., 1981). Segregation of morphogenetic determinants to the different blastomeres can currently only be determined experimentally by isolation or deletion of specific blastomeres, whose fate in normal development is known from cell-lineage studies.

A. Development of Blastomeres Isolated at the Two- or Four-Cell Stage

In eggs with polar lobes the lobe is segregated to the CD blastomere at first cleavage. It is not surprising, therefore, that after isolation of the two blastomeres at first cleavage only the CD blastomere forms a polar lobe at the next cleavage and develops lobe-dependent structures, whereas the AB blastomere cleaves equally without forming a polar lobe and develops as a lobeless embryo.

In *Dentalium* (Wilson, 1904a; Cather and Verdonk, 1979) and *Mytilus* (Rattenbury and Berg, 1954), the AB halves gastrulate and form partial larvae with velar cells, but without an apical tuft or a posttrochal region. CD larvae form an apical tuft and a posttrochal region, but the differentiation of adult structures such as foot and shell is defective. In *Ilyanassa* (Clement, 1956) AB halves gastrulate and develop into embryos with velar cells, pigment, muscle cells, and a ciliated enteron. CD halves form in addition the adult structures, foot, shell, and eyes (Fig. 6). According to Hess (1956a), AB halves of *Bithynia* first exogastrulate, but later part of the endoderm may invaginate. In his view only the AB halves would form a stomodaeum, whereas both AB and CD halves would be able to form a shell gland. From an extensive histological study of 25 pairs of *Bithynia,* Verdonk and Cather (1973) concluded that AB halves formed larval head cells, ganglia, muscle cells, and larval liver. Stomodaeal structures and

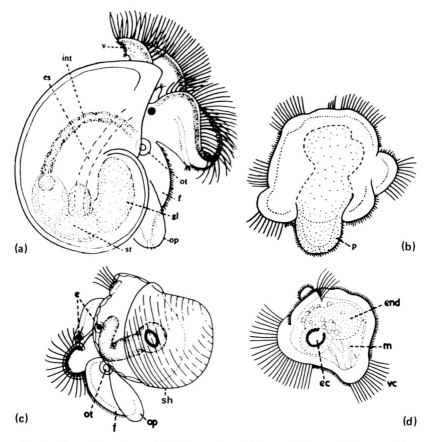

Fig. 6. Comparison of normal, lobeless, and partial larvae of *Ilyanassa*. In the normal larva (**a**) several adult structures can be observed, which are also present in the CD-half larva (**c**). In the lobeless embryo (**b**) and the AB larva (**d**) adult structures are almost completely absent. e, Eye; ec, enteric cavity; end, endoderm; es, esophagus; f, foot; gl, digestive gland; int, intestine; m, muscle; op, operculum; ot, statocyst; p, posterior protrusion (everted stomodaeum); sh, shell; st, stomach; v, velum; vc, velar cells. (From Clement, 1952 (**a** and **b**), from Clement, 1956 (**c** and **d**).)

shell were always restricted to the CD halves, as were eyes, tentacles, foot, operculum, shell, etc. After second cleavage in *Dentalium, Mytilus,* and *Il-yanassa,* isolated A, B or C blastomeres develop as lobeless embryos, whereas the D blastomere develops lobe-dependent structures. It is surprising that in *Dentalium* (Wilson, 1904a) the isolated C blastomere does not develop an apical tuft because in normal development this structure is formed by descendants of both the C and the D quadrant (van Dongen and Geilenkirchen, 1974) and after removal of the polar lobe at second cleavage an apical tuft does develop even though the posttrochal region is missing. Cather and Verdonk (1979) observed the presence of an apical tuft in ABC (−D) embryos of *Dentalium*. In the gastropod *Ilyanassa* AD halves, obtained by puncturing the B and C blastomere at the four-cell stage, develop into partial larvae with one or two eyes, shell, and foot. ABC combinations with the D quadrant removed develop as lobeless embryos (Clement, 1956, 1962). Although other combinations were not studied by Clement, one may assume that in *Ilyanassa* combinations of blastomeres in which the D blastomere is missing develop as lobeless embryos. The situation in another gastropod, *Bithynia*, is quite different. Verdonk and Cather (1973) found that both AD and BC combinations could form adult structures such as eyes, tentacles, foot, and shell. Gills were the only structure restricted to one half (BC). Cather et al. (1976a) studied the development of the various types of three-quarter embryos, obtained by deletion of one known blastomere at the four-cell stage. When the A blastomere was removed, BCD embryos appeared to be normal except for the head region where the left eye and tentacle were absent. Also ACD embryos with the B blastomere removed were normal in appearance except in the region of the head, where often one cyclopic eye and a single tentacle were present. When the C blastomere was killed, ABD embryos most often developed as chaotic monsters. In some cases, however, a snaillike embryo with a reduced shell and with the right eye and tentacle missing were produced. Gills were always absent in these embryos. With the D blastomere removed, ABC embryos developed a turret-like shell. The mantle cavity was missing or everted so the gills were exposed. The foot was always reduced. The head may be more or less normal with eyes, tentacles, and velar structures.

From these isolation and deletion experiments one may conclude that the polar lobe in most molluscs segregates the morphogenetic potential nearly exclusively to the D quadrant. In *Bithynia*, however, the morphogenetic determinants are divided over both the C and the D quadrant. This may be related to the behavior of the vegetal body, in which the morphogenetic determinants are contained. This body, present in the uncleaved egg and in the first polar lobe, cannot be detected in histological sections of the CD blastomere after a period just before the second cleavage. We have assumed that the contents of the vegetal body are distributed to the C and the D blastomere at second cleavage (Verdonk and Cather, 1973). However, in electron-microscopic preparations a large aggregate of vesicles is still present at the vegetative side of the CD blastomere. The individual vesicles are apparently unchanged but the cup shape of the vegetal body is lost. Whereas in earlier stages the vesicles are arranged in dense clusters with vesicle-free areas in between, this pattern disappears just before second cleavage (Dohmen and Verdonk, 1979). At second cleavage a large mass of these vesicles is segregated into the D blastomere, the vesicles disappear from that area during the four-cell stage. Whether or not some vesicles or their contents are segregated to the C blastomere is still uncertain.

Although in eggs with polar lobes one blastomere of the two-cell stage receives the essential morphogenetic determinants and after isolation only this blastomere can form adult structures, it never develops into a complete well-organized larva. The reason may be that an isolated blastomere of a polar-lobe forming egg always cleaves as if still part of a whole embryo. Isolated CD blastomeres form a polar lobe at the next cleavage, which fuses with one of the daughter cells. Subsequently both blastomeres split off micromeres at the animal side. One might assume that a situation of two macromeres giving off duets of micromeres at the subsequent cleavages leads to a configuration of cells in which the normal course of cell interactions is hardly possible or in any case greatly abnormal. This is in agreement with the results obtained by deletion of one blastomere at the two-cell stage of equally cleaving eggs. In the gastropod *Lymnaea palustris,* Morrill et al. (1973) killed one blastomere at the two-cell stage and obtained completely normal embryos from the surviving blastomere. An absolute condition, however, was that the uninjured blastomere cleaved as a whole egg, passing again through two equal cleavages before four micromeres were formed. If the isolated blastomere cleaved as though still part of a whole embryo, forming two macromeres and two micromeres, development was always abortive. Similar results were obtained by Verdonk (1979) using eggs of *Lymnaea stagnalis, L. ovata, Physa acuta,* and *Succinea putris.* In all these gastropods, completely normal embryos (Fig. 7) may be formed provided that the uninjured blastomere follows the cleavage pattern of a normal egg. These embryos can develop into young snails, which hatch and form adults.

Morrill et al. (1973) also obtained normal snails after destroying one of the

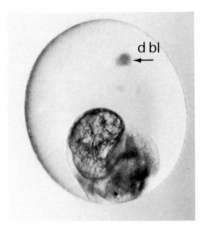

Fig. 7. Normal embryo of *Lymnaea stagnalis,* which developed from one of the blastomeres of the two-cell stage. The other blastomere (d bl), killed at the two-cell stage, is still visible in the egg capsule.

blastomeres at the four-cell stage of *L. palustris*. In this case the remaining blastomeres cleaved as though still part of a normal embryo. Apparently, a configuration of three macromeres with triplets of micromeres is close enough to the normal situation to enable normal development. As previously mentioned, because of the presence of a polar lobe in *Bithynia,* the blastomeres can be identified at the four-cell stage and removal of the A or B blastomere at this stage has no great effect on development. Also, in this species three-quarter embryos may form fairly normal embryos, provided that a blastomere is removed which does not get part of the polar-lobe material. Unfortunately, the yield of normal embryos after deletion of one blastomere at the two- or four-cell stage of *Lymnaea* and other pulmonates is too low (4–30%) to decide whether or not morphogenetic determinants are differentially segregated during first and second cleavage of equally cleaving eggs. Due to the presence of a very tough vitelline membrane in the pulmonate egg, attempts to separate blastomeres and to rear both halves in order to see whether identical twins can be obtained have been unsuccessful. This *experimentum crucis* might decide whether or not the theory of van den Biggelaar and Guerrier (Chapter 5, this volume) that in equally cleaving eggs the blastomeres of the four-cell stage have identical developmental potential is correct.

B. Developmental Potential of Micromeres

According to cell-lineage studies, a strict relationship exists between specific blastomeres and the formation of distinct organs. This has been demonstrated most precisely for the cells of the first three quartets of micromeres, whose fate can be followed easily because they give rise to the ectodermal structures. Deletion of micromeres, whose prospective significance in normal development is known, may provide information as to whether the developmental potential is segregated exclusively to specific cells during cleavage.

In *Dentalium,* the apical tuft is formed by derivatives of 1c and 1d (van Dongen and Geilenkirchen, 1974). Removal of both 1c and 1d at the eight-cell stage results in embryos without an apical tuft (Cather and Verdonk, 1979). In *Ilyanassa* (Clement, 1967), deletion of the 1a micromere at the eight-cell stage results in absence of the left eye and a slight velar reduction, whereas removal of 1c leads to loss of the right eye and slight velar reduction. Removal of 1b tends to reduce the distance between the eyes, whereas loss of 1d does not produce any significant defect (Fig. 8). Van Dam and Verdonk (1982) removed the whole first quartet of micromeres in the gastropod *Bithynia* and observed that eyes and tentacles were always missing, although other adult structures such as shell, foot, heart, and gut were present. These data are in accordance with cell-lineage studies on the development of the head region in gastropods (Verdonk, 1965).

Clement (1963, 1971) also determined the effect of deletions in the second

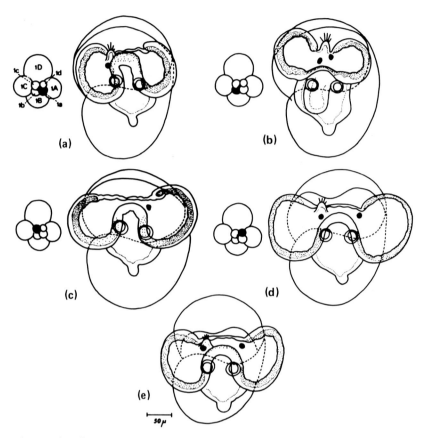

Fig. 8. The effect of deletion of a first quartet micromere on the development of *Ilyanassa*. The dark cell in the diagram (to the left of the resultant veliger) depicts the cell deleted at the eight-cell stage. (**a**) left eye missing after 1a deletion; (**b**) synophthalmia after 1b deletion; (**c**) right eye missing after 1c deletion; (**d**) no detectable abnormality after 1d deletion; (**e**) control larva. (Modified from Clement, 1967.)

quartet of micromeres of *Ilyanassa*. Deletion of 2a affects the development of the left velar lobe and may also influence the development of the left eye and statocyst. Loss of 2b has no detectable effect on development. Deletion of 2c leads to a reduction of the shell, absence of the larval heart, and eversion of the stomodaeum. Loss of 2d causes a reduction of the shell or its absence. Data of Cather (1967), which will be discussed later more extensively, show that the shell in *Ilyanassa* is only absent when both 2c and 2d are deleted. Apparently the shell in gastropods does not originate exclusively from the first somatoblast, 2d, as in other molluscs. In *Dentalium*, removal of 2d results in absence of the shell, although a reduced foot with one statocyst may be present. This is in agreement with the cell-lineage studies of van Dongen and Geilenkirchen (1974), which show that the shell and one half of the foot originate from 2d, whereas the other half of the foot comes from 3d. Accordingly, after removal of 3d the foot is reduced and in nearly all cases only one statocyst is present. In *Ilyanassa* (Clement, 1971), 3a and 3b do not seem to be of great embryonic value, because after their removal only part of the velar lobes are reduced. After deletion of 3d, the left half of the foot and the left statocyst are missing, and after loss of 3c the right statocyst and the right half of the foot.

In molluscs, the 4d blastomere is the mesentoblast, which gives rise to the mesoderm bands from which heart and intestine originate. After removal of this cell in *Ilyanassa* (Clement, 1960) these structures are missing. Apart from these deficiencies, the embryo looks completely normal. After removal of 4d in *Dentalium*, defects in the larva are not seen (Cather and Verdonk, 1979). This may be due to the extensive development of the ectomesoderm in this species, which obscures observations on the endomesoderm (van Dongen, 1977).

The results from blastomere deletion experiments indicate an early specification of cell values in molluscs. In nearly all cases, removal of a micromere leads to the absence of structures in the embryo, which according to cell-lineage studies are derived from that cell. Morphogenetic potential of the egg is apparently distributed selectively during the formation of the quartets of micromeres. This does not mean that morphogenetic determinants, segregated in the quartets of micromeres, are already definitively localized in the egg before the first cleavage and passively distributed by the formation of cleavage planes. If this were the case, removal of the animal parts of an egg would lead to distinct defects because in most molluscs the quartets of micromeres are formed from the animal half. Clement (1968) observed that vegetal halves, obtained by centrifugation of uncleaved eggs of *Ilyanassa* are able to form veligers with eyes. In *Dentalium*, the vegetal half of an unfertilized egg may form a normal embryo with an apical tuft present (Wilson, 1904a; Verdonk et al., 1971). Because the eyes in gastropods and the apical tuft in scaphopods originate from the first quartet of micromeres, which in normal development is formed from the region around the animal pole, one has to assume that this region was certainly absent in the vegetal fragments. The quartets of micromeres are thus formed from egg regions, which they do not receive in normal development and yet form their normal structures.

It is known in *Dentalium,* where the determinants for apical-tuft formation are localized in the uncleaved egg. Removal of the vegetal-pole region of the egg or of the polar lobe at first cleavage always leads to an embryo without an apical tuft (Wilson, 1904a; Geilenkirchen et al., 1970; Verdonk et al., 1971). However, when the lobe is removed at second cleavage an apical tuft may be formed. Apparently these determinants, originally situated at the vegetal pole, are segregated through the polar lobe of the first cleavage to the CD blastomere. During the two-cell stage they move up in the animal direction so that they are not incorporated in the polar lobe at second cleavage. At third cleavage they must be present at the animal pole of the C and D blastomere as they are segregated to 1c and 1d.

Freeman (1976a,b, 1978, 1979) has convincingly shown that in the eggs of the ctenophore *Mnemiopsis* and the nemertean *Cerebratulus* morphogenetic determinants, although already present in the uncleaved egg, are not strictly pre-localized. Cleavage appears to play an active, directional role in establishing where localization of determinants will occur. In molluscs, cleavage may play a similar role in the segregation of morphogenetic potential in the micromeres. The behavior of the apical-tuft factor in *Dentalium* points in this direction.

The segregation of morphogenetic potential in the micromeres leads to a commitment of these cells to the formation of specific structures at an early stage of development. Although the consequence of removal of a specific blastomere is the absence of a distinct structure in the embryo, this does not prove that this micromere can differentiate independently into this structure. Direct evidence for independent differentiation of the micromeres is rather limited in molluscs. In *Dentalium* one might suppose that, after isolation of individual first quartet micromeres, 1c and 1d would form an apical tuft, because it is only after removal of both these cells that the apical tuft is missing. Wilson (1904a), however, found that only the isolated 1d micromere can form an apical tuft. Before adopting a complicated theory such as that proposed by van Dongen and Geilenkirchen (1974), one would do well to repeat some of Wilson's experiments. In *Patella* (Wilson, 1904b), each isolated first-quartet cell forms four primary trochoblasts, two secondary trochoblasts, one quarter of the apical tuft, and a group of undifferentiated ectoderm cells. However, they never produce adult structures like eyes and tentacles, which in the normal development of snails originate from 1a and 1c. Wilson's statement that isolated first-quartet cells in *Patella* form the same structures that they would have produced if still part of a whole embryo is only true as far as the trochoblasts are concerned. Self-differentiation in molluscs is restricted to larval structures and there is no direct evidence that micromeres from which adult structures originate are able to differentiate independently. With respect to the determination of adult structures, morphogenetic determinants only bias the differentiation of competence of the cells to which they are segregated.

V. The D Quadrant as an Organizer in Development

The limited developmental capacity of isolated blastomeres and especially of the micromeres indicates that the segregation of morphogenetic potential during cleavage is in no way sufficient for the final differentiation into adult structures.

Differentiation of cells in molluscs as in other animals is dependent on interactions with other cells. It seems legitimate to suppose that in cellular interactions in molluscs the D quadrant plays a dominant role, because micromeres fail to differentiate not only in isolation, but also in the absence of a D quadrant (as is evident from the development of lobeless embryos in which no D quadrant is specified (see p. 219). Differentiation is not only dependent upon, but also strictly related to the D quadrant, which acts as an organizer in molluscan development (Cather, 1971).

A. Segregation of Morphogenetic Potential in the D Quadrant

The morphogenetic determinants present in the polar lobe at first cleavage are shunted exclusively or for the major part to the D quadrant at second cleavage. Their segregation through the blastomeres derived from the D quadrant during further cleavage has been studied by deleting macromeres (1D, 2D, etc.) at successive cleavage stages of *Ilyanassa* (Clement, 1962) and *Dentalium* (Cather and Verdonk, 1979). For *Ilyanassa,* the experiments are diagrammatically represented in Fig. 9.

The results of removing the 1D macromere in *Dentalium* are similar to the deletion of the D quadrant at the four-cell stage. The embryo has an apical tuft, but the posttrochal region is missing. No shell or foot is formed. In *Ilyanassa* ABC + 1d ($-$1D) embryos do not form adult structures and develop like lobeless embryos. Although the first quartet of micromeres is formed before the macromere 1D is removed, this quartet fails to develop eyes and tentacles, which according to cell-lineage studies originate from this quartet.

After deletion of the macromere 2D in *Dentalium,* ABC + 1d + 2d ($-$2D) embryos form a small foot with a single statocyst but do not form a shell. The absence of a shell gland or shell may be due to a partial exogastrulation of these embryos. In *Ilyanassa* the majority of these $-$2D embryos develop as lobeless embryos and do not form adult structures, but occasionally a small external shell is present. Although in both *Dentalium* and *Ilyanassa* the micromeres that normally form the shell are present, generally in $-$2D embryos they do not form an external shell. After removal of 3D in *Dentalium,* ABC + 1d + 2d + 3d embryos are structurally complete with foot, statocyst, and shell. In *Ilyanassa,* the development of the $-$3D embryos may vary from a disorganized lobeless larva without adult structures to a fairly normal veliger with eyes, shell, foot, and statocysts; only heart and intestine are absent. After removal of 4D, the embryos of both *Dentalium* and *Ilyanassa* are structurally complete and differ only in size from the controls. Consequently the macromere 4D is only nutritive in nature.

These results indicate that the morphogenetic determinants segregated to the D quadrant exert their influence mainly during the interval between the fifth and the sixth cleavage when the macromere 3D is present. Deletion of the D quadrant

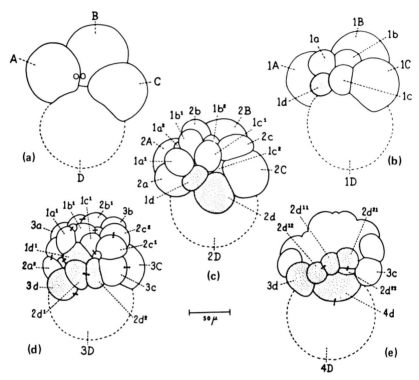

Fig. 9. (a–c) Successive cleavage stages of *Ilyanassa* when the D quadrant macromere was deleted. (**a**) D blastomere removed; ABC combination. (**b**) 1D macromere removed, ABC + 1d combination. (**c**) 2D macromere removed; ABC + 1d + 2d combination. (**d**) 3D macromere removed; ABC + 1d + 2d + 3d combination. (**e**) 4D macromere removed; ABC + 1d + 2d + 3d + 4d combination. (From Clement, 1962.)

macromere (2D) prior to fifth cleavage has the same effect as removal of the polar lobe. After the division of 3D into the macromere 4D and the mesentoblast 4d only the latter cell appears to be of some importance for the development as removal of 4d leads to a veliger without heart and intestine (Clement, 1960). Both in *Dentalium* and *Ilyanassa*, deletion of 3D after fifth cleavage may result in well-organized larvae with most of the adult structures present. By varying the time of removal of 3D from 30 min after its appearance to just before the formation of 4d, Clement (1962) discovered a strong indication that *Ilyanassa* embryos deprived of the 3D macromere in the latter part of the interval between the divisions showed a better differentiation than the embryos operated on just after fifth cleavage. Because several adult structures (e.g., eyes, tentacles, shell) observed in well-organized −3D larva, originate from micromeres that are already present before 3D is formed, one may conclude that in the interval between fifth and sixth cleavage the macromere 3D exerts an inductive action on these micromeres.

Also, in equally cleaving eggs the interval between the fifth and the sixth cleavage appears to be of special importance for the development of the embryo, because in this interval by an interaction between the animal micromeres with one of the macromeres, this macromere is determined to the dorsal macromere 3D (see Chapter 5, this volume). Several observations indicate that this macromere 3D, once determined, immediately starts to influence other blastomeres of the embryo. As mentioned already (p. 228) in *Lymnaea,* the ectosomes, RNA-rich granules present in the four macromeres, start to behave differently in 3D and the other macromeres. In 3D they are dispersed, whereas in 3A, 3B, and 3C they become more condensed and are finally extruded in the cleavage cavity (van den Biggelaar, 1976b). Up to and including the fifth cleavage, both the cleavage pattern and the chronology of the divisions in the quadrants are similar. Already at the sixth cleavage, immediately after the determination of 3D, differences in cleavage between the dorsal and ventral quadrants become evident in various tiers of micromeres. In the third quartet of micromeres the division in the dorsal cells 3c and 3d is different from the division in the ventral cells 3a and 3b. In the first quartet of micromeres the division of the dorsal cells $1d^1$, $1c^2$, and $1d^2$ is retarded with respect to the corresponding ventral cells (van den Biggelaar, 1971a,b, 1976a). These observations point to the important role that cell interactions play in the interval between the fifth and sixth cleavage of equally cleaving eggs. Cell interactions may be mediated by communication channels which, as is evident from injection of lucifer yellow CH in *Patella,* become functional after the determination of 3D (de Laat et al., 1980). Whether such communication channels between 3D and adjacent cells also exist in eggs with polar lobes is unknown.

The above data indicate that both in eggs with polar lobes and in equally cleaving eggs the 3D macromere is of paramount importance for the organization of the embryo. Even if the way in which 3D is determined is different in polar-lobe eggs (by segregation of morphogenetic determinants) from equally cleaving eggs (by an interaction of the animal micromeres with one of the macromeres) the end result is the same. Van den Biggelaar and Guerrier (1979) have suggested that similar interactions play a role in eggs with polar lobes and eggs with equal cleavage. They suppose that the morphogenetic determinants are segregated in an inactive form during the early cleavages and are activated in 3D by an interaction of this cell with the micromeres. Deletion of these micromeres, which in *Patella* suppresses the differentiation of one of the macromeres into 3D and concomitantly the origin of dorsoventral organization in the embryo, does not have the same effect in the polar-lobe forming egg of *Bithynia.* Van Dam and Verdonk (1982) showed that after this operation embryos are formed that show a clear dorsoventral axis indicated by shell and foot.

B. Influence of the D Quadrant on Morphogenesis of the Head Region

In all spiralian molluscs the head region is derived from the first quartet of micromeres, set apart at third cleavage. Cell-lineage studies have shown that the adult head structures, including eyes, tentacles, and cerebral ganglia, originate from these micromeres. Deletion experiments have corroborated this view, for loss of first quartet micromeres always affects the head region. After removal of all micromeres (1a–1d) at the eight-cell stage of *Bithynia,* the head with eyes and tentacles is missing. Thus, these micromeres can be regarded as the only cells in the embryo competent to form head structures (van Dam and Verdonk, 1982). From experiments of Clement (1967) one might even conclude that in *Ilyanassa* the fate of each micromere within the first quartet is already determined at the eight-cell stage, because removal of a specific blastomere always had the same effect on the differentiation of the head region (see p. 233). Cather et al. (1976b), after deletion of individual blastomeres in *Lymnaea* and *Bithynia,* concluded that the head region in gastropods develops a rather rigidly fixed, cell-limited system. Experiments with eggs of *Bithynia* (van Dam and Verdonk, 1982) have shown, however, that regulation in the head region may occur. After removal of 1a or 1c, the left or right eye and tentacle were missing at first. Prolonged culture of these operated embryos showed, however, that in most cases the missing eye and tentacle eventually appeared. The question therefore arises whether in *Ilyanassa* a regulation might also occur if the embryos were kept alive somewhat longer than in Clement's experiments. Also in *Ilyanassa* regulation is quite obvious in experiments, in which blastomeres were isolated at the four-cell stage (Clement, 1956). CD halves sometimes formed two eyes, although only one could be expected. Even D quarters sometimes formed a single eye, despite the fact that an eye does not belong to the D lineage in snails.

In *Lymnaea stagnalis* the cephalic plates, from which eyes, tentacles, and cerebral ganglia originate, are formed by descendants of 1a and 1c with a small contribution of 1b (Fig. 10a,c,e). The cephalic plates are separated by seven large cells of the apical plate. Ventrally and laterally the cephalic plates are bordered by the prototroch or velum, and the dorsal side of the head region is occupied by large cells of the head vesicle (Figs. 11a,c). Raven (1952) found that after lithium treatment various types of head malformations appeared, including cyclopia, synophthalmia, asymmetric monophthalmia and anophthalmia. Later studies in Raven's laboratory showed that at the trochophore stage all future head malformations are very similar (Verdonk, 1965). They are characterized by a girdle of small cells connecting the cephalic plates at the dorsal side of the apical plate (Fig. 11b). Cell-lineage studies showed that this bridge of small cells originates by an abnormal division of a single cell, $1d^{121}$, which is the basal cell of the dorsal arm of the cross. In normal development this cell divides in a plane

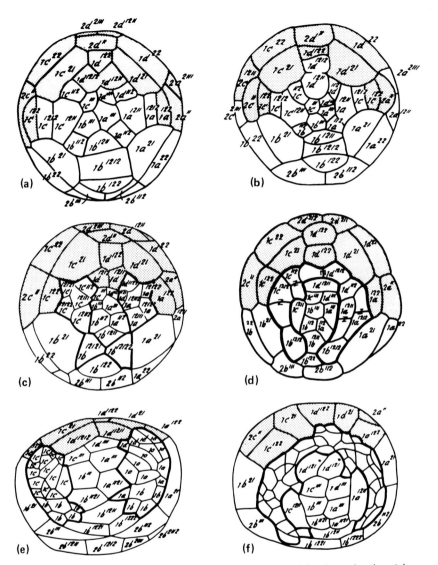

Fig. 10. (a–f) Embryos of *Lymnaea stagnalis*. The cells of the future head vesicle are stippled. (**a**) Normal embryo at the early cross stage with the basal cell of the dorsal arm (1d^{121}) dividing typically. (**b**) Lithium-treated embryo at the same stage; the basal cell of the dorsal arm is dividing at right angles to the normal direction. (**c**) Normal embryo at a late cross stage. The cells forming the cephalic plates are outlined. (**d**) Lithium-treated embryo in which the inner median cell of the dorsal arm 1d^{1212} has divided; the basal cells of the right arm 1c^{1211} and of the dorsal arm 1d^{1211} undivided. Future cephalic plate cells outlined. (**e**) Normal embryo at an early trochophore stage. Division has continued in the future cephalic plate cells (outlined). (**f**) Lithium-treated embryo at the early trochophore stage. Through continued division of 1d^{1212} the cephalic plates (outlined) are connected at the dorsal side. Supernumerary cells (1a^{1211}; 1c^{1211}; 1d^{121}*) have joined the apical plate. (From Verdonk, 1965.)

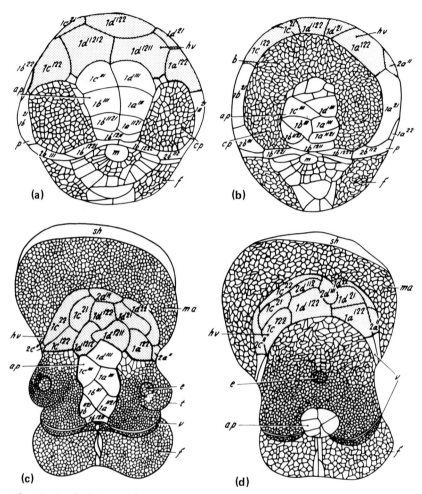

Fig. 11. (a–d). Embryos of *Lymnaea stagnalis*. (a) Normal trochophore, showing the small-celled cephalic plates separated by the seven large cells of the apical plate. (b) Lithium-treated embryo, trochophore stage. The cephalic plates are connected by a bridge (b) of small cells above the apical plate. (c) Normal veliger. From each cephalic plate an eye and tentacle have been formed. (d) Cyclopic embryo after lithium treatment. The eye is located in the bridge of small cells connecting the cephalic plates above the apical plate. ap, Apical plate; cp, cephalic plates; e, eye; f, foot; hv, head vesicle; m, mouth; ma, mantle; sh, shell; t, tentacle. (From Verdonk, 1965.)

parallel to the long axis of the dorsal arm, so that the two daughter cells ($1d^{1211}$ and $1d^{1212}$) become situated side by side (Fig. 10a). Subsequently these cells stop dividing and become the basal cells of the head vesicle (Fig. 10a,c,e and Fig. 11a,c). In lithium-treated embryos the basal cell of the dorsal arm ($1d^{121}$) divides in a plane at right angles to the long axis of the dorsal arm so that the two daughter cells become aligned in the direction of the arm (Fig. 10b). As a consequence, the configuration in the dorsal arm is exactly the same as in the lateral arms, also its further development is identical to that of the lateral arms. Both daughter cells of $1d^{121}$ may continue dividing and form the transverse bridge of small cells connecting the cephalic plates. In this way a horseshoe-shaped girdle of small cells is formed surrounding the apical plate laterally and dorsally (Fig. 10b,d,f and Fig. 11b,d). In this girdle of small cells, eyes and tentacles may be formed at various places. The most common malformations observed after lithium treatment are cyclopia and synophthalmia with one or two eyes present in the bridge of small cells connecting the cephalic plates (Fig. 11d). Apparently the basal cells of the dorsal arm, which in normal development stop dividing and differentiate into larval cells, may continue dividing under experimental conditions and differentiate into adult structures.

Whereas in the normal embryo the apical plate always consists of seven cells, in lithium-treated embryos a higher number of apical plate cells is often observed. These additional apical-plate cells appear to be derived from the basal cells of the lateral arms of the cross. In normal development these cells ($1a^{1211}$ and $1c^{1211}$) continue dividing and form a part of the cephalic plates, whereas after lithium treatment they may stop dividing and become apical-plate cells (Fig. 10f). This is exactly what, in normal development, is observed in the ventral B arm of the cross. Apparently in embryos with head malformations lithium may change a dorsal arm of the cross into a lateral arm and a lateral arm into a ventral arm.

Although treatment of *Lymnaea* eggs with lithium chloride at lower concentrations (about 2.5×10^{-5} M) causes malformations of the head in otherwise normal embryos, which may be reared to adult snails, at higher concentrations it causes exogastrulae. By comparing the development of the first quartet of micromeres in embryos with head malformations and in exogastrulae, certain similarities were observed (Verdonk, 1968c). In some exogastrulae, the cephalic-plate cells are arranged in a horseshoe-shaped area as in embryos with head malformations, whereas in other exogastrulae the cephalic-plate cells surround the apical plate completely. The apical plate always consists of more than seven cells. By following the cell lineage of these exogastrulae it appears that the cleavage pattern in the arms of the cross is the same; each arm behaves as a ventral arm in normal development. The development of the head region is radially symmetrical.

To explain the above data one may assume that the anlage of the head region in

Lymnaea originally has a radial symmetry, which is subsequently altered into a bilateral symmetry during development. Lithium then suppresses the change from radial into bilateral symmetry either partially (embryos with head malformations) or completely (exogastrulae).

The effect of lithium on *Lymnaea* resembles the effect of polar-lobe removal in *Ilyanassa* (Clement, 1952) and *Dentalium* (van Dongen and Geilenkirchen, 1974). In both molluscs, delobing results in a suppression of the bilaterally symmetrical organization. The cell lineage of the first quartet in *Dentalium* was followed far enough to support the interpretation that in lobeless embryos all arms of the cross develop as a ventral arm does in normal development, just as happens in *Lymnaea* after treatment with higher concentrations of lithium chloride. In normal embryos of *Dentalium* the dorsal arm deviates from the lateral arms and these in turn from the ventral arm. As this cannot be due to a direct segregation of polar-lobe material, the development of the first quartet in *Dentalium* is controlled by the D quadrant. One may assume that in *Lymnaea,* where no polar lobe is present, the D quadrant also controls the development of the first quartet of micromeres, which forms the head region. This is in agreement with the view that in equally cleaving gastropods the quadrants of the future embryo are not determined before the interval between the fifth and the sixth cleavage. After the determination of one of the quadrants to be the dorsal quadrant, the micromeres of the first quartet, formed at third cleavage, differentiate in agreement with their position to the dorsal quadrant.

The above data indicate that the cells that are competent to form the head region are set apart at third cleavage. Differentiation of the first quartet cells into a bilaterally symmetrical head pattern is dependent on an interaction with the D quadrant. In *Bithynia* the role of the D quadrant can be taken over by the C quadrant, according to Cather et al. (1976a). ABC embryos, obtained by removal of the D blastomere at the four-cell stage, may develop a more or less normal head with two eyes and tentacles.

C. Cellular Interactions in Shell Development

The first somatoblast 2d has generally been considered as the origin of the shell gland in molluscs since Lillie (1895) traced the shell gland to this cell in his study on the development of the Unionidae. Except for the lamellibranchs and the scaphopods (*Dentalium,* van Dongen and Geilenkirchen, 1974), where the first somatoblast is a very conspicuous cell, the lineage of 2d cannot be followed far enough to ascertain its relationship to the shell gland. Nevertheless, Lillie's view on the origin of the shell gland was generally accepted to hold for other molluscs as well (e.g., *Crepidula,* Conklin, 1897).

Cather (1967) investigated the origin of the shell in the gastropod *Ilyanassa* by marking cells with carbon particles. When 2d was marked, the particles were found in and around the invaginating shell gland, but when 2c was marked the carbon was recovered on the mantle of the right side and in the mantle cavity of the young veliger. Apparently the shell gland originates normally from 2d, whereas the progeny of 2c are incorporated into the mantle edge of the growing shell.

After removal of 2d a shell was formed in more than 90% of the embryos. The shell was always reduced in size and usually abnormal in shape. Deletion of 2c resulted in embryos with a cup-shaped shell of reduced size. If, however, both 2c and 2d were removed, a shell was always missing. Apparently the micromeres 2c and 2d are the only cells in the embryo that form the shell.

In his study on the development of isolated half and quarter embryos of *Ilyanassa*, Clement (1956) reported shell as developing only in AD, CD, and D isolates. Later (1962) he reported that A, B, C, or AB isolates could form internal shell masses. From a histological study of these embryos Cather (1967) concluded that these masses were formed by typical shell-gland cells arranged in closed sacs.

Apparently all quadrants have the capacity to form shell. One might even suppose that this capacity is situated in the second quartet of micromeres, because in normal development shell formation is always restricted to cells of this quartet. Internal shell develops only in the absence of the D blastomere, whereas in its presence external shell is formed. This indicates that the D quadrant suppresses the formation of internal shell material and promotes the formation of external shell. This view is corroborated by Clement's (1962) experiments, described on p. 237 in which he removed the D quadrant at successive stages. In −D (ABC) and −1D (ABC + 1d) an external shell was missing, but internal shell material was present. In the −2D embryos, which have the cells 2c and 2d, a small external shell was occasionally observed. In the −3D embryos various intermediate states were observed between the condition of internal shell masses and the presence of an external shell. After removal of 4D a normal shell is formed and internal shell masses are absent.

The D quadrant apparently exerts its influence on shell formation mainly during the stage when 3D is present in the embryo and in fact some of the −3D embryos show a single, well-developed shell. Therefore, one might suppose that ectoblasts obtained by removing all macromeres (3A–3D) after the first three quartets of micromeres had been formed, would be able to form external shell. Cather (1967) made 1013 of these ectoblast embryos. They all formed hollow, balllike structures with a circumferential ring of ciliated cells and with an apical tuft, but shell material either externally or internally was never observed. However, when any one of the macromeres (3A, 3B, 3C, or 3D) or the mesentoblast 4d was left in combination with this ectoblast a larva with a small and sometimes normally shaped shell may be formed. Even a combination of a mesentoblast, obtained by killing AB at the two-cell stage, followed by a deletion of 3C and

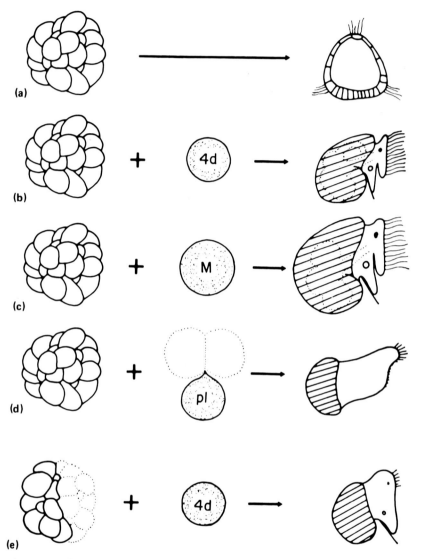

Fig. 12. A diagrammatic summary of the interactions preceding shell formation in *Il-yanassa*. (**a**) isolated ectoblast, (**b**) isolated ectoblast plus mesentoblast (4d), (**c**) isolated ecto-blast plus macromere (M), (**d**) isolated ectoblast plus polar lobe (pl), (**e**) posterior-half ectoblast plus mesentoblast (4d). (Modified after Cather, 1967.)

4D, formed an embryo with an external shell. Finally, if a detached polar lobe is held in contact with the vegetal side of an isolated ectoblast, internal or external shell may be formed. The various interactions that result in the formation of shell are shown in Fig. 12.

Raven (1952) observed that in embryos of *Lymnaea,* treated with lithium, the archenteron is often displaced and that a shell gland always forms at the place where the tip of the archenteron touches the inner side of the ectoderm. Reviewing the literature on shell-gland formation in gastropods, he noted that the shell gland always forms in the ectoderm at the site of contact between the small-celled endoderm of the archenteron and the ectoderm. He concluded that contact induction of the shell gland by the archenteron wall was a general phenomenon in gastropods. Hess (1956a,b) observed that in half embryos or exogastrulae of *Bithynia* a belated invagination of digestive gland entoderm may induce shell gland at the place of contact between endoderm and ectoderm. Clement (1962) and Collier and McCann-Collier (1964) have pointed out that an interaction between the archenteron and the ectoderm cannot occur in animals with epibolic gastrulation such as *Ilyanassa* because the shell gland forms prior to the differentation of endoderm in the proximity of the shell gland. In these species, nevertheless, an interaction comparable to the induction exerted by the endoderm in *Lymnaea* and *Bithynia* is needed, as isolated ectoblasts in Cather's experiments do not form shell. This interaction seems not to be specific for any particular quadrant, as any macromere or even an isolated polar lobe have the ability to induce shell formation.

From the experimental data on shell formation in *Ilyanassa* the following conclusions can be drawn.

1. All quadrants have the potency to form shell material.
2. In normal development only micromeres 2c and 2d form shell.
3. This restriction is effected by the D quadrant, due to the segregation of the polar lobe to this quadrant.
4. In order to function as shell-forming cells, the descendants of 2c and 2d need an inductive influence, which can be exerted by any macromere, the mesentoblast, or the archenteron.

One may suppose that this scheme is also applicable to other molluscs. In *Dentalium,* 2d is the only cell competent to form shell and in *Bithynia* the C quadrant can replace the D quadrant.

VI. Conclusions

In spiralian molluscs morphogenetic determinants, situated at the vegetal pole of the uncleaved eggs, play an important role in establishing the primordial organization of the future embryo. These morphogenetic determinants may be localized in special cytoplasmic structures or they may be bound to the cortex of the vegetal region. In this region, unique surface structures are observed, consisting of ridges, folds, or a special type of microvilli. Their relationship to the localization of morphogenetic determinants is still unclear. The biochemical nature and the mode of action of morphogenetic determinants at a molecular level is discussed in Chapter 7.

In unequally cleaving eggs, with or without the formation of a polar lobe, the morphogenetic determinants, present at the vegetal region of the egg, are segregated exclusively or for the main part into one of the blastomeres at the four-cell stage, and in this way determine it to form the dorsal quadrant (D) of the embryo. In equally cleaving eggs the vegetal region is divided among the four blastomeres of the four-cell stage. The D quadrant is determined after the fifth cleavage by an interaction between the animal micromeres and one of the vegetal macromeres.

In all molluscs the D quadrant, once determined, acts as an organizer, which by cellular interactions controls the development of the other quadrants. Relatively little is known about the development of cellular contacts, which must play an important role in these cell interactions.

Segregation of morphogenetic potential in the successive quartets of micromeres leads to a commitment of these cells to the formation of special structures. These determinants specify only the competence of these cells, they do not fix the terminal differentiation. Differentiation is dependent on and related to the D quadrant. The molluscan embryo is in no way a mosaic of self-differentiating blastomeres.

References

Atkinson, J. W. (1971). Organogenesis in normal and lobeless embryos of the marine prosobranch gastropod *Ilyanassa obsoleta*. *J. Morphol.* **133**, 339–352.

Cather, J. N. (1967). Cellular interactions in the development of the shell gland of the gastropod, *Ilyanassa*. *J. Exp. Zool.* **166**, 205–224.

Cather, J. N. (1971). Cellular interactions in the regulation of development in annelids and molluscs. *Adv. Morphog.* **9**, 67–125.

Cather, J. N., and Verdonk, N. H. (1974). The development of *Bithynia tentaculata* (Prosobranchia, Gastropoda) after removal of the polar lobe. *J. Embryol. Exp. Morphol.* **31**, 415–422.

Cather, J. N., and Verdonk, N. H. (1979). Development of *Dentalium* following removal of D-quadrant blastomeres at successive cleavage stages. *Wilhelm Roux's Arch. Dev. Biol.* **187**, 355–366.

Cather, J. N.,Verdonk, N. H., and Dohmen, M. R. (1976a). Role of the vegetal body in the regulation of development in *Bithynia tentaculata* (Prosobranchia, Gastropoda). *Am. Zool.* **16**, 455–468.

Cather, J. N., Verdonk, N. H., and Zwaan, G. (1976b). Cellular interactions in the early development of the gastropod eye, as determined by deletion experiments. *Malacol. Rev.* **9**, 77–84.

Clement, A. C. (1952). Experimental studies on germinal localization in *Ilyanassa*. I. The role of the polar lobe in determination of the cleavage pattern and its influence in later development. *J. Exp. Zool.* **121,** 593–626.

Clement, A. C. (1956). Experimental studies on germinal localization in *Ilyanassa*. II. The development of isolated blastomeres. *J. Exp. Zool.* **132,** 427–446.

Clement, A. C. (1960). Development of the *Ilyanassa* embryo after removal of the mesentoblast cell. *Biol. Bull. (Woods Hole, Mass.)* **119,** 310.

Clement, A. C. (1962). Development of *Ilyanassa* following removal of the D-macromere at successive cleavage stages. *J. Exp. Zool.* **149,** 193–216.

Clement, A. C. (1963). Effects of micromere deletion on development in *Ilyanassa*. *Biol. Bull. (Woods Hole, Mass.)* **125,** 375.

Clement, A. C. (1967). The embryonic value of the micromeres in *Ilyanassa obsoleta,* as determined by deletion experiment. I. The first quartet cells. *J. Exp. Zool.* **166,** 77–88.

Clement, A. C. (1968). Development of the vegetal half of the *Ilyanassa* egg after removal of most of the yolk by centrigual force, compared with the development of animal halves of similar visible composition. *Dev. Biol.* **17,** 165–186.

Clement, A. C. (1971). *Ilyanassa. In* "Experimental Embryology of Marine and Freshwater Invertebrates" (G. Reverberi, ed.), pp. 188–214. North-Holland Publ., Amsterdam.

Clement, A. C. (1976). Cell determination and organogenesis in molluscan development: A reappraisal based on deletion experiments in *Ilyanassa*. *Am. Zool.* **16,** 447–453.

Collier, J. R., and McCann-Collier, M. (1964). Shell gland formation in the *Ilyanassa* embryo. *Exp. Cell Res.* **34,** 512–514.

Conklin, E. G. (1897). The embryology of *Crepidula*. *J. Morphol.* **13,** 1–226.

Crampton, H. E. (1896). Experimental studies on gastropod development. *Arch. Entwicklungs mech. org.* **3,** 1–19.

Davidson, E. H., and Britten, R. J. (1971). Note on the control of gene expression during development. *J. Theor. Biol.* **32,** 123–130.

de Laat, S. W., Tertoolen, L. G. J., Dorresteijn, A. W. C., and van den Biggelaar, J. A. M. (1980). Intercellular communication patterns are involved in cell determination in early molluscan development. *Nature (London)* **287,** 546–548.

Dohmen, M. R., and Lok, D. (1975). The ultrastructure of the polar lobe of *Crepidula fornicata* (Gastropoda, Prosobranchia). *J. Embryol. Exp. Morphol.* **34,** 419–428.

Dohmen, M. R., and van der Mast, J. M. A. (1978). Electron microscopical study of RNA-containing cytoplasmic localizations and intercellular contacts in early cleavage stages of eggs of *Lymnaea stagnalis* (Gastropoda, Prosobranchia). *Proc. K. Ned. Akad. Wet., Ser. C* **81,** 403–414.

Dohmen, M. R., and Verdonk, N. H. (1979). The ultrastructure and role of the polar lobe in development of molluscs. *In* "Determinants of Spatial Organization" (S. Subtelny and I. R. Konigsberg, eds.), pp. 3–27. Academic Press, New York.

Freeman, G. (1976a). The role of cleavage in the localization of developmental potential in the ctenophore *Mnemiopsis leidyi*. *Dev. Biol.* **49,** 143–177.

Freeman, G. (1976b). The effects of altering the position of cleavage planes on the process of localization of developmental potential in ctenophores. *Dev. Biol.* **51,** 332–337.

Freeman, G. (1978). The role of asters in the localization of factors that specify the apical tuft and the gut of the nemertine *Cerebratulus lacteus*. *J. Exp. Zool.* **206,** 81–108.

Freeman, G. (1979). The multiple roles which cell division can play in the localization of developmental potential. *In* "Determinants of Spatial Organization" (S. Subtelny and I. R. Konigsberg, eds.), pp. 53–76. Academic Press, New York.

Geilenkirchen, W. L. M., Vedonk, N. H., and Timmermans, L. P. M. (1970). Experimental studies on morphogenetic factors localized in the first and second polar lobe of *Dentalium* eggs. *J. Embryol. Exp. Morphol.* **23**, 237–243.

Guerrier, P. (1970a). Les caractères de la segmentation et la détermination de la polarité dorso-ventrale dans le développement de quelques Spiralia. I. Les formes à premier clivage égal. *J. Embryol. Exp. Morphol.* **23**, 611–637.

Guerrier, P. (1970b). Les caractères de la segmentation et la détermination de la polarité dorso-ventrale dans le développement de quelques Spiralia. III. *Pholas dactylus* et *Spisula subtruncata* (Mollusques Lamellibranches) *J. Embryol. Exp. Morphol.* **23**, 667–692.

Guerrier, P., van den Biggelaar, J. A. M., van Dongen, C. A. M., and Verdonk, N. H. (1978). Significance of the polar lobe for the determination of dorsoventral polarity in *Dentalium vulgare* (da Costa). *Dev. Biol.* **63**, 233–242.

Hess, O. (1956a). Die Entwicklung von Halbkeimen bei dem Süsswasser-Prosobranchier *Bithynia tentaculata* L. *Wilhelm Roux's Arch. Entwicklungsmech. org.* **148**, 336–361.

Hess, O. (1956b). Die Entwicklung von Exogastrulakeimen bei dem Süsswasser-Prosobranchier *Bithynia tentaculata* L. *Wilhelm Roux's Arch. Entwicklungsmech. Org.* **148**, 474–488.

Hörstadius, S. (1937). Investigations as to the localization of the micromere-, the skeleton- and the entoderm-forming material in the unfertilized egg of *Arbacia punctulata*. *Biol. Bull. (Woods Hole, Mass.)* **73**, 295–316.

Jaffe, L. A., and Guerrier, P. (1981). Localization of electrical excitability in the early embryo of *Dentalium*. *Dev. Biol.* **83**, 370–373.

Jaffe, L. F. (1966). Electrical currents through the developing *Fucus* egg. *Proc. Natl. Acad. Sci. U.S.A.* **56**, 1102–1109.

Lillie, F. R. (1895). The embryology of the Unionidae. *J. Morphol.* **10**, 1–100.

Minganti, A. (1950). Acidi nucleici e fosfatasi nello sviluppo della *Limnaea*. *Riv. Biol.* **42**, 295–319.

Morgan, T. H. (1933). The formation of the antipolar lobe in *Ilyanassa*. *J. Exp. Zool.* **64**, 433–467.

Morgan, T. H. (1935). The separation of the egg of Ilyanassa into two parts by centrifuging. *Biol. Bull. (Woods Hole, Mass.)* **68**, 280–295.

Morrill, J. B. (1964). The development of fragments of *Limnaea stagnalis* eggs centrifuged before second cleavage. *Acta Embryol. Morphol. Exp.* **7**, 5–20.

Morrill, J. B., Blair, C. A., and Larsen, W. J. (1973). Regulative development in the pulmonate gastropod *Lymnaea palustris* as determined by blastomere deletion experiments. *J. Exp. Zool.* **183**, 47–56.

Nuccitelli, R. (1978). Ooplasmic segregation and secretion in the *Pelvetia* egg is accompanied by a membrane-generated electrical current. *Dev. Biol.* **62**, 13–33.

Rattenbury, J. C., and Berg, W. E. (1954). Embryonic segregation during early development of *Mytilus edulis*. *J. Morphol.* **95**, 393–414.

Raven, C. P. (1952). Morphogenesis in *Limnaea stagnalis* and its disturbance by lithium. *J. Exp. Zool.* **121**, 1–77.

Raven, C. P. (1963). The nature and origin of the cortical morphogenetic field in *Limnaea*. *Dev. Biol.* **7**, 130–143.

Raven, C. P. (1967). The distribution of special cytoplasmic differentiations of the egg during early cleavage in *Limnaea stagnalis*. *Dev. Biol.* **16**, 407–437.

Raven, C. P. (1970). The cortical and subcortical cytoplasm of the *Lymnaea* egg. *Int. Rev. Cytol.* **28**, 1–44.

Raven, C. P. (1974). Further observations on the distribution of cytoplasmic substances among the cleavage cells in *Lymnaea stagnalis*. *J. Embryol. Exp. Morphol.* **31**, 37–59.

Raven, C. P. (1976). Morphogenetic analysis of spiralian development. *Am. Zool.* **16**, 395–403.

Tyler, A. (1930). Experimental production of double embryos in annelids and mollusks. *J. Exp. Zool.* **57**, 347–407.

van Dam, W. I., and Verdonk, N. H. (1982). The morphogenetic significance of the first quartet micromeres for the development of the snail *Bithynia* tentaculata. *Wilhelm Roux's Arch. Dev. Biol.* **191**, 112–118.

van den Biggelaar, J. A. M. (1971a). Timing of the phases of the cell cycle during the period of asynchronous division up to the 49-cell stage in *Lymnaea*. *J. Embryol. Exp. Morphol.* **26**, 367–391.

van den Biggelaar, J. A. M. (1971b). Development of division asynchrony and bilateral symmetry in the first quartet of micromeres in eggs of *Lymnaea*. *J. Embryol. Exp. Morphol.* **26**, 393–399.

van den Biggelaar, J. A. M. (1976a). Development of dorsoventral polarity preceding the formation of the mesentoblast in *Lymnaea stagnalis*. *Proc. K. Ned. Akad. Wet., Ser. C* **79**, 112–126.

van den Biggelaar, J. A. M. (1976b). The fate of maternal RNA containing ectosomes in relation to the appearance of dorsoventrality in the pond snail, *Lymnaea stagnalis. Proc. K. Ned. Akad. Wet., Ser. C* **79**, 421–426.

van den Biggelaar, J. A. M., and Guerrier, P. (1979). Dorsoventral polarity and mesentoblast determination as concomitant results of cellular interactions in the mollusk *Patella vulgata. Dev. Biol.* **68**, 462–471.

van den Biggelaar, J. A. M., Dorresteijn, A. W. C., de Laat, S. W., and Bluemink, J. G. (1981). The role of topographical factors in cell interaction and determination of cell lines in molluscan development. *In* "International Cell Biology 1980–1981" (H. G. Schweiger, ed.), pp. 526–538. Springer-Verlag, Berlin and New York.

van Dongen, C. A. M. (1976a). The development of *Dentalium* with special reference to the significance of the polar lobe. V and VI. Differentiation of the cell pattern in lobeless embryos of *Dentalium vulgare* (da Costa) during late larval development. *Proc. K. Ned. Akad. Wet., Ser. C* **79**, 245–266.

van Dongen, C. A. M. (1976b). The development of *Dentalium* with special reference to the significance of the polar lobe. VII. Organogenesis and histogenesis in lobeless embryos of *Dentalium vulgare* (da Costa) as compared to normal development. *Proc. K. Ned. Akad. Wet., Ser. C* **79**, 454–465.

van Dongen, C. A. M. (1977). Mesoderm formation during normal development of *Dentalium dentale. Proc. K. Ned. Akad. Wet., Ser. C* **80**, 372–376.

van Dongen, C. A. M., and Geilenkirchen, W. L. N. (1974). The development of *Dentalium* with special reference to the significance of the polar lobe. I, II and III. Division chronology and development of the cell pattern in *Dentalium dentale* (Scaphoda). *Proc. K. Ned. Akad. Wet., Ser. C* **77**, 57–100.

van Dongen, C. A. M., and Geilenkirchen, W. L. M. (1975). The development of *Dentalium* with special reference to the significance of the polar lobe. IV. Division chronology and development of the cell pattern in *Dentalium dentale* after removal of the polar lobe at first cleavage. *Proc. K. Ned. Akad. Wet., Ser. C* **78**, 358–375.

Verdonk, N. H. (1965). Morphogenesis of the head region in *Limnaea stagnalis* L. Ph.D. Thesis, Rijksuniversiteit Utrecht.

Verdonk, N. H. (1968a). The effect of removing the polar lobe in centrifuged eggs of *Dentalium. J. Embryol. Exp. Morphol.* **19**, 33–42.

Verdonk, N. H. (1968b). The relation of the two blastomeres to the polar lobe in *Dentalium. J. Embryol. Exp. Morphol.* **20**, 101–105.

Verdonk, N. H. (1968c). The determination of bilateral symmetry in the head region of *Limnaea stagnalis. Acta Embrhol. Morphol. Exp.* **10**, 221–227.

Verdonk, N. H. (1979). Symmetry and asymmetry in the embryonic development of molluscs. *In* "Pathways in Malacology" (S. van der Spoel, A. C. van Bruggen, and J. Lever, eds.), pp. 25–45. Bohn, Scheltema & Holkema, Utrecht.

Verdonk, N. H., and Cather, J. N. (1973). The development of isolated blastomeres in *Bithynia tentaculata* (Prosobranchia, Gastropoda). *J. Exp. Zool.* **186**, 47–62.

Verdonk, N. H., Geilenkirchen, W. L. M., and Timmermans, L. P. M. (1971). The localization of morphogenetic factors in uncleaved eggs of *Dentalium*. *J. Embryol. Exp. Morphol.* **25**, 57–63.

Wierzejski, A. (1905). Embryologie von *Physa fontinalis* L. *Z. Wiss. Zool.* **83**, 502–706.

Wilson, E. B. (1904a). Experimental studies on germinal localization. I. The germ regions in the egg of *Dentalium*. *J. Exp. Zool.* **1**, 1–72.

Wilson, E. B. (1904b). Experimental studies on germinal localization. II. Experiments on the cleavage-mosaic in *Patella* and *Dentalium*. *J. Exp. Zool.* **1**, 197–268.

7

The Biochemistry of Molluscan Development[1]

J. R. COLLIER

Biology Department
Brooklyn College of the City University of New York
Brooklyn, New York

I. Introduction

The molluscan embryo is exemplary of those embryos that have a spiral pattern of cleavage, which is intimately associated with the determinative segregation of the egg cytoplasm into the early blastomeres. The biochemistry of these eggs is of special importance to studies of the role of egg cytoplasm in establishing embryonic determination.

The biochemistry of molluscan embryos has been most extensively studied among the Protostomia and provides another basis, in addition to their basic body plans and modes of embryonic development, for comparison with the Deuterostomia. In addition to their different strategies for formation of the enteron

[1]The preparation of this chapter and the author's original data included in it were supported by grant 13558 from the City University of New York PSC–BHE Research Award Program and by grant PCM78-02773 from the National Science Foundation.

THE MOLLUSCA, VOL. 3
Development

and origin of the mesoderm, the role of cytoplasmic localization in determination is more prominent among protostomian than deuterostomian embryos. Is the early evolutionary divergence of the Protostomia and Deuterostomia reflected in the chemistry of their development, particularly, as might be expected from the prevalence of determinate cleavage among the Protostomia, in the modes of regulation of gene expression employed during embryogenesis?

In this chapter the available information on the biochemistry of molluscan development is reviewed, brought to focus on some general problems of developmental biology, and used as a basis for comparison with the developmental chemistry of the Deuterostomia. Earlier reviews devoted partially or entirely to the developmental biochemistry of molluscs include those of Collier (1965, 1966), Brachet (1950, 1967), Raven (1972), Brahmachary (1973), Collier (1976a), Kidder (1976b), and Morrill et al. (1976).

II. The Molluscan Genome

Hinegardner (1974) measured the DNA content of 110 molluscan species and found that the haploid genome size was symmetrically distributed around 9.6×10^{11} daltons with some tailing toward higher values. This genome size is very close to the median genome size of 1.2×10^{12} daltons for the entire animal kingdom. Hinegardner established a positive correlation between DNA content, chromosome number, and body size. He also emphasized that among the bivalves, the more generalized species such as the mussels had a higher DNA content than the more specialized pectens and oysters. The even higher amounts of DNA among the neogastropods, which had an average haploid DNA content of 1.9×10^{12} daltons, was correlated by Hinegardner with the recent origin of this group and its evolutionary radiation.

Table I shows the genome sizes of those molluscs frequently used in developmental studies and for which there are data on the melting temperature, density, and base composition. Among the species listed in Table I, the DNA of the bivalves has a lower $G + C$ content than the single gastropod represented, however, it is certainly not clear from the limited data that this is a general trend among these organisms. Satellite DNA has been detected only in *Mulinia* (Kidder, 1976a); this satellite represents 3–5% of the genome and, on the basis of hybridization experiments, Kidder (1976b) suggests that it probably contains the rDNA sequences. Oocyte DNA, of which the excess of the DNA content of the tetraploid oocyte nucleus is mitochondrial DNA, has been reported for only two species, a bivalve and a gastropod. The gastropod egg contains disproportionately more mitochondrial DNA by volume and genome size than the bivalve egg.

Table II lists kinetic components, each characterized by its rate constant, abundance, reiteration frequency, and complexity, of several molluscan gen-

TABLE I

DNA Content and Base Composition of the Molluscan Genome

Organism	Haploid DNA content		T_m, °C (G + C)	Density, g/cm³ (G + C)	References
	Daltons	Base pairs			
Bivalves					
Crassostrea gigas	7.6×10^{11}	1.2×10^9	82.5 (32.2)	1.694 (33.6)	McLean and Whitely (1973)
C. virginica	4.2×10^{11}	6.3×10^8	84.0 (36.6)	—	Goldberg et al. (1975)
Mytilus edulis	9.6×10^{11}	1.5×10^9	—	—	Hinegardner (1974)
Mulinia lateralis	5.0×10^{11}	7.7×10^8	—	1.693 (34.0)	Kidder (1976a)
Spisula solidissima	7.2×10^{11}	1.1×10^9	—	—	Hinegardner (1974)
Gastropods					
Aplysia californica	1.1×10^{12}	1.7×10^9	—	—	Angerer et al. (1975)
Ilyanassa obsoleta	1.9×10^{12}	2.9×10^9	86.0 (40.8)	1.701 (42.0)	Collier and McCann-Collier (1962) Collier and Yuyama (1969)
Lymnaea stagnalis	1.2×10^{12}	1.8×10^9	—	—	van den Biggelaar (1971b)

omes. More kinetic components and higher levels of reiteration of DNA sequences were found among gastropod than bivalve DNAs, especially when optical methods were used to follow the reassociation kinetics (Collier and Tucci, 1980). [Compare these with the fewer kinetic components measured earlier by Davidson et al. (1971) using hydroxyapatite chromatography.] Additional data may show a relationship between reiteration frequency of DNA sequences and the evolutionary radiation of the gastropods.

Among those molluscs in which the organization of single copy and repeated sequences of DNA have been investigated, *Spisula* (Goldberg et al., 1975), *Crassostrea* (Goldberg et al., 1975), and *Aplysia* (Angerer et al., 1975), it was found that 70, 75, and 66%, respectively, of the single copy DNA was interspersed with 2–400 nucleotides of repeated DNA sequences. This sequence organization of the molluscan genome is similar to the *Xenopus* model of organization (Davidson et al., 1975) and dissimilar to the *Drosophila* genome (Manning et al., 1975).

Genomic redundancy of ribosomal genes has been found in the three species of bivalves and the single gastropod that has been investigated (Table III). The level of redundancy of rRNA genes among these molluscs is linearly correlated with

TABLE II

Kinetic Components of the Molluscan Genome

Organism	Reassociation rate constant, k_2 (l/mol/s)	Proportion of kinetic component in genome (%)	Reiteration frequency	Complexity (ntp)	References
Bivalves					
Crassostrea virginica	1.3×10^{-3}	38	1.0	2.5×10^8	Goldberg et al. (1975)
	5.0×10^{-2}	53	40	8.7×10^8	
	100				
Crassostrea gigas	1.2×10^{-3}	70	1.0		McLean and Whiteley (1973)
	2.5×10^{-2}	23	20		
	1.0	7	830		
Spisula solidissima	7.7×10^{-4}	32	1.0	3.7×10^8	Goldberg et al. (1975)
	2.5×10^{-2}	40	30	1.6×10^7	
	2.8×10	18	350	6.0×10^5	
Gastropods					
Aplysia californica	5.6×10^{-4}	50	1.0	9.0×10^8	Angerer et al. (1975)
	4.8×10^{-2}	20	85	3.7×10^6	
	2.5	25	4,600	5.1×10^4	
	4.0×10^3	10	7×10^6		
		7			
Ilyanassa obsoleta	2.2×10^{-4}	46	1.0	1.3×10^9	Collier and Tucci (1980)
	2.1×10^{-3}	16	10	4.4×10^8	
	1.2×10^{-2}	8	55	2.5×10^8	
	5.7×10^{-2}	9	260	2.5×10^8	
	2.0×10^{-1}	4	910	7.8×10^{10}	
	1.0	6	4,500	1.1×10^{11}	
	5.0	2	23,000	3.0×10^{10}	
		10		1.8×10^{11}	

TABLE III

Redundancy and Amplification of Ribosomal Genes in the Molluscan Genome

Organism	Number of rRNA genes per haploid genome in somatic cells	Percentage of the genome	Times amplified in oocytes	References
Bivalves				
Mulinia lateralis	120	0.1	2	Kidder (1976a)
Mytilus edulis	220	0.10	—	Vincent (see Collier, 1971)
Spisula solidissima	195	0.12	—	Vincent (see Collier, 1971)
			5	Brown and David (1969)
Gastropods				
Ilyanassa obsoleta	395	0.09	—	Collier (1971); see also Kidder (1976a)

genomic complexity, a correlation established for a larger group or organisms by Birnstiel et al. (1971).

The extent of amplification of rDNA in the molluscan oocyte is somewhat less clear than is the case for rDNA reiteration. The only detailed study is Kidder's (1976a) investigation of the ribosomal cistrons in the coot clam *Mulinia lateralis* in which he found a twofold amplification in the oocyte. Brown and Dawid (1968) estimated, from the ratio of rDNA to 4-S DNA in eggs and sperm, that there is at least a fivefold amplification of rDNA in the egg of the surf clam *Spisula solidissima*. Kidder (1976b) considers the biological significance of such low levels of amplification and suggests that the production of a large amount of ribosomal RNA may not be necessary in the molluscan oocyte—the low level of amplification in *Mulinia* and *Spisula* may have arisen and persisted in the absence of selective pressure for limiting the level of amplification.

III. Ribosomal and Transfer RNA Synthesis

Koser and Collier (1971, 1972) have determined, by polyacrylamide-gel electrophoresis, molecular weights of 1.37×10^6 and 0.71×10^6 for the 26- and 18-S ribosomal RNA of *Ilyanassa obsoleta* and 2.4 to 2.5×10^6 for the precursor to these rRNAs. The 26- and 18-S rRNA of *Mytilus edulis* and *Spisula solidissima* were also measured and found to have the same molecular weights (J. R. Collier and R. B. Koser, unpublished observations). The 26-S rRNA of *Ilyanassa* was labile to heat (60°C), urea, and Mg^{2+} deprivation (Koser and Collier, 1971); the rRNAs of both *Mytilus* and *Spisula* were also heat labile (J. R. Collier and R. B. Koser, unpublished observations). This lability of 26-S rRNA accounts for the degradation of 26-S rRNA frequently seen in sucrose gradients of RNA from *Ilyanassa* (Newrock and Raff, 1975; Naus and Kidder, 1982). The base composition of *Ilyanassa* rRNA has also been determined (Collier, 1976a).

Table IV shows the earliest stages reported for the detection of rRNA synthesis among molluscan embryos. In most cases rRNA synthesis may have begun much earlier than indicated but went undetected because of low levels of synthesis and/ or cofractionation with other RNAs. This is clearly possible for *Spisula* (Firtel and Monroy, 1970), in which a heterodisperse pattern of radioactivity was associated on a sucrose gradient with the 28-S and 18-S RNAs extracted from polysomes of four-cell embryos pulsed with [³H]uridine. Similarly, a heterodisperse distribution of tritiated RNA is seen in sucrose-gradient profiles of RNA from early and late cleavage embryos of *Mulinia* (Kidder, 1972a). The report of rRNA synthesis in the four-cell embryo of *Ilyanassa* (Collier, 1976a) is also subject to the artifact of co-elution with RNAs other than rRNA.

Van den Biggelaar (1971a, b, c) pointed out a relationship between the length of the cell cycle and the amount of RNA synthesis in *Lymnaea*, and Kidder

TABLE IV

The Beginning of RNA Synthesis during Development of Molluscan Embryos

Organism	Ribosomal RNA		Transfer RNA	Heterogeneous RNA	Messenger RNA	References
	5S	28 and 18S				
Bivalves						
Mulinia lateralis	Cleavage	Gastrula	Middle to late cleavage	Cleavage	2-cell	Kidder (1972a,b, 1976b)
Spisula solidissima		Trochophore			2-cell	Firtel and Monroy (1970)
Gastropods						
Lymnaea stagnalis		24-cell		16-cell		van den Biggelaar (1971a)
Acmaea scutum		—		16-cell		Karp (1973)
Helix aspersa		Early blastula		8-cell		Dauwalder (1963)
Ilyanassa obsoleta	4-cell	4-cell Late cleavage Late cleavage	4-cell	4-cell	4- to 16-cell	Collier (1976a) Newrock and Raff (1975) Naus and Kidder (1982)

(1976b) has expanded this topic relative to rRNA synthesis. Kidder (1976b) suggested that a rapid cell cycle does not permit enough time for the formation of functional nucleoli, which are probably required for rRNA synthesis. This reasonable suggestion is not, however, entirely supported by the evidence cited by Kidder. For example, Kidder (1976b, Table 3) cites the data of Sconzo et al. (1970) for the initiation of rRNA synthesis in the mesenchyme blastula of the sea urchin, although he later discusses the observations of Emerson and Humphreys (1971), which support the synthesis of rRNA during early cleavage. The uncertainties in regard to Emerson and Humphreys's (1971) interpretation of the status of the nucleolus during early cleavge in the sea urchin embryo, as pointed out by Kidder (1976b), do not alter the detailed biochemical evidence presented by Emerson and Humphreys (1970) for the synthesis of rRNA during early cleavage of the sea urchin embryo. The hypothesis that rRNA is synthesized continuously during sea urchin embryogenesis but at an undetectably low level during early cleavage (Emerson and Humphreys, 1970) has been confirmed by Surrey et al. (1979) who demonstrated the synthesis of rRNA by the early blastula in two species of sea urchins; in one species, *Strongylocentrotus purpuratus*, cells of both the early blastula and the gastrula synthesized 40 molecules of ribosomal precursor RNA per minute. This observation shows that there is no enhancement of rRNA transcription at gastrulation in this species and that the measured level of transcription would not be detectable during early cleavage when each embryo consists of only a few cells. Further, the times for the initiation of rRNA synthesis for *Lymnaea, Acmaea, Oncopeltus,* and the rabbit (cited in Table 3, Kidder, 1976b) do not support a correlation between the length of the cell cycle and the onset of rRNA synthesis.

Naus and Kidder (1982) have completed a detailed study of the ultrastructure and cytochemistry of fibrillar nucleoli, which they designate as nucleolar-like bodies (NLBs), that appear during early cleavage of the *Ilyanassa* egg. They have compared these NLBs to mature nucleoli and noted the absence of a granular component and detectable incorporation of uridine into the former. By late cleavage these NLBs become associated with chromatin, develop a granular component, and incorporate tritiated uridine. Naus and Kidder (1982) favor their suggestion that the incorporation of uridine into nucleoli during later cleavage (Fig. 1a,b) is the result of the activation of rDNA transcription, although they recognize the possibility that a low level of rRNA synthesis may occur throughout cleavage. They suggest that such a low level of synthesis is not associated with NLBs and that the rRNA precursor rapidly decays instead of being processed into 18- and 28-S ribosomal RNAs.

The electron microscopic autoradiographs of Naus and Kidder (1982) (Fig. 1a,b) and the number of nucleoli observed appear to support the suggestion that rDNA transcription in *Ilyanassa* is activated during late cleavage; however, it is possible that low levels of rRNA synthesis could go undetected in the ultrathin sections required for ultrastructural autoradiographs. Schmekel and Fioroni

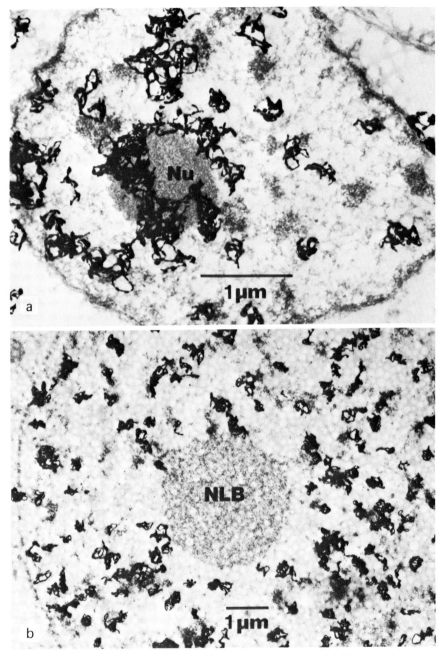

Fig. 1. Ribosomal RNA synthesis in *Ilyanassa* embryogenesis. (**a**) Electron microscope autoradiograph of a postgastrula embryo labeled for 2 h with [5-³H]uridine. Both nucleolus (Nu) and nucleoplasm are labeled. (**b**) Electron microscope autoradiograph of a midcleavage embryo (12-h posttrefoil) labeled for 2 h. The nucleolar-like body (NLB) is unlabeled, despite the heavy labeling of the nucleoplasm. (From Naus and Kidder, 1982 with permission.)

(1976) reported the presence of mature nucleoli, that is, nucleoli with a granular component, in the eight-cell embryo of *Nassarius reticulatus,* an observation which, although in another species, suggests an earlier appearance of functional nucleoli than seen by Naus and Kidder (1982). It is important that this question be resolved because erroneous implications about the activation of rDNA transcription would be misleading in regard to the regulation of transcription. The answer to this question may be that a low level of rRNA synthesis occurs at all stages of development but the rate of rRNA synthesis is modulated at various stages of embryogenesis. Indeed, recalling the observations of Humphreys (1973) on the difference in the rate of rRNA synthesis in fed and unfed sea urchin larvae, it may be that the rate of rRNA synthesis is modulated in response to cellular requirements as well as by a developmental program. The more sensitive and specific biochemical assay for the methylation of rRNA used by Surrey et al. (1979) may be useful in resolving this problem. Also a distinction must be made between rate changes in transcription and in rRNA processing as Surrey et al. found just before gastrulation.

During early cleavage of the annelid embryo, *Ophryotrocha labronica,* Emanuelsson (1973) described the regular occurence of karyomeres (chromosomal vesicles) at all cleavages up to the 16-cell stage. Karyomeres appear as separate units, which later coalesce to form a nucleus, that all contain one or more NLBs. Nucleoli with granular components that incorporate uridine do not appear until the 32-cell stage. The NLBs of *Ophryotrocha* are pronase sensitive and incorporate [³H]leucine and [³H]myoinositol. Emanuelsson suggested that the NLBs are probably composed of lipoproteins and that their function may be to provide reserve membranous material for subsequent divisions. These observations on an annelid embryo may be pertinent to our understanding the role of similar NLBs in molluscan embryos. These NLBs may be totally unrelated to nucleoli. (Emanuelsson poses the counterpoint that they are nucleoli whose rRNA synthesis is repressed.) Is rRNA transcribed from the nucleolar organizing region(s) of the chromosome(s) during early cleavage and only later from rDNA in the nucleolus? Do the NLBs described by Emanuelsson (1973) and Naus and Kidder (1982) contain DNA?

Ribosomal RNA synthesis has been shown to continue until the trochophore stage in *Spisula* (Firtel and Monroy, 1970) and *Mulinia* (Kidder, 1972a), and until the veliger stage in *Ilyanassa* (Collier, 1976a). In *Ilyanassa,* uridine incorporation into rRNA has been reported as early as the four-cell stage (Collier, 1976a) and by late cleavage (Newrock and Raff, 1975; Naus and Kidder, 1982), however, the first major increase in RNA accumulation that is attributable to the synthesis of rRNA occurs only after three days of development (Collier, 1975a). Thus, the *Ilyanassa* egg has sufficient ribosomes to support development for the first three days, which is one day beyond gastrulation. Mirkes (1972) has reported a significant increase in polyribosomes as early as the eight-cell stage in

Ilyanassa and a fourfold increase by the time a veliger larva has been formed. The occurrence of an initial increase in polyribosomes as early as the third cleavage, a developmental age that precedes a major synthesis of ribosomes, indicates the existence of preformed monosomes and their mobilization into polyribosomes by mRNAs either synthesized or activated during early cleavage.

Data for the synthesis of 5-S ribosomal RNA is available only for *Mulinia* (Kidder, 1972a) and *Ilyanassa* (Collier, 1976a). In *Mulinia*, radioactive label probably accumulates in 5-S rRNA during cleavage and is not in synchrony with the onset of 18- and 28-S rRNA synthesis (Kidder, 1972a, 1976a,b). In *Ilyanassa*, the synthesis of 5-S RNA and the high-molecular-weight rRNAs begins during early cleavage (Collier, 1976a); there is a 1:1 molar ratio of 5-S RNA to high-molecular-weight rRNAs, which is maintained throughout embryogenesis.

Kidder (1972a) detected the synthesis of tRNA from mid to late cleavage in *Mulinia,* and Collier (1976a) found the synthesis of this RNA as early as the four-cell stage in *Ilyanassa*. The molar ratio of tRNA to rRNA in *Ilyanassa* was 20:1 in the egg and 10:1 in the postgastrular embryo (Collier, 1976a). This ratio changes when the first major increase in total RNA occurs in *Ilyanassa* and then remains constant during later stages of embryogenesis. This suggests a coordinate synthesis of these two RNAs during oogenesis that undergoes a proportional change during later embryogenesis. Transfer RNA accounts for 18% of the total RNA of the *Ilyanassa* egg, a percentage that remains high until after gastrulation when it drops to an average of 12%. This level of tRNA is similar to, but somewhat lower in actual value than, the high tRNA content of the pre-vitellogenic *Xenopus* oocyte (Wegnez and Denis, 1979).

The uncleaved egg of the freshwater pulmonate *Lymnaea* incorporates ^{32}P into a NaOH-labile fraction, presumably RNA; this incorporation increases at the two-cell stage and reaches a maximum level in the trochophore (Brahmachary et al., 1968, 1971, 1972). Brahmachary and Palchoudhury (1971) reported a 10-S fraction of [^{32}P]RNA, presumed to be mRNA, isolated on a sucrose gradient from RNA extracted from *Lymnaea*. The radioactivity in this fraction reached a peak in the morula, was present and actinomycin D sensitive in the trochophore, but insensitive to actinomycin D in the veliger of *Lymnaea*. The authors presume this to be mRNA; however, in the absence of RNase treatment, and from the atypical pattern of rRNA displayed in their sucrose gradients the nature of this 10-S component remains unclear. Tapaswi (1972, 1974) reported the incorporation of ^{32}P into NaOH-sensitive components by the morula and blastula of *Lymnaea*. These labeled components; presumably RNA, were fractionated by agarose-gel electrophoresis. No electrophoregrams were presented in either of these papers and it is therefore difficult to evaluate this work.

This work on the incorporation of ^{32}P into alkali-stable materials of the *Lymnaea* egg suggests that RNA synthesis begins in the uncleaved egg and continues during cleavage, as shown autoradiographically with [^{3}H]uridine by van den

Biggelaar (1971a), and throughout other stages of development. These observations suffer from lack of basic data in some papers (e.g., Tapaswi, 1972, 1974, as previously indicated); caution is urged in interpreting alkali-labile ^{32}P radioactivity to have resulted from incorporation into RNA, because ^{32}P in proteins, except those containing phosphotyrosine (Cooper and Hunter, 1981), is also alkaline labile.

IV. Synthesis and Processing of Messenger RNA

The presence, synthesis, and translation of mRNAs can be detected by a variety of biochemical methods. Most of the observations on molluscan embryos do not involve the isolation of mRNAs; rather, they are measures of polyribosomes, incorporation of radioactive precursors into polyribosomes, heterodisperse RNAs, *in vivo* and *in vitro* synthesis of proteins, and *in vitro* translation of RNAs from which the presence and translation of mRNAs may be inferred. In this section the evidence for the synthesis of mRNAs during molluscan development and the existence of maternal mRNAs in oocytes and mature eggs will be reviewed. The localization and translation of maternal mRNAs will be reviewed in subsequent sections.

Among bivalves, the incorporation of amino acids and nucleosides into polyribosomes, the repression by actinomycin D of nonribosomal RNA synthesis in polyribosome fractions, and the mobilization of ribosomes into polyribosomes have been followed throughout the development of *Spisula solidissima* by Firtel and Monroy (1970). The [^{3}H]RNA extracted from polyribosomes of four-cell embryos and analyzed on sucrose gradients was heterodisperse with a main peak that had a sedimentation coefficient of 11 S. This result supports the interpretation that the short pulse time (30 min) used detected mainly mRNA during early cleavage and in later stages of *Spisula* embryogenesis. There is a continuous increase in the incorporation of [^{3}H]uridine into this class of RNA during early and late cleavage, blastulation, and until the formation of the trochophore larva, which was the latest stage studied.

These authors also observed a linear increase in the proportion of total ribosomes converted into polyribosomes from a time prior to the maturation of the oocyte until the nonswimming blastula stage; after this stage there was an abrupt increase in the rate of mobilization of monosomes into polyribosomes, which continued until the differentiation of the trochophore larva. That these were functional polyribosomes was demonstrated by their incorporation of [^{3}H]leucine into nascent proteins.

Firtel and Monroy (1970) also found that actinomycin D repressed the amount of polyribosomes by 26% with a proportionally greater repression of ''light polyribosomes'' than ''heavy polyribosomes.'' They suggested that the light polyribosomes, which were more active in the incorporation of amino acids than were the heavy polyribosomes, were engaged in the synthesis of histones required for cleavage. This interpretation is consistent with the finding of Kedes and Gross (1969) that histones are synthesized on light polyribosomes in the sea urchin embryo and is directly supported by the observation of Gabrielli and Baglioni (1975) that only the H1 histone is coded for by RNA extracted from the *Spisula* egg. The 26% reduction in the number of polyribosomes by actinomycin-D treatment implies that 74% of the normal increase in polyribosomes is unaffected and may therefore be attributed to the availability of maternal mRNAs for the mobilization of a major proportion of the monosomes into polyribosomes.

Kidder (1972b) measured the increase in active polyribosomes and the appearance of newly synthesized RNA on polyribosomes during early and late cleavage of the coot clam, *Mulinia lateralis*. His results support the early synthesis of mRNA, that is, nonribosomal RNA associated with active polyribosomes, and the continuous increase in the proportion of total ribosomes in polyribosomes during the development of bivalves. Kidder's study of *Mulinia* adds the important demonstration that there is an influx of newly synthesized mRNA into the polyribosomes as early as the two-cell stage as well as during late cleavage.

McLean and Whitely (1974) have shown that the rate of RNA synthesis in the unfertilized egg of the Pacific oyster (*Crassostrea gigas*) does not change until the embryo begins to swim, which is the time they detected by RNA–DNA hybridization experiments the appearance of new RNA species. Unfortunately, both the measurements of RNA synthesis and the hybridization experiments depict changes in total RNA, and it can only be surmised that they accurately reflect the synthesis of mRNAs.

Karp and Whiteley (1973), from DNA–RNA hybridization experiments with RNA from the limpet *Acmaea scutum,* found a similarity of the most abundant RNA sequences in all of several stages of development from the unfertilized egg to the midveliger stage. The RNA from the adult limpet was a poor competitor compared to midveliger reference RNA, that is, the labeled RNA used in the competition experiment. This observation suggests that a large proportion of the RNA sequences produced during embryonic and larval development are not shared in common with the tissues of the adult. This result, which implies the presence of a distinct set of RNA sequences for each state of differentiation, is similar to, though not refined to the detection of single-copy DNA transcripts, the observations of Galau et al. (1976) for the embryonic and adult tissues of the sea urchin.

In addition to the evidence for the existence of maternal mRNA in the *Spisula* egg from Firtel and Monroy's (1970) actinomycin-D experiments, Gabrielli and Baglioni (1975, 1977) have isolated mRNAs, as shown by their ability to support *in vitro* protein synthesis, from both postribosomal supernatants and polyribosomes of the *Spisula* egg. The mRNAs extracted from each of these egg fractions were translated into H1 histone and a variety of other proteins, which were resolved by one-dimensional electrophoresis on sodium dodecyl sulfate acrylamide gels. The mRNAs for the other histones were not detected in the RNAs extracted from the *Spisula* egg. Similar evidence for maternal mRNAs in the *Spisula* egg has also been reported by Rosenthal et al. (1980). These results are the most direct evidence for the presence of maternal mRNAs in molluscan eggs and will be considered in greater detail in a later section.

Histone mRNAs synthesized during cleavage of the *Spisula* embryo have been identified by electrophoresis in acrylamide gels by Steele et al. (1978). These authors also measured the reiteration frequency of the histone genes from the reassociation kinetics of trace amounts of labeled histone mRNA with a large excess of DNA. The rate of reassociation of histone mRNA with DNA was compared with the rate of reassociation of single copy DNA; it was estimated from this comparison that the histone genes were reiterated less than 100 copies per haploid genome. This level of reiteration is considerably lower than that observed in sea urchin embryos and in view of the absence of all but the H1 mRNA in the oocyte (Gabrielli and Baglioni, 1975) it, as pointed out by Steele et al., emphasizes that there are a variety of strategies used by different organisms to provide the histones required for cell division during early embryogenesis.

Is the absence of all histone mRNAs, except H1, in the egg (Gabrielli and Baglioni, 1975) of special significance? Perhaps the other histones are synthesized and stored during oogenesis, as has been shown for *Xenopus* (Woodland and Adamson, 1977), and only the continued synthesis of H1, the nucleosome linker, is required to provide the complement of histones required for rapid cell division. That this may be so is supported by the use of embryonal mRNAs for the synthesis of 80% of H1 histone in interordinal hybrids of echinoid embryos (Easton and Whiteley, 1979). The alternate forms of H1 histones and their changing pattern during embryogenesis (in *Spisula,* Gabrielli and Baglioni, 1975; *Ilyanassa,* Collier, 1981b; sea urchins, Ruderman and Gross, 1974; Arceci et al., 1976; Newrock et al., 1977; Hieter et al., 1979; *Xenopus,* Woodland et al., 1979; Risley and Eckhardt, 1981) suggest that the continued synthesis of mRNAs for this class of histones, as opposed to the use of stored mRNAs, may provide a mechanism for activating different genomic elements (the chromatin remodeling of Newrock et al., 1977; the *variegated chromosomes* of Weintraub et al., 1977) that are responsible for changes in the developmental program that occur during early embryogenesis. The observations of Ruderman and Gross (1974) that the changing pattern of histone synthesis is actinomycin D insensitive, and therefore independent of the embryonal genome, suggests that the egg autonomously regulates the changing pattern of histone synthesis. Posttranslational modifications of histones may also play a role in activating those genomic elements underlying the changes in the developmental program that characterize embryogenesis.

Among the gastropods it is evident from autoradiographic determination of radioactivity in pulse-labeled embryos that nonnucleolar RNA synthesis begins during early cleavage (Table IV); however, although some of these RNAs may be transcripts of structural genes, this is not an unequivocal conclusion. Similarly, the isolation of heterogeneous RNAs by MAK chromatography from the *Ilyanassa* embryo during early cleavage and later stages of development (Collier, 1976a) is suggestive, but inconclusive, evidence for the synthesis of mRNA, as is the resolution of a 10–15-S RNA by acrylamide electrophoresis during early and late development of this embryo (Koser and Collier, 1976).

The best evidence for mRNA synthesis during early cleavage and all later stages of development among the gastropods is from the polyadenylation of nascent RNAs by the *Ilyanassa* embryo (Collier, 1975b). In these experiments the total RNA was labeled with [^{3}H]uridine and fractionated by the ribonuclease T_2-labile binding of polyadenylated RNAs to Millipore filters and by the hybridization of these RNAs to poly(U). About 40% of the newly synthesized RNA during early cleavage (4- to 16-cell stage) was polyadenylated, 32% by the mensentoblast embryo, that is, the 25-cell embryo, and approximately 9–10% during all later stages of development. These are not absolute values but the percentage of the total nascent RNA that was polyadenylated.

Low levels of actinomycin D (2.5–25 µg/ml), which preferentially inhibit rRNA synthesis, increased the levels of polyadenylated RNAs in the 3-day embryo from 8 to 16%. This proportional change in the polyadenylated RNA indicates that at least 16% of the newly synthesized nonribosomal RNA at this stage was polyadenylated mRNA. These observations establish that most of the RNAs made during early cleavage are polyadenylated and are therefore probably poly(A)$^+$ mRNAs. The relation of these embryogenic mRNAs to the maternal mRNAs in the egg at this stage will be discussed later.

Vitually nothing is known about the actual processing of mRNAs, such as *capping*, presence of introns and exons, nucleolytic cleavage of mRNAs, and so forth, in molluscan embryos; however, a detailed study of poly(A) and oligo(A) tracts in RNAs of *Ilyanassa* has been made in Kidder's laboratory (Clark and Kidder, 1977; Kidder et al., 1977; Gerdes and Kidder, 1979).

Clark and Kidder (1977) have shown that the poly(A) content of the *Ilyanassa* embryo, as measured by hybridization with [^{3}H]poly(U), changed very little during early cleavage and gastrulation, but increased by threefold in the veliger larva. This increase in poly(A) in the veliger probably began in the postgastrula embryo during organogenesis. They also measured, by electrophoresis in polyacrylamide gels that were carefully calibrated with 4-S and 5-S RNA markers, the lengths of poly(A) tracts attached to RNAs extracted from embryos of several different stages of development. The mean tract length of 30 bases found through gastrulation increased to 39 in the veliger. The authors pointed out that tracts of poly(A) are considerably shorter than those reported for most eukaryotes. Even more striking is the large array of eight size classes of poly(A) tracts, six of which were observed repeatedly in all five stages of embryogenesis investigated. Clark and Kidder suggest that these size classes of poly(A) might function in the identification and utilization of distinct mRNAs at different periods of development.

From the mean tract lengths of poly(A) and the RNA content of the *Ilyanassa* embryo (Collier, 1975b), Clark and Kidder (1977) estimated the number of poly(A) tracts per embryo during several stages of development, presuming each poly(A) tract to correspond to an individual mRNA. The uncleaved egg was estimated to contain 25×10^7 tracts, and the average for the blastula and gastrula was 37×10^7. To the extent that these values represent individual mRNAs they show that the bulk of these mRNAs is present in the uncleaved egg and that embryogenic mRNA synthesis accounts for an increase of about one-third during the first two days of development.

Kidder et al. (1977) subsequently established that the poly(A) tracts in *Ilyanassa* were not of mitochondrial origin and that they could be classified into two groups according to their rate of incorporation of [^{3}H]adenosine. One class was 27–30 bases long and incorporated [^{3}H]adenosine very slowly; the other class was smaller (11–14 and 16–18 bases), incorporated label rapidly, and was stable for at least 20 h. Special care was taken to rule out the possibility that these kinetic classes arose from degradation of poly(A) during its isolation.

Gerdes and Kidder (1979) have also isolated oligo(A) tracts (oligo(A)$_8$ and oligo(A)$_{18}$ of 8 and 18 bases respectively) from *Ilyanassa* embryos, which are rapidly synthesized and stable during a 10-h chase. These oligo(A) tracts were shown to exist independently of poly(A), that is, they are neither precursors nor degradation products of poly(A), and are not attached to the 3' end of RNA. Both of these oligo(A) tracts are covalently attached to longer RNA molecules, but only the oligo(A)$_8$ parallels the distribution of mRNA. Gerdes and Kidder (1979) suggest that the oligo(A)$_{18}$ RNA may have a regulatory function similar to that postulated by Bester et al. (1975) for an oligo(U) RNA in chick embryo muscle. In light of recent evidence for translational control of protein synthesis in the *Ilyanassa* embryo (Brandhorst and Newrock, 1981; Collier and McCarthy, 1981), which will be discussed later, this is a very pertinent suggestion. These two classes of oligo(A) RNAs may provide a means of isolating and characterizing the processes regulating protein synthesis in this embryo.

V. Protein Synthesis

Some general aspects of protein synthesis in molluscan embryos are reviewed here; observations on protein synthesis specifically related to the localization of mRNAs and control of their translation are reviewed in a later section.

The work reviewed earlier on the presence of functional polyribosomes in the eggs of *Spisula* and *Mulinia* (Firtel and Monroy, 1970; Kidder, 1972a) showed a continuous pattern of protein synthesis during the development of these lamellibranch embryos. Bell and Reeder (1967) measured the relative rates of protein synthesis before and after fertilization of the *Spisula* egg and found a low level of protein synthesis in the unfertilized egg, which increased three- to fourfold after fertilization. An increase in permeability to amino acids occurred about 50 min after fertilization, and the relative rate of protein synthesis continued to increase with a concomitant increase in polyribosomes throughout early development. Similarly, Mirkes (1970, 1972) has shown that the unfertilized egg of *Ilyanassa* contains polyribosomes active in protein synthesis and that both uptake and incorporation of labeled amino acids into protein increase within 15 min after fertilization. Mirkes demonstrated by preloading the unfertilized egg with radioactive amino acids that the 2.5-fold increase in incorporation of amino acids into protein reflected a real increase in the rate of protein synthesis. Mirkes (1972) also found that by the eight-cell stage the number of active polyribosomes in the unfertilized egg had increased by 80% and by fivefold in the veliger larva. Collier and Schwartz (1969) measured the total amino acid content of the acid-soluble pool of *Ilyanassa* and found that the amino acid pool remained relatively constant during the first four days of development and showed some sharp fluctuations during later stages. Although fluctuations in the size of the amino acid pool were less than 50% throughout development, the uptake of leucine varied over a tenfold range at different stages of development. Taking into account these factors, the best estimate of the relative rate of protein synthesis was made from the mol % of labeled amino acids incorporated, and this value showed a steady increase during the first three days of *Ilyanassa* development, leveled off during organogenesis, and finally decreased during later stages of development.

The observations on the increase in polyribosomes and the estimates of relative rates of amino acid incorporation into protein show, not surprisingly, that the molluscan embryo responds to fertilization by an increase in the relative rate of protein synthesis and in the mobilization of ribosomes into polyribosomes. The rate of protein synthesis continues to increase throughout most stages of development. Because adequate measurements of the amino acid pool size and the absolute rate of protein synthesis have not been determined for any molluscan embryo, the available data on the rates of protein synthesis are less precise than for other embryos, nonetheless, the evidence at hand supports the conclusions made here and is sufficient for most purposes.

The data thus far reviewed have been from marine species that undergo indirect development and produce pelagic larvae that metamorphose into adults. Many freshwater gastropods, typified by *Lymnaea* and other pond snails, undergo direct development, that is, they do not have a free-living larval stage but develop in an egg capsule containing a perivitellin–albuminous fluid. In a review

of protein synthesis in pulmonates, Morrill et al. (1976), in whose laboratory
much of the biochemical work on this group has been done, pointed out that the
capsule fluid of *Lymnaea* is pinocytotically ingested by the embryo and presumably serves as a nutrient after the consumption of the egg yolk by the end of the
gastrula stage. Morrill (1964) emphasizes that removal of most of the egg yolk
by centrifugal fragmentation does not prevent normal development (see also
Clement, 1938), whereas eggs removed from their egg capsules do not develop
beyond the gastrula. From these points and a parallel increase of embryonic
proteins with a decrease in capsular fluid protein that begins during the second
day of development (see Fig. 2) when organ primoridia first appear, Morrill
(1964; Morrill et al., 1976) concluded that the capsular fluid, which also contains
galactogen and glycogen, serves as a major source of nutrient for *Lymnaea*. [See
Morrill (1973) for the increase in protein content during the development of
Physa fontinalis.] The transport of galactogen by macro- and micropinocytosis
from the perivitelline fluid into various cells of the embryo of *Lymnaea stagnalis*
has been shown by Arni (1974). Arni detected galactogen both in cellular vacuoles and in the cytoplasm of embryonic cells. Horstmann (1956) had shown that

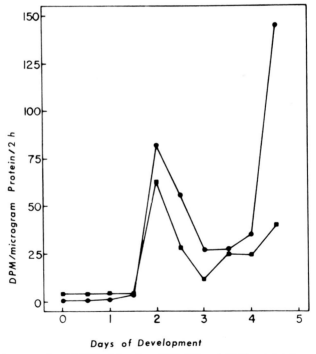

Fig. 2. Incorporation of [3H]uridine into RNA (●) and [3H]leucine into protein (■) of
Lymnaea palustris during development (24°C). (From Morrill et al., 1976, with permission.)

the galactogen content of *Lymnaea stagnalis* decreased during development, and Goudsmit (1976) later demonstrated the catabolism of galactogen by embryonic homogenates of another freshwater snail, *Bulimnaea megasoma*. Goudsmit also found that adults of this freshwater snail used [^{14}C]glucose for the synthesis of the galactogen deposited in the egg. These observations show that the galactogen of the capsule fluid is taken up and catabolized by developing embryos of these freshwater snails.

The egg-capsule fluid of *Lymnaea* was shown by Morrill et al. (1964) to contain 11 protein bands as resolved by one-dimensional starch-gel electrophoresis and to be species specific by immunoelectrophoretic analysis among *Lymnaea palustris, L. stagnalis,* and *L. columella.* The origin of the capsular-fluid proteins from the adult albumen gland was indicated by the identical electrophoretic pattern of proteins from these sources. Morrill et al. (1964) found no evidence for the selective utilization of any of the capsular proteins by the *Lymnaea* embryo. Presumably some of the protein absorbed from the capsule fluid by the embryo is degraded and contributes to the amino acid pool. [See Morrill (1963) for an amino acid analysis of the capsular protein.] The possibility that some of the proteins of the capsular fluid are functionally important to the developing embryo should not be overlooked. Norris and Morrill (1964) have briefly reviewed other metabolic events, such as enzyme localization, respiration, and glycogen and galactogen metabolism in freshwater pulmonates.

Jockusch (1968) autoradiographically measured the incorporation of [^{3}H]leucine into *Lymnaea* eggs and found that by the eight-cell stage the rate of incorporation had increased 25-fold over the rate observed prior to syngamy. The maximal rate of puromycin-sensitive incorporation of leucine, presumably protein synthesis, occurred during interphase of the cell cycle and dropped to a minimum during metaphase. Two interesting aspects of these findings are that the increase in protein synthesis does not immediately follow fertilization and the implication that fusion of pronuclei may provide the stimulus for change in protein synthesis.

Van der Wal (1976) has autoradiographically detected the incorporation of [^{3}H]leucine into components of the egg cytoplasm during the first, second, and third cleavages of *Lymnaea stagnalis.* From an analysis of the grains per square centimeter in electron microscopic autoradiographs van der Wal has shown a maximal incorporation into mitochondria, significant incorporation into three varieties of yolk granules and fat vacuoles, and a moderate level of incorporation into the cytoplasm. Those grains 100 nm beyond a specific structure were scored as cytoplasmic. The high level of incorporation into mitochondria is surprising and quite different from the 97% inhibition of protein synthesis by cycloheximide during early cleavage of *Ilyanassa,* an observation which indicates that mitochondria synthesize less than 3% of the proteins made during early cleavage (Collier, 1981b). Van der Wal's autoradiographs also show extensive labeling of yolk granules, which presumably arises from translocation of nascent proteins into these granules; he suggests that this radioactivity represents newly synthesized proteins that are engaged in the disintegration and transformation of the yolk granules.

Boon-Niermeyer (1975) has measured by autoradiography of paraffin sections the incorporation of [^{3}H]lysine during early cleavage in normal and puromycin-treated *Lymnaea* eggs. Puromycin treatment for 10–30 min retarded cleavage, was most effective during S phase of the cell cycle, was reversible, and repressed incorporation of [^{3}H]lysine by 50%. Boon-Niermeyer suggests that cleavage delay is caused by the inhibition of synthesis of specific proteins associated with cleavage rather than by the effect of puromycin on total protein synthesis and that the head malformations that occurred in puromycin-treated embryos arose from perturbations in the division chronology, specifically the cleavage of the first quartet of micromeres, rather than from a differential effect on protein synthesis.

Rebhun et al. (1973) have suggested that the inhibition of cleavage in *Spisula* and sea urchin eggs by puromycin may arise, in part, from the direct action of the purine component of puromycin (6-dimethylaminopurine), which is an effective inhibitor of cleavage, rather than or in addition to the effect on protein synthesis. This suggestion is consistent with the interpretation of Boon-Niermeyer of the effect of puromycin on differentiation of *Lymnaea*.

J. B. Morrill and his students have extensively studied protein synthesis and the activities of hydrolases in adult organs, normal embryo, and actinomycin D-treated embryos of the freshwater pulmonates *Lymnaea* and *Physa*. Morrill's (Morrill et al., 1976) data for the incorporation of uridine into RNA and leucine into protein of *Lymnaea palustris* embryos is reproduced in Fig. 3. The synthesis of both RNA and protein, as judged from incorporation data, reaches a peak in the trochophore, declines thereafter, and increases as adult organ primordia are formed. Norris and Morrill (1964) measured the activity of nine electrophoretically mobile hydrolases, including acid and alkali phosphatases, two aminopeptidases, galacto- and glucosidases, in 16 adult organs and during embryogenesis of *Lymnaea palustris*. [See Morrill et al. (1976) for a similar analysis of phosphatases in the development of *Physa fontinalis*]. Among these nine hydrolases 42 multiple bands were resolved when starch-gel electrophoreograms were assayed for hydrolase activity.

The distribution of these enzyme activities defines organs and stages of development (see Fig. 4). Each of 16 adult organs and 7 stages of development was shown to have a distinct pattern of bands with respect to their presence, absence, and intensity of staining. Prior to the onset of organogenesis, embryos contained only five bands (two of acid phosphatase, one band each of two peptidases, and one β-galactosidase band), which the authors suggest may be associated with the metabolism of yolk and perivitelline fluid during early development. As development proceeded, the number and complexity of electrophoretically mobile bands increased to a total of 28 in the hatched snail.

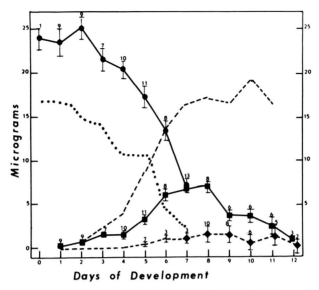

Days of Development

Fig. 3. Total protein, galactogen, and glycogen of embryo and capsule fluid of Lymnaea *palustris*. (●———●), Capsule fluid galactogen; (••••), capsule fluid protein; (■———■), embryo galactogen; (◆- -◆), embryo glycogen; (- - - -), embryo protein. (From Morrill et al., 1976 with permission.)

Moon and Morrill (1979) have followed by polyacrylamide gel electrophoresis the appearance of acid phosphatase isozymes during the development of *Lymnaea palustris*. The multiple forms of this enzyme show a more heterogeneous pattern during development than previously reported by Norris and Morrill (1964). This difference is attributable to the improved resolution of acrylamide gel electrophoresis.

Other examples of isozymes have been reported in embryos of *Ilyanassa* (malate dehydrogenase, Goldberg and Cather, 1963; aminopeptidase, Morrill and Norris, 1965; catalase and aminopeptidase, Nelson and Scandalios, 1977; alkaline phosphatase and esterases, Freeman, 1971); and *Argobuccinum oregonense* (lactic dehydrogenase, Goldberg and Cather, 1963). Gooch et al. (1972) and Berger (1977) have evaluated genetic polymorphism and reviewed isozyme studies in the genera *Nassarius* (*Ilyanassa*)[1] and *Littorina*.

The isozymes of catalase and aminopeptidase in *Ilyanassa* embryos and adults

[2]*Nassarius obsoletus* was removed from the genus *Nassarius* (Dumeril, 1806) by Stimpson and placed into the new genus *Ilyanassa* (Stimpson, 1865) on the basis that *Ilyanassa obsoleta* has a different operculum and lacks the caudal pedal cirri or metapodial tentacles characteristic of the genus *Nassarius*. Although embryologists tend to use the correct term *Ilyanassa obsolete* (Say), scientists in other fields often still use the now outmoded designation *Nassarius obsoletus*.

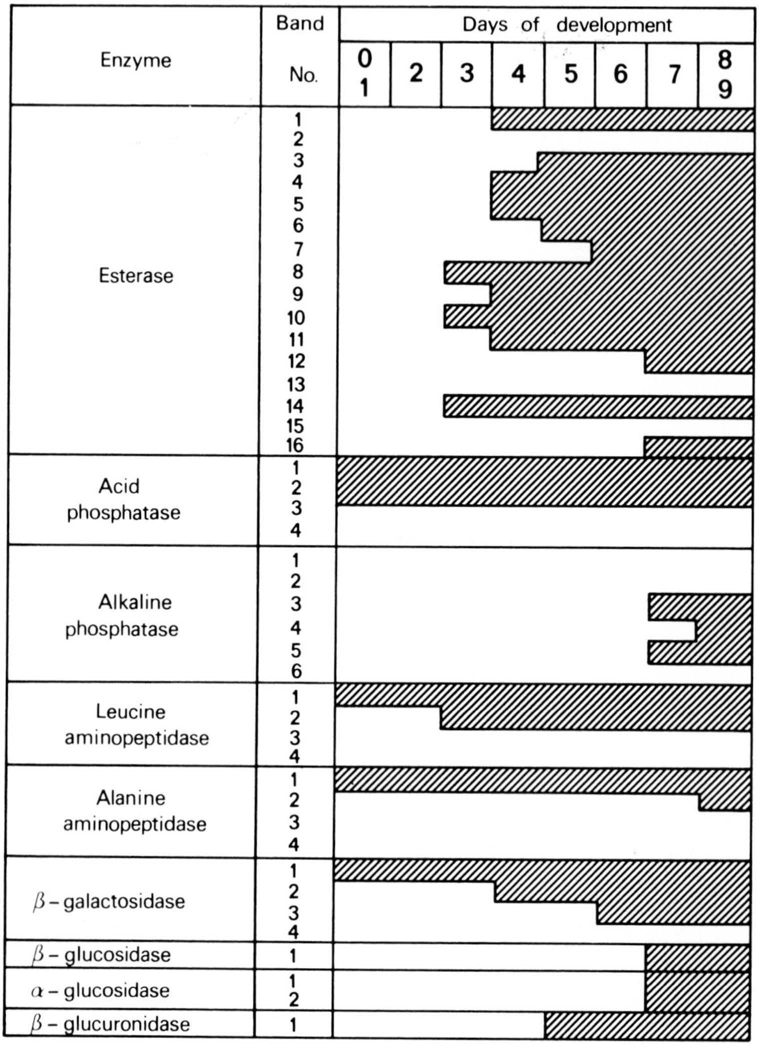

Enzyme	Band No.	Days of development
		0 1 / 2 / 3 / 4 / 5 / 6 / 7 / 8 9
Esterase	1 2 3 4 5 6 7 8 9 10 11 12 13 14 15 16	
Acid phosphatase	1 2 3 4	
Alkaline phosphatase	1 2 3 4 5 6	
Leucine aminopeptidase	1 2 3 4	
Alanine aminopeptidase	1 2 3 4	
β – galactosidase	1 2 3 4	
β – glucosidase	1	
α – glucosidase	1 2	
β – glucuronidase	1	

Fig. 4. Appearance of electrophoretically mobile hydrolases during the development of *Lymnaea palustris*. (From Norris and Morrill, 1964 with permission.)

have been thoroughly characterized by Nelson and Scandalios (1977). They have suggested that the two isozymes of catalase and the two major forms of aminopeptidase, both of which are invariant during development, are each the products of two genetic loci and that the observed variations in their activity are caused by differential rates of synthesis and/or degradation. The isozymes in *Ilyanassa* described by Freeman (1971) are related to the removal of the polar lobe and will be discussed later.

Whether this changing pattern of multiple forms of hydrolases represents changes in enzyme activity, degradation, and/or synthesis of polypeptides is an important question, which could probably be resolved, in part, by inhibition experiments with actinomycin D, α-amanitin, and puromycin. Whatever the answer to this question, it should be appreciated that these studies do provide a metabolic description of each adult organ and show that each stage or state of differentiation has a unique set of metabolic properties. However, if isozymes are to be of further value in characterizing differentiation a detailed study of their genetics, activation, and degradation is required.

Are isozymes significant to embryogenesis because of differences among conformers in kinetics, intracellular localization, or response to regulators? Or are they products of related multigene families, such as the vitellogenin gene family (Wahli et al., 1981), whose variable appearance in development could be caused by the operation of polycistronic regulators that control large groups of genes? According to this view, which is ventured as a trial hypothesis, different isozymes would exist as a part of genetic heterozygosity and their variable appearance during development would be a consequence of being included into or excluded from the domain of polycistronic gene regulators that exercise a broad level of control over gene transcription. Several structural genes coding for a group of conformers and their associated spacer regions, that is, a multigene family, would occupy several thousand DNA bases, which could be variably selected or excluded from a battery of adjacent genes as outlined above.

Stage-specific changes in polypeptide synthesis have been demonstrated by two-dimensional electrophoresis of proteins produced between early cleavage and the mesentoblast stage of development in the *Ilyanassa* embryo (Brandhorst and Newrock, 1981; Collier and McCarthy, 1981). Collier and McCarthy (1981) observed the disappearance of 12 peptides, the appearance of 24 new peptides and changes in the relative rate of accumulation of 38 peptides between the 8- and 29-cell *Ilyanassa* embryo, these changes represented approximately 4, 6, and 9%, respectively, of the polypeptides resolved at these stages of development.

VI. Biochemistry of Ooplasmic Segregation

The Mollusca is the largest phylum among a group of organisms whose embryos have spiral cleavage, the Spiralia. Spiral cleavage is the left to right alternation of cleavage planes that are oblique to the egg axis; this succession of cleavage planes describes a spiral. From experiments of classical embryology it has long been known that, in most cases, isolated blastomeres of the Spiralia do not produce normal embryos, thus cleavage of the spiralian embryo is determinate. The striking pattern of spiral cleavage and its association with the determination of the embryonic fate of early blastomeres emphasizes the role of the egg cytoplasm in differentiation. This determinative role of the egg cytoplasm, which is a general feature of most animal development, has been variously described as germinal localization (from His's "Organbildende Keimbezirke"), cytoplasmic localization, prelocalization (Wilson, 1928), precocious segregation (Lankester, 1877), and, a derivative of the former, ooplasmic segregation (Costello, 1948). Lankester (1877) was remarkably astute when he wrote "Thus, since the fertilized egg already contained hereditarily acquired molecules, . . . , invisible though differentiated, there would be a possibility that these two kinds of hereditarily acquired molecules should part company, not after the egg-cell had broken up into many cells as a morula, but at the very first step in the multiplication of the egg-cell [p. 410]." Written just more than 100 years ago this is indeed a prescient statement as we think of "hereditarily acquired molecules" of maternal messenger RNA, messenger ribonucleoprotein particles, and so forth. Yet we know little about the localization of the "physiological molecules" of Lankester and their segregation into the blastomeres of the young embryo.

The biochemistry of ooplasmic segregation has been studied by an analysis of the contents and synthetic capabilities of regions of ooplasm or isolated blastomeres and by determining the effects of deleting or inactivating part of the ooplasm on macromolecular syntheses, Both of these approaches will be reviewed.

Many molluscan eggs, such as those of the gastropods *Bithynia, Crepidula,* and *Ilyanassa,* the scaphopod *Dentalium,* and the bivalve *Mytilus,* contain a vegetal protrusion of cytoplasm, the polar lobe, that is essential for differentiation (Clement, 1952, 1971; Cather and Verdonk, 1974). This polar lobe can be readily isolated (Clement, 1952; Collier, 1981a) and its biochemistry (see review by Collier, 1976a) and role in development (see Chapter 6, this volume; and Dohmen and Verdonk, 1979) have been extensively studied. Obviously, this anucleate region of the egg provides a useful approach for studying the role of the ooplasm in differentiation. Most specifically, it is suited for studying the role of the vegetal ooplasm, which in most animal eggs contains determinative information. The cytomechanics and biochemistry of polar-lobe formation have been studied extensively in *Ilyanassa* and are reviewed in Chapter 3 of this volume.

Early investigations on the biochemistry of ooplasmic segregation were those of Berg (1954) and Collier (1957, 1960a,b). Berg (1954) found an equal distribution of alanylglycine peptidase activity between the AB and CD blastomeres of *Mytilus edulis;* Collier (1957) showed that the activity of this enzyme was distributed in proportion to the measured volume of hyaline cytoplasm among the lobeless egg, third polar lobe, AB, CD, C, and D blastomeres, and the ecto- and mesentoblast of the *Ilyanassa* embryo. Collier (1960a) found a slightly greater concentration of total RNA in the AB cell than in the CD blastomere and the polar lobe of *Ilyanassa.* Dohmen and Verdonk (1974) have demonstrated a RNase labile methyl green-pyronin staining component, which is probably RNA, in the polar lobe of *Bithynia tentaculata*. This staining reaction is confined to the vegetal body, a cup-shaped mass of small vesicles, contained in the polar lobe of this species.

Table V summarizes the distributions of five molecular components among the egg, lobeless egg, and polar lobe of *Ilyanassa*. The most significant point to emerge from these data is the higher concentration per cubic millimeter of cytoplasm of acid-soluble precursors, specifically ATP (Collier and Garone, 1975), in the polar lobe. This distribution of ATP, acid-soluble precursors (as measured by absorbancy at 260 nm), and acid-soluble phosphorus, is correlated with the distribution of yolk in this egg. Yolk platelets that are isolated in nonaqueous media and which have lost the amorphous cortex of cytoplasm and the limiting membrane associated with yolk platelets in *Ilyanassa* (M. M. Collier, 1975) do not contain ATP (J. R. Collier and M. E. McCarthy, unpublished observations). These observations suggest that ATP and perhaps other acid-soluble nucleic acid precursors are sequestered in the cortex of the yolk platelet. The cytoplasmic cortex and the limiting membrane of the yolk platelet could sequester other molecules and thereby segregate them along with the yolk, which in the *Ilyanassa* egg is differentially distributed into the D quadrant (Collier, 1957), and indeed a small amount of yolk is passed into the primary mesentoblast (4d). However, the segregation of yolk per se does not appear to oe a morphologenetically significant event as normal development occurs after deletion of the 4D macromere (Clement, 1962) and many lobe-dependent structures differentiate in yolk-free merogones of *Ilyanassa* (Clement, 1968).

Van Dongen et al. (1981) have analyzed, by isotachophoresis, the nucleotide content of the polar lobe and lobeless egg of *Nassarius reticulatus*. By the use of this sensitive method they demonstrated that the amount of di- and triphosphate nucleotides, especially GTP, was greater in the polar lobe than in the lobeless egg. This detailed extension of the earlier observations reviewed above on the nucleotides and ATP in the polar lobe of *Ilyanassa obsoleta,* and the possibility of further studies by isotachophoresis, promises to be valuable in resolving the long-standing problem of cytoplasmic localization in polar lobes.

Whether the higher concentration of ATP in the polar lobe of *Ilyanassa* is morphogenetically significant has not been established. The ability of the lobeless embryo to synthesize DNA, proteins, and all classes of RNA (Collier, 1975a, 1981b; Collier and McCarthy, 1981; Brandhorst and Newrock, 1981) at rates comparable to normal embryos implies that the loss of ATP by the lobeless embryo may not be morphogenetically significant. (For an exception to DNA synthesis in later development, see Collier, 1975a.) However, the emergence of specialized intercellular junctions (gap junctions) and the demonstration of their operation in the transport of low-molecular-weight components in certain cell lines during early development, as described in *Patella vulgata* by de Laat et al. (1980) and reviewed by van der Biggelaar et al. (1981), suggests an attractive

TABLE V
Biochemical Lineage of the *Ilyanassa* Egg

Characteristic	Egg	Lobeless egg	Polar lobe	References
Volume, 10^{-4} mm^3				
Total	21.8	16.9	4.9	Collier (1957)
Cytoplasmic	13.3	11.2	2.2	
Yolk	7.0	4.3	2.7	
RNA				
Nanograms	6.60	6.70	0.90	Collier (1960a, 1975a)
ng/μl cytoplasm	5.00	5.10	4.10	
Acid soluble precursors				
Picomoles of nucleotide	158.00	99.50	58.50	Berg and Kato (1959); J.
pM/10^{-4} mm^3 cytoplasm	11.90	8.90	26.60	R. Collier (unpublished
pM/10^{-4} mm^3 yolk	22.60	23.10	21.70	results)
ATP				
Picomoles	25.40	15.30	9.60	Collier and Garone (1975)
pM/10^{-4} mm^3 cytoplasm	1.70	1.20	4.30	
pM/10^{-4} mm^3 yolk	3.60	3.60	3.50	
Acid soluble phosphorus				
Nanograms	5.30	3.50	2.20	Collier (1960b)
μg/μl cytoplasm	2.50	3.10	10.00	
μg/μl yolk	7.60	8.10	8.20	
Phospholipid phosphorus				
Nanograms	2.60	1.70	1.10	Collier (1960b)
μg/μl total volume	1.20	1.00	2.10	
μg/μl yolk	3.71	3.95	4.07	

mechanism for segregating low-molecular-weight molecules such as ATP into specific cells. The transfer of ATP from the D macromere, which is enriched in ATP by having received the polar-lobe cytoplasm, into specific micromeres could increase the cleavage rates of these cells at a later time so that the number of stem cells required for the differentiation of an organ would be produced. In the absence of the polar lobe these stem cells would not be formed in sufficient number, or stem cells would not divide rapidly enough, to form a tissue or organ. [See Collier (1975a) for an earlier discussion of this point and van der Biggelaar (1971b) for a discussion of the energy requirements for DNA synthesis and cell division.]

The ability of isolated polar lobes to incorporate labeled amino acids into an acid-insoluble fraction, presumably protein, has been shown in *Mytilus* (Abd-El-Wahab and Pantelouris, 1957) and in *Ilyanassa* (Clement and Tyler, 1967; Geuskens and de Jonghe d'Ardoye, 1971). Geuskens and de Jonghe d'Ardoye (1971) observed by electron microscope autoradiography the incorporation of ribonuclease-resistant [³H]uridine into mitochondria, the periphery of yolk platelets, and multimembranous vesicles in isolated polar lobes of *Ilyanassa*; [³H]thymidine was also incorporated into yolk platelets of the polar lobe. These authors suggested that there is DNA in yolk platelets and that RNase-resistant uridine may have been incorporated into mitochondrial RNA–DNA hybrids or into polysaccharides.

The effect of removal of the polar lobe at first cleavage on nucleic acid synthesis by the *Ilyanassa* embryo has been studied by Collier (1975a,b, 1976b, 1977), Koser and Collier (1976), and reviewed by Collier (1976a). Collier (1975a) found that the first major increase in total RNA, principally rRNA and tRNA as measured in this study, occurred in the normal embryo during the fourth day of development and was delayed until the sixth day in the lobeless embryo, although the rate of accumulation of total RNA in the lobeless embryo after day six was the same as in the normal embryo. This delay cannot be attributed to differences in cell number as both the normal and lobeless embryos have the same cell number until the end of the fourth day of development. A second observation was that the rate of accumulation of DNA by lobeless embryos was significantly less after the fifth day.

The delay in RNA accumulation may have resulted from the absence of the shell gland, which contains large columnar cells rich in RNA, and mesodermal bands in the lobeless embryo. Thus, this defect in RNA synthesis may be a result, rather than a cause, of the abnormal differentiation of the lobeless embryo. It was suggested above and earlier (Collier, 1975a) that many of the morphogenetic deficiencies of the lobeless embryo result from the failure of stem cells of the lobe-dependent organs to proliferate and/or from the formation of an insufficient number of stem cells. This is certainly indicated by the lower rate of DNA synthesis in the older lobeless embryo and is consistent with Atkinson's

(1971) findings that the lobeless embryo produces cell types of many of the lobe-dependent organs, such as digestive gland, style sac, esophagus, and stomach, that otherwise fail to differentiate in the lobeless embryo.

Measurements of RNA synthesis in normal and lobeless embryos by the incorporation of radioactive precursors have been suggestive of differences in the rates of RNA synthesis in the normal and lobeless embryo (Davidson et al., 1965), but were uncertain until the size of the adenosine triphosphate pool was measured (Collier and Garone, 1975). This permitted measurements of absolute rates of RNA synthesis (Collier, 1976b, 1977) and the demonstration that at the mesentoblast and gastrula stages the synthesis of RNA was less in the lobeless embryo. There was no significant difference in the rate of RNA synthesis between these embryos during early cleavage.

Koser and Collier (1976) fractionated, by polyacrylamide electrophoresis, RNA extracted from normal and lobeless embryos of *Ilyanassa* after a 24 h incubation with [^{3}H]uridine. Replicate analyses of both classes of embryos were made at daily intervals throughout development. The percentage distribution of the total radioactivity in each of several electrophoretically identified fractions was compared by a three-factor analysis of variance. This analysis established that removal of the polar lobe at first cleavage (1) did not alter the pattern of RNA synthesis, that is, the proportional distribution of radioactivity in RNA classes, during the early determinant cleavages of this spiralian embryo; (2) resulted in proportionally less radioactivity in 34-S RNA during postgastrular development, presumed to result from less incorporation into hnRNA as there was no effect on the accumulation of radioactivity in rRNA; and (3) caused the accumulation of proportionally more radioactivity in a 12–16-S RNA component in the postgastrular lobeless embryo. These observations were interpreted as follows: (1) the morphogenetic defects caused by polar lobe extirpation are not reflected by detectable changes in the pattern of RNA synthesis before gastrulation, (2) the inferred decrease in hnRNA production may be a transcriptional defect caused by the loss of the polar lobe, and (3) the proportional increase in radioactivity in the 12–16-S RNA resulted from the accumulation of mRNAs that were not translated by the lobeless embryo. It is not known whether the possible effect of the polar lobe on hnRNA transcription is a cause or a result of the failure of the lobeless embryo to differentiate.

Collier (1975b) measured the proportion of [^{3}H]uridine-labeled RNA polyadenylated by the *Ilyanassa* embryo throughout its development. About 35–40% of the [^{3}H]RNA was polyadenylated during early cleavage and the mesentoblast stage; at all later stages only about 10% of [^{3}H]RNA was polyadenylated. The high proportion of polyadenylated RNA seen in early development reflected the preponderance of heterogeneous RNA made during this time of development and confirms earlier observations (Collier, 1976a). It was also shown that removal of the polar lobe at first cleavage did not alter the proportion of poly(A) RNA synthesized by the 25-cell embryo.

These studies have shown that the morphogenetic failure of the lobeless embryo is not caused by a major defect in RNA synthesis. It cannot be ruled out that the lobeless embryo does not transcribe some specific regulatory and/or mRNAs. This aspect of the problem has been investigated by studies of polypeptide synthesis and will be reviewed later.

Another significant finding of this work was the observation that the rate of accumulation of DNA decreased in the lobeless embryo during later stages of development (Collier, 1975a). The relationship of these findings to the proliferation of stem cells has already been outlined and will be considered in greater detail later.

In addition to the preceding observations on the effect of the polar lobe on nucleic-acid synthesis, information is now available on the influence of the polar-lobe cytoplasm on protein synthesis. The more recent of these studies, which resolved labeled polypeptides extracted from embryos by two-dimensional electrophoresis, detected the expression of a few to several hundred individual structural genes. The ability to follow the final product of this many structural genes permits reliable inferences about the source and localization of mRNAs and the regulation of their expression.

Early studies with the embryo of *Ilyanassa obsoleta* showed that the absence of the polar lobe reduced the level of incorporation of labeled amino acids into the proteins of the lobeless egg (Collier, 1961) and that the isolated polar lobe incorporated amino acids into protein (Clement and Tyler, 1967). These early observations were extended by a series of studies using one-dimensional electrophoresis to fractionate proteins produced by the lobeless embryo and the polar lobe (Teitelman, 1973), normal and actinomycin D–treated normal and lobeless embryos (Newrock and Raff, 1975), and isolated AB and CD blastomeres (Donohoo and Kafatos, 1973). These studies, each of which detected different patterns of radioactivity among gel slices of polypeptides from normal and lobeless embryos, were severely limited by the acknowledged restriction of the resolving power of one-dimensional electrophoresis and do not permit any significant conclusions about the influence of the polar lobe on protein synthesis.

Freeman (1971) has resolved the enzyme activities of three isozymes of alkaline phosphatase and seven isozymes of esterase in the normal and lobeless egg and embryo of *Ilyanassa obsoleta*. There were no differences between the normal and lobeless egg in the activity of any of these isozymes, but the 10-day lobeless embryo was deficient in all but one of the phosphatases and two of the esterases. These results are significant because they detect differences in activities of individual proteins, and, to this extent, avoid some of the vagaries of one-dimensional electrophoresis. However, the embryological and physiological meaning of these results are not clear.

Protein synthesis in the *Ilyanassa* embryo has been analyzed by two-dimensional electrophoresis (Brandhorst and Newrock, 1981; Collier and McCarthy, 1981). Collier (1981b) has shown that the relative rates of protein synthesis are the same in the egg, lobeless egg, and the polar lobe of *Ilyanassa*, and that mitochondrial contribution to protein synthesis during early cleavage is negligible.

The data from the two laboratories (Brandhorst and Newrock, 1981; Collier and McCarthy, 1981) that have investigated protein synthesis during early development of *Ilyanassa* are in remarkable agreement. Both groups find that the normal egg, lobeless egg, and the isolated polar lobe make the same set of [^{35}S]methionine-labeled polypeptides during early cleavage (see Figure 5), and that the removal of the polar lobe at first cleavage causes some quantitative, but no qualitative changes in the stage-specific proteins synthesized by the mesentoblast-stage embryo, that is, 25-cell embryo. Similarly, both laboratories find that the 24-h-isolated polar lobe synthesizes some, but not all, of the proteins made by the mesentoblast-stage embryo. Both groups concur in the interpretation of their results: the lobeless egg and the polar lobe contain the same set of mRNAs and the translation of a new set of polypeptides by the 24-h polar lobe demonstrates transcriptionally independent regulation of protein synthesis by the *Ilyanassa* embryo. Similarly, the stage-specific disappearance of some proteins in the mesentoblast embryo and the 24-h polar lobe suggests the operation of translational repressors or, as said by Brandhorst and Newrock (1981), the selective inactivation of mRNAs.

Collier and McCarthy (1981) investigated protein synthesis by actinomycin D–treated embryos and found that the translocation of 98% of the 350 polypeptides detected by two-dimensional electrophoresis was insensitive to actinomycin D, and that the stage-specific changes that appear in the mesentoblast embryo were qualitatively insensitive to continuous actinomycin D treatment. From this result they concluded that most of the polypeptides made during early embryogenesis of *Ilyanassa* are translated from maternal mRNAs (mmRNA) and that the regulation of most, but not all, of the new polypeptides synthesized by the mesentoblast embryo are made from mmRNAs, presumably stored as messenger ribonucleoprotein particles (mRNP) that are activated by transcriptionally independent events.

Cheney and Ruderman (1978) have resolved by two-dimensional electrophoresis polypeptides translated *in vitro* from RNA extracted from the AB and CD blastomeres of *Spisula solidissima*. They report that "A few maternal mRNA sequences appear to be segregated at first cleavage [p. 349]." This work has been published only as an abstract and cannot be evaluated at this time.

In addition to measuring the relative rates of protein synthesis in *Ilyanassa*, Collier (1981b) found that two variants of H1 histone, histones H2A, H2B, H3, and H4, and high mobility group peptides 14 and 17 were synthesized at approximately the same rate in the egg, lobeless egg, and isolated polar lobe. As these proteins were made by the isolated polar lobe their synthesis does not require concurrent gene transcription. Because the chromosomes of the *Ilyanassa* egg

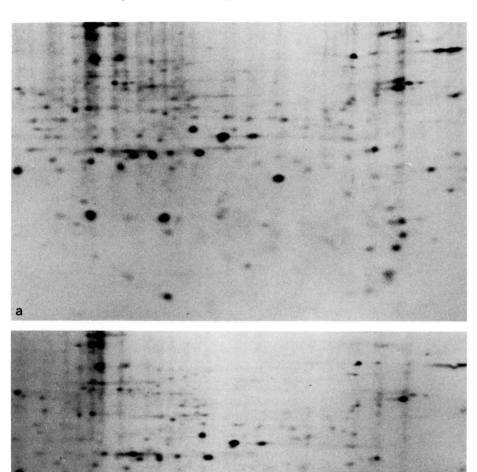

Fig. 5. Two-dimensional electrophoreograph of [35S]methionine *in vitro* labeled proteins extracted from (**a**) normal eggs labeled for 4 h from trefoil to the 12-cell stage and (**b**) isolated polar lobes from the same population of eggs labeled as in (**a**) and for the same time interval.

were in a nontranscribing stage before isolation of the polar lobe, it is probable that these peptides were translated from mmRNAs.

MacKay and Newrock (1982) have identified protein A24 among [^{3}H]leucine-labeled proteins extracted from *Ilyanassa* embryos. Newrock separated *Ilyanassa* proteins by SDS and Triton–urea (9 M urea and 6 mM Triton X-100) poly-acrylamide-gel electrophoresis. He identified the A24 protein in *Ilyanassa* by using purified mouse A24 protein as a marker. The A24 protein separated by Newrock on SDS gels occupies approximately the same position as the presumed high-mobility group peptides 14 and 17 (HMG 14 and 17) reported by the author (Collier, 1981b). In the latter case, there were two bands of [^{35}S]methionine-labeled proteins rather than the single band observed by Newrock. Because we have used different methods for labeling, protein extraction, and electrophoresis, further work will be required to establish unequivocally that we have resolved three polypeptides (A24, HMG 14, and 17). A detailed examination of the histones and related polypeptides in *Ilyanassa* is of special interest in view of the relationship of these peptides to chromatin structure (Newrock et al., 1977; Weintraub et al., 1977) and the synthesis of HMG 14 and 17 (and possibly the A24 protein) by the isolated polar lobe (Collier, 1981b).

A major inference to be drawn from these observations on polypeptide synthesis during early embryogenesis are (1) most of the mRNAs used during this time are maternal in origin, (2) mRNAs do not appear to be differentially localized in the polar lobe, and (3) some of the stage-specific changes in protein synthesis during early development are not under the immediate control of the nuclear genome, that is, they are translationally controlled by cytoplasmic factors.

VII. Regulation of Gene Expression

Gene expression may be regulated by controlling either the transcription of mRNAs or a variety of posttranscriptional events, the last of which is translation. In this section, we will try to determine what levels of control are operating in molluscan embryos and to review the available evidence for the regulation of gene expression so as to learn as much as possible about the mechanism by which these levels of control function.

Verdonk (1973) has made a useful estimate of gene expression during early embryogenesis of *Lymnaea stagnalis*. He scored the developmental stages when X-ray induced mutations interfered with development. He found no perturbations during cleavage, the induction of 14 lethal factors in the gastrula, 63 in the early trochophore, 24 in the late trochophore, 12 in the early veliger, and none in the late veliger. From these observations it was inferred that embryonic genes targeted by X rays were inactive during cleavage and that their activity at other stages was proportional to the number of induced lethal factors. It is significant that the first evidence for transcription of the embryonic genome is in the gastrula, reaches a peak in the early trochophore when organ formation has begun, decreases thereafter, and is not evident in the late veliger. The latter observations suggest that those genes required for the final differentiation of the veliger larva are precociously transcribed. (This precocity is similar to the finding of Collier (1966) that gene transcription for organogenesis in the *Ilyanassa* embryo precedes by about 24 h the actual differentiation of a give organ.)

These observations of Verdonk are especially valuable because they extend over the entire developmental history of the *Lymnaea* egg and, in general, confirm the findings of Feigenbaum and Goldberg (1965) and Collier (1966) with the *Ilyanassa* embryo. Verdonk correctly points out that it has not been demonstrated that enough actinomycin D enters the cells of early embryos, specifically *Ilyanassa*, to block RNA synthesis. This point has subsequently been answered in part by Newrock and Raff (1975), who showed the inhibition of RNA synthesis by actinomycin D in the 18-h *Ilyanassa* embryo.

The failure of lethal factors to accumulate during cleavage of *Lymnaea*, the cleavage and continued protein synthesis by actinomycin D-treated *Ilyanassa* (Feigenbaum and Goldberg, 1965; Collier, 1966; Collier and McCarthy, 1981) and *Lymnaea* embryos (Morrill et al., 1976) all emphasize that early development of the molluscan embryo is, as are most other animal embryos investigated, supported by the precocious transcripts of the maternal genome.

That transcription of the embryonic genome makes an essential contribution to development, at least from the gastrula onwards, has been demonstrated in the *Ilyanassa* embryo (Collier, 1966) by suppression of the differentiation of specific organs by sequential treatment of older embryos with actinomycin D. In this work it was shown, as mentioned above, that those genes required for the differentiation of a specific organ were transcribed about 24 h before its organogenesis.

Although the evidence points to a major role for maternal gene transcripts during early embryogenesis it is important to emphasize that there are some structural genes transcribed by the embryonic genome during early development. This point was clearly shown when Collier and McCarthy (1981) found that six polypeptides failed to be synthesized in *Ilyanassa* embryos treated continuously with actinomycin D until the 29-cell stage. This relatively small number of proteins, which has been shown to be slightly larger by more recent work (J. R. Collier, unpublished results), could be essential for the later development.

Gabrielli and Baglioni (1977), in their study of the translation of mmRNAs in *Spisula* embryos, found that (1) histone mRNAs bind to oligo(T) cellulose and therefore probably contain poly(A) sequences on their 3' termini, (2) *Spisula* eggs do contain mRNAs for all histones but very small amounts of all except histone H1 (contrary to an earlier report by Gabrielli and Baglioni, 1975), and (3) some of the histone mRNAs are not associated with polysomal RNA. They also reported their unpublished observations that *Spisula*-egg mRNA contains the 5' cap of M^7Gppp and is therefore probably not activated by capping. Gabrielli and Baglioni state that ''The presence of histone mRNA among the nonpolysomal mRNA species suggests that the supply of mRNA exceeds the capacity of the eight-cell embryos to translate this mRNA [p. 531].'' From this point they suggest that translational regulation in the *Spisula* egg may arise from the successful competition of histone mRNA for the rate-limiting factors that initiate translation. This may be a mode of translational regulation but it does not follow that (1) the localization of histone mRNAs in a nonpolysomal fraction means that their supply exceeds the translational capacity of the embryo (indeed, it is reasonable for mRNAs to occur at nonpolysomal sites, i.e. messenger ribonucleoprotein particles, regardless of the supply of mRNA) nor that (2), in the absence of specific data, this mode of translational regulation is to be preferred over other possibilities.

Rosenthal et al. (1980) have shown, by one-dimensional electrophoresis of proteins translated in a cell-free system from phenol-extracted RNA, that *Spisula* oocytes and embryos (10–70 min postfertilization) contained the same set of mRNAs, yet different sets were associated with oocyte and embryo ribosomes. *In vivo* stage-specific changes in polypeptide synthesis that occurred following fertilization were shown to be accurately reflected by the cell-free translation of a crude phenol-free supernatant of oocyte or embryo homogenates. The stage-specific translation of proteins was not retained when oocyte or embryo homogenates were deproteinized by phenol extraction. From these observations Rosenthal et al. suggested that the translation control of protein synthesis in this embryo involves the association of protein with mRNA (presumably as mRNP) and that these proteins are responsible for the failure of specific mRNA to be translated. The fidelity of the *in vitro* translation system for the stage-specific changes of protein synthesis of *Spisula* embryos demonstrated by Rosenthal et al. (1980) is a useful experimental approach for deciphering the methods of translational control.

Collier and McCarthy (1981) observed, by two-dimensional electrophoresis of *in vivo* labeled proteins, an increased rate of accumulation of 14 polypeptides by *Ilyanassa* embryos that had been continuously treated with actinomycin D until the mesentoblast stage of development. Other experiments showed that actinomycin D did not selectively alter the rate of accumulation of any polypeptides synthesized during early cleavage. Subsequent control experiments (J. R. Collier, unpublished results) have shown that polypeptide synthesis in isolated polar lobes is unaffected by actinomycin D. These controls, especially the latter, strongly indicate that the effect of actinomycin D on protein synthesis during the early development of *Ilyanassa* occurs by the repression of transcription and not as a side effect of actinomycin D on protein synthesis. Thus, the increase in the accumulation of specific polypeptides by actinomycin D treatment is interpreted as resulting from a transcriptionally dependent event. Collier and McCarthy (1981) have postulated the operation of regulatory genes that produce repressors that decrease the rate of translation of specific polypeptides. Whether these repressors are RNAs or proteins is not specified. Translational control RNAs, such as those postulated by Bester et al. (1975), are possible, as are protein controllers of the type envisioned by Rosenthal et al. (1980); indeed, either an RNA or a protein would be capable of releasing a mRNA from an mRNP. It is assumed by Collier and McCarthy (1981) that mmRNAs in the *Ilyanassa* egg do exist as mRNPs and their observation, as well as those of Brandhorst and Newrock (1981), that the isolated polar lobe produces some peptides that are specific for the mesentoblast embryo leads to the suggestion that there are cytoplasmic factors (contained in the isolated polar lobe) that are capable of activating mmRNAs that were presumably present in mRNPs. Thus, there is good evidence for the translational control of mmRNAs by concurrent transcription of the embryonic genome and by some of the maternal transcripts, that is, those activating factors in the isolated polar lobe that were made during oogenesis. That regulatory gene transcripts may be produced during early development is consistent with the amount of heterogeneous RNA made during early cleavage of the *Ilyanassa* embryo (Collier, 1976a).

Raff et al. (1975) have found that the synthesis of microtubule proteins begins in the gastrula of *Ilyanassa* and that their continued synthesis is insensitive to actinomycin D, although microtubule assembly was sensitive to actinomycin D. These authors suggested that there is a mechanism for the selection of tubulin mRNAs for translation at a specific time in development. This mechanism may be similar to the cytoplasmic factor(s) described above for the activation of mmRNAs in the isolated polar lobe. It is of interest that the production of tubulins is not under immediate genomic control whereas actinomycin D-sensitive events, presumably gene transcription, are required for the assembly of microtubules.

Maul and Avdalovic (1980) have developed a technique for the mass isolation of nuclear envelopes from germinal vesicles of *Spisula solidissima* oocytes. They have identified eight major proteins contained in the fibrous lamina pore complex. Three of these proteins are phosphorylated. Are they involved in the transport of mmRNAs into the cytoplasm and regulated by phosphorylation?

The evidence reviewed in this section demonstrates that oogenic mRNAs play a major role in early molluscan development. Some of these mmRNAs are differentially selected for translation by cytoplasmic factors, some of which are contained in the polar lobe. The presence of activating factors in the polar lobe implies that these factors were produced during oogenesis and, if this is so, that the maternal genome not only supplies and stabilizes structural gene transcripts but also provides a mechanism for their subsequent activation. Thus, the preparations made during oogenesis for early development are extensive *and* are inclusive of some of the control mechanisms used for the regulation of mmRNA translation. The integration of the activities of the maternal genome with the transcription of the embryonic genome is emphasized by the role of regulatory genes in controlling the rate of translation of some of the mmRNAs.

Direct evidence for transcriptional control of gene expression during both early and late embryogenesis has also been reviewed. Differential gene transcription is a major feature of organogenesis (Collier, 1966) and is precocious in respect to the differentiation of individual organs of the embryo.

The transcription of a small number of mRNAs (Collier and McCarthy, 1981) by the embryonic genome during early cleavage has been emphasized, not only because these mRNAs may have an important role in embryogenesis, but also because the classical actinomycin D experiments, in the absence of a method for detecting changes in a small number of structural genes, may have led us to overlook the importance of transcription during this early and crucial shape of development.

VIII. Concluding Remarks

It was asked earlier whether the evolutionary divergence of the Protostomia and the Deuterostomia would be reflected in the biochemistry of their development. This does not appear to be the case. Both groups have similar genomes in terms of the presence and organization of single copy and repeated sequences of DNA. Indeed, in the two bivalves and one gastropod investigated (Goldberg et al., 1975; Angerer et al., 1975), the sequence organization of the molluscan genome is similar to the model observed in *Xenopus* (a deuterostome) and not to the organization of the *Drosophila* (a protostome) genome (Manning et al., 1975). The relatively low level of rDNA amplification (Kidder, 1976a; Brown and Dawid, 1969) and reiteration of histone genes (Steele et al., 1978) suggests a difference between the molluscan genome and that of some deuterostomes. However, rDNA amplification is not universal among the deuterostomes, and the data on the reiteration of histone genes is not extensive in either group.

The initiation and duration of the synthetic periods for various classes of RNA during molluscan development are similar to that of the sea urchin and other animals where RNA synthesis has been sufficiently studied to permit a comparison. The role of maternal mRNAs and embryonic transcription in the development of molluscan embryos is basically the same as reported for other animal embryos.

Each state of differentiation of the molluscan embryo is characterized by a unique set of biochemical properties, such as sequences of DNA transcribed in *Acmaea* (Karp and Whiteley, 1973), hydrolase activities in *Lymnaea* (Norris and Morrill, 1964), and the transcription of structural genes in *Ilyanassa* (Collier and McCarthy, 1981; Brandhorst and Newrock, 1981). This situation is similar to the unique patterns of single copy DNA transcription at different developmental stages in the sea urchin embryo (Galau et al., 1976).

The effect of polar lobe removal on RNA synthesis in *Ilyanassa* was reviewed and the evidence suggests that the morphogenetic influence of the polar lobe does not arise from a major defect in RNA synthesis. Although polar lobe removal does reduce the absolute rate of total RNA synthesis (Collier, 1976a, 1977), it is not clear to what extent this is a cause or a result of the morphogenetic effect of the polar lobe.

From the studies of polypeptide synthesis by the lobeless embryo (Brandhorst and Newrock, 1981; Collier and McCarthy, 1981), it is clear that the loss of the polar lobe does not qualitatively alter the transcription of structural genes that are detectable by two-dimensional electrophoresis of [^{35}S]methionine-labeled proteins. These two studies provide the best available evidence on the distribution of *in vivo* translated mRNA [see discussion in preceding sections for the basis of rejecting earlier data from one-dimensional electrophoresis of proteins; also, see review by Jeffery (1983) for a different evaluation of this work], and it is clear that there is no indication of an exclusive localization of *in vivo* translated mRNA in the polar lobe. These data do not rule out the possible localization of mRNAs that are not detected by *in vivo* translation. The only available evidence on this point (Cheney and Ruderman, 1978) has not been published in detail.

J. S. Peterson and W. R. Jeffery (unpublished data, 1982; reviewed and illustrated by Jeffery, 1983) used poly(U) and cloned histone DNA sequences for hybridization probes to show the enrichment of histone mmRNA in the germinal vesicle of *Spisula*. When the germinal vesicle breaks down, this mmRNA is dispelled into the cytoplasm. These observations indicate that at least some of the mmRNAs are stored not in the cytoplasm of the oocyte but in its nucleus. Most significantly, this experimental approach using poly(U) and cloned genes as hybridization probes is well suited for observing the *in situ* localization of mRNAs. The complex pattern of polyadenylated RNAs described in the *Ilyanassa* egg by Gerdes and Kidder (1979) may be useful in conjunction with this approach to establish the localization of different categories of mRNAs in this egg.

The relative role of maternal mRNAs and gene transcription by the embryo has been discussed in Section VII and will not be repeated except to reiterate the possible significance of the restricted transcription of the embryonic genome in early cleavage, the probable regulation of mmRNA translation by activators produced during oogenesis and localized in the cytoplasm, and finally to note that precocious transcription plays a significant role in all phases of embryogenesis.

The precocity seen in gene transcription, both in the production of maternal mRNAs and in the transcription of organ-specific genes (Collier, 1966) prior to organogenesis, is a basic feature of embryogenesis. Precocious transcription, at whichever stage of development it occurs, is a mechanism that results in the production of sufficient mRNAs to produce enough proteins for an anticipated developmental event, that is, it is a temporal form of amplification. This precocity of mRNA production, which is seen in the development of all animals so far investigated, implies that there is an anticipatory feature written into the genomic program for embryogenesis.

The increased rate of accumulation of polypeptides in actinomycin D–treated embryos, which led to the postulation of regulatory genes that repress the translation of specific mRNAs (Collier and McCarthy, 1981), has been confirmed in an extended series of experiments using both actinomycin D and α-amanitin as inhibitors of RNA transcription (J. R. Collier, unpublished results, 1981). The discovery of these regulatory genes, which control the rate of translation of maternal mRNAs, promises to be a significant step in understanding the regulation of gene expression during early embryogenesis.

On the basis of a lower rate of DNA accumulation by the lobeless embryo of *Ilyanassa* during later stages of development and the high concentration of ATP in the polar lobe, this author has suggested that the morphogenetic deficiencies of the lobeless embryo result from the failure of stem cells of lobe-dependent organs to proliferate and/or from the formation of an insufficient number of stem cells. (See pp. 261–262 and Collier, 1975a.) I call this the stem-cell hypothesis and relate to it the findings of de Laat et al. (1980) on the appearance during early development of functional intercellular gap junctions in certain cell lines as a mechanism for distributing ATP and other low-molecular-weight components into specific cells. This stem-cell hypothesis is consistent with Atkinson's (1971) findings that the lobeless embryo produces cell types of many lobe-dependent organs that otherwise fail to differentiate in the lobeless embryo. This hypothesis is designed to explain a systemic or generalized effect of the polar lobe, as previously discussed (Collier, 1976a), and is not intended to exclude the possibility that the polar lobe may contain specific instructions, or elements that invoke specific instructions, for the differentiation of lobe-dependent organs.

Acknowledgments

I thank Miss Arlene Hughes and Miss Helena Rubenstein for their assistance. I thank Dr. G. M. Kidder for making available his unpublished data and Dr. J. B. Morrill for sharing his knowledge of the literature and for many helpful discussions. I especially thank Dr. Marjorie McCann-Collier for her critical reading of the manuscript and for her invaluable editorial assistance.

References

Abd-El-Wahab, A., and Pantelouris, E. M. (1957). Synthetic processes in nucleated and non-nucleated part of *Mytilus* eggs. *Exp. Cell Res.* **13**, 78–82.

Angerer, R. C., Davidson, E. H., and Britten, R. J. (1975). DNA sequence organization in the mollusc *Aplysia californica*. *Cell* **6**, 29–39.

Arceci, R. J., Senger, D. R., and Gross, P. R. (1976). The programmed switch in lysin-rich synthesis at gastrulation. *Cell* **9**, 171–178.

Arni, P. (1974). Licht- und elektronenmikroskopische Untersuchungen an Embryonen von *Lymnaea stagnalis* L. (Gastropoda, pulmonata) mit besonderer Berücksichtigung der fruhembryonalen Ernährung. *Z. Morphol. Tiere* **78**, 299–323.

Atkinson, J. W. (1971). Organogenesis in normal and lobeless embryos of the marine prosobranch gastropod *Ilyanassa obsoleta*. *J. Morphol.* **133**, 339–352.

Bell, E., and Reeder, R. (1967). The effect of fertilization on protein synthesis in the egg of the surf clam *Spisula solidissima*. *Biochim. Biophys. Acta* **142**, 500–511.

Berg, W. E. (1954). Peptidases in isolated blastomeres of *Mytilus edulis*. *Proc. Soc. Exp. Biol. Med.* **85**, 606–608.

Berg, W. E., and Kato, Y. (1959). Localization of polynucleotides in the egg of *Ilyanassa*. *Acta Embryol. Morphol. Exp.* **2**, 227–233.

Berger, E. (1977). Gene-enzyme variation in three sympatric species of *Littorina*. II. The Roscoff population, with a note on the origin of North American *L. littorea*. *Biol. Bull. (Woods Hole, Mass.)* **153**, 255–264.

Bester, A. J., Kennedy, D. S., and Heywood, S. M. (1975). Two classes of translational control RNA: Their role in the regulation of protein synthesis. *Proc. Natl. Acad. Sci. U.S.A.* **72**, 1523–1527.

Birnstiel, M. L., Chipchase, M., and Speirs, J. (1971). The ribosomal RNA cistrons. *Prog. Nucleic Acid Res. Mol. Biol.* **11**, 351–385.

Boon-Niermeyer, E. K. (1975). The effect of puromycin on the early cleavage cycles and morphogenesis of the pond snail *Lymnaea stagnalis*. *Wilhelm Roux's Arch. Dev. Biol.* **177**, 29–40.

Brachet, J. (1950). "Chemical Embryology." Wiley-Interscience, New York.

Brachet, J. (1967). Behaviour of nucleic acids during early development. *Compr. Biochem.* **28**, 23–54.

Brahmachary, R. L. (1973). Molecular embryology of invertebrates. *Adv. Morphog.* **10**, 115–174.

Brahmachary, R. L., and Palchoudhury, S. R. (1971). Further investigations on transcription and translation in *Limnaea* embryos. *Can. J. Biochem.* **49**, 926–932.

Brahmachary, R. L., Banerjee, K. P., and Basu, T. K. (1968). Investigations on transcription in *Limnaea* embryos. *Exp. Cell Res.* **51**, 177–184.

Brahmachary, R. L., Ghosal, D., and Tapaswi, P. K. (1971). Rhythmic incorporation of ^{32}P and ^{14}C-uracil in early mitotic cycles of *Limnaea* (Mollusc) eggs. *Z. Naturforsch., B: Anorg. Chem., Org. Chem., Biochem., Biophys., Biol.* **26B**, 822–824.

Brahmachary, R. L., Mallik, B., and Tapaswi, P. K. (1972). Transcription and degradation of heavy (pre 28 S) RNA and 4 S RNA during embryogenesis in *Limnaea*. *Z. Naturforsch., B: Anorg. Chem., Org. Chem., Biochem., Biophys., Biol.* **27B**, 1261–1266.

Brandhorst, B. P., and Newrock, K. M. (1981). Post-transcriptional regulation of protein synthesis in *Ilyanassa* embryos and isolated polar lobes. *Dev. Biol.* **83**, 250–254.

Brown, D. D., and Dawid, I. B. (1969). Specific gene amplification in oocytes. *Science* **160**, 272–279.

Cather, J. N., and Verdonk, N. H. (1974). The development of *Bithynia tentaculata* (Prosobranchia, Gastropoda) after removal of the polar lobe. *J. Embryol. Exp. Morphol.* **31**, 415–422.

Cheney, C. M., and Ruderman, J. V. (1978). Segregation of maternal mRNA at first cleavage in embryos of the mollusc *Spisula solidissima*. *J. Cell Biol.* **79**, 349a.

Clark, A. J., and Kidder, G. M. (1977). Polyadenylic acid in *Ilyanassa*: Estimates of the number and mean length of poly(A) tracts in embryonic and larval stages. *Differentiation* **8**, 113–122.

Clement, A. C. (1938). The structure and development of centrifuged egg fragments of *Physa heterostropha*. *J. Exp. Zool.* **79**, 435–460.

Clement, A. C. (1952). Experimental studies on germinal localization in *Ilyanassa*. I. The role of the polar lobe in determination of the cleavage pattern and its influence in later development. *J. Exp. Zool.* **121,** 593–626.

Clement, A. C. (1962). Development of *Ilyanassa* following removal of the D macromere at successive cleavage stages. *J. Exp. Zool.* **149,** 193–216.

Clement, A. C. (1968). Development of the vegetal half of the *Ilyanassa* egg after removal of most of the yolk by centrifugal force, compared with the development of animal halves of similar visible composition. *Dev. Biol.* **17,** 165–186.

Clement, A. C. (1971). *Ilyanassa*. In "Experimental Embryology of Marine and Fresh-water Invertebrates" (G. Reverberi, ed.), pp. 188–214. North-Holland Publ., Amsterdam.

Clement, A. C., and Tyler, A. (1967). Protein-synthesizing activity of the anucleate polar lobe of the mud snail *Ilyanassa obsoleta*. *Science* **158,** 1457–1458.

Collier, J. R. (1957). A study of alanylglycine dipeptidase activity during the development of *Ilyanassa obsoleta*. *Embryologia* **3,** 243–260.

Collier, J. R. (1960a). The localization of ribonucleic acid in the egg of *Ilyanassa obsoleta*. *Exp. Cell Res.* **21,** 126–136.

Collier, J. R. (1960b). The localization of some phosphorus compounds in the egg of *Ilyanassa obsoleta*. *Exp. Cell Res.* **21,** 548–555.

Collier, J. R. (1961). The effect of removing the polar lobe on the protein synthesis of the embryo of *Ilyanassa obsoleta*. *Acta Embryol. Morphol. Exp.* **4,** 70–76.

Collier, J. R. (1965). Morphogenetic significance of biochemical patterns in mosaic embryo. In "The Biochemistry of Animal Development" (R. Weber, ed.), Vol. 1, pp. 203–244. Academic Press, New York.

Collier, J. R. (1966). The transcription of genetic information in the spiralian embryo. *Curr. Top. Dev. Biol.* **1,** 39–59.

Collier, J. R. (1971). Number of ribosomal cistrons in the marine mud snail *Ilyanassa obsoleta*. *Exp. Cell Res.* **69,** 181–184.

Collier, J. R. (1975a). Nucleic acid synthesis in the normal and lobeless embryo of *Ilyanassa obsoleta*. *Exp. Cell Res.* **95,** 254–262.

Collier, J. R. (1975b). Polyadenylation of nascent RNA during the embryogenesis of *Ilyanassa obsoleta*. *Exp. Cell Res.* **95,** 263–268.

Collier, J. R. (1976a). Nucleic acid chemistry of the *Ilyanassa* embryo. *Am. Zool.* **16,** 483–500.

Collier, J. R. (1976b). The interconversion of purine nucleotides during *Ilyanassa* embryogenesis. *Exp. Cell Res.* **101,** 438–440.

Collier, J. R. (1977). Rates of RNA synthesis in the normal and lobeless embryo of *Ilyanassa obsoleta*. *Exp. Cell Res.* **106,** 390–394.

Collier, J. R. (1981a). Methods of obtaining and handling eggs and embryos of the marine mud snail *Ilyanassa obsoleta*. In "Marine Invertebrates," pp. 217–232. Report of the Committee on Marine Invertebrates, Institute of Laboratory Animal Resources, National Research Council—National Academy Press, Washington, D.C.

Collier, J. R. (1981b). Protein synthesis in the polar lobe and lobeless egg of *Ilyanassa obsoleta*. *Biol. Bull. (Woods Hole, Mass.)* **160,** 366–375.

Collier, J. R., and Garone, L. M. (1975). The localization of ATP in the polar lobe of the *Ilyanassa* egg. *Differentiation* **4,** 195–196.

Collier, J. R., and McCann-Collier, M. (1962). The deoxyribonucleic acid content of the egg and sperm of *Ilyanassa obsoleta*. *Exp. Cell Res.* **27,** 553–559.

Collier, J. R., and McCarthy, M. E. (1981). Regulation of polypeptide synthesis during early embryogenesis of *Ilyanassa obsoleta*. *Differentiation* **19,** 31–46.

Collier, J. R., and Schwartz, R. (1969). Protein synthesis during *Ilyanassa* embryogenesis. *Exp. Cell Res.* **54,** 403–406.

Collier, J. R., and Tucci, J. (1980). The reassociation kinetics of the *Ilyanassa* genome. *Dev., Growth Differ.* **22,** 741–748.

Collier, J. R., and Yuyama, S. (1969). Characterization of DNA-like RNA in the *Ilyanassa* embryo. *Exp. Cell Res.* **56,** 281–291.

Collier, M. M. (1975). Doctoral Dissertation, City University of New York, New York.

Cooper, J. A., and Hunter, T. (1981). Changes in protein phosphorylation in Rous sarcoma virus-transformed chicken embryo cells. *Mol. Cell. Biol.* **1,** 165–178.

Costello, D. P. (1948). Ooplasmic segregation in relation to differentiation. *Ann. N.Y. Acad. Sci.* **49,** 663–683.

Dauwalder, M. (1963). Initiation of RNA synthesis and nucleolar modification during cleavage in *Helix aspersa. J. Cell Biol.* **19,** 19A.

Davidson, E. H., Haslett, G. W., Finney, R. J., Allfrey, V. G., and Mirsky, A. E. (1965). Evidence for prelocalization of cytoplasmic factors affecting gene activity in early embryogenesis. *Proc. Natl. Acad. Sci. U.S.A.* **54,** 696–704.

Davidson, E. H., Hough, B. R., Chamberlin, M. E., and Britten, R. J. (1971). Sequence repetition in the DNA of *Nassaria (Ilyanassa) obsoleta. Dev. Biol.* **25,** 445–463.

Davidson, E. H., Galau, G. A., Angerer, R. C., and Britten, R. J. (1975). Comparative aspects of DNA organization in metazoa. *Chromosoma* **51,** 253–259.

de Laat, S. W., Tertoolen, L. G. J., Dorresteijn, A. W. C., and van den Biggelaar, J. A. M. (1980). Intercellular communication patterns involved in cell determination in early molluscan development. *Nature (London)* **287,** 546–548.

Dohmen, M. R., and Verdonk, N. H. (1974). The structure of a morphogenetic cytoplasm, present in the polar lobe of *Bithynia tentaculata* (Gastropoda, Prosobranchia). *J. Embryol. Exp. Morphol.* **31,** 423–433.

Dohmen, M. R., and Verdonk, N. H. (1979). The ultrastructure and role of the polar lobe in development of molluscs. *In* "Determinants of Spatial Organization" (S. Subtelny and I. R. Konigsberg, eds.), pp. 3–27. Academic Press, New York.

Donohoo, P., and Kafatos, F. C. (1973). Differences in the proteins synthesized by the progeny of the first two blastomeres of *Ilyanassa*, a "mosaic embryo." *Dev. Biol.* **32,** 224–229.

Easton, D. P., and Whiteley, A. H. (1979). The relative contribution of newly synthesized and stored messages to H1 histone synthesis in interspecies hybrid echinoid embryos. *Differentiation* **12,** 127–133.

Emanuelsson, H. (1973). Karyomeres in early cleavage embryos of *Ophryotrocha labronica* LaGreen and Bacci. *Wilhelm Roux's Arch. Entwicklungsmech. Org.* **173,** 27–45.

Emerson, C. P., Jr., and Humphreys, T. D. (1970). Regulation of DNA-like RNA and the apparent activation of RNA synthesis in sea urchin embryos: Quantitative measurements of newly synthesized RNA. *Dev. Biol.* **23,** 86–112.

Emerson, C. P., Jr., and Humphreys, T. D. (1971). Ribosomal RNA synthesis and the multiple, atypical nucleolus in cleaving embryo. *Science* **171,** 898–901.

Feigenbaum, L., and Goldberg, E. (1965). Effect of actinomycin D on morphogenesis in *Ilyanassa. Am. Zool.* **5,** 198.

Firtel, R. A., and Monroy, A. (1970). Polysomes and RNA synthesis during early development of the surf clam *Spisula solidissima. Dev. Biol.* **21,** 87–104.

Freeman, S. B. (1971). A comparison of certain isozyme patterns in lobeless and normal embryos of the snail, *Ilyanassa obsoleta. J. Embryol. Exp. Morphol.* **26,** 339–349.

Gabrielli, F., and Baglioni, C. (1975). Maternal messenger RNA and histone synthesis in embryos of the surf clam *Spisula solidissima. Dev. Biol.* **43,** 254–263.

Gabrielli, F., and Baglioni, C. (1977). Regulation of maternal mRNA translation in developing embryos of the surf clam *Spisula solidissima. Nature (London)* **269,** 529–531.

Galau, G. A., Klein, W. H., Davis, M. M., Wold, B. J., Britten, R. J., and Davidson, E. H. (1976). Structural gene sets active in embryos and adult tissues of the sea urchin. *Cell* **7**, 487–505.

Gerdes, P. A., and Kidder, G. M. (1979). Metabolism of oligo(A) tracts and association of oligo(A)(+)RNA with polysomes in molluscan embryos. *Can. J. Biochem.* **57**, 1305–1314.

Geuskens, M., and de Jonghe d'Ardoye, V. (1971). Metabolic patterns in *Ilyanassa* polar lobes. *Exp. Cell Res.* **67**, 61–72.

Goldberg, E., and Cather, J. N. (1963). Molecular heterogeneity of lactic dehydrogenase during development of the snail *Argobuccinum oregonense* Redfield. *J. Cell. Comp. Physiol.* **61**, 31–38.

Goldberg, R. B., Crain, W. R., Ruderman, J. V., Moore, G. P., Barnett, T. R., Higgins, R. C., Gelfand, R. A., Galau, G. A., Britten, R. J., and Davidson, E. H. (1975). DNA sequence organization in the genomes of five marine invertebrates. *Chromosoma* **51**, 225–251.

Gooch, J. L., Smith, B. S., and Knupp, D. (1972). Regional survey of gene frequencies in the mud snail *Nassarius obsoletus*. *Biol. Bull. (Woods Hole, Mass.)* **142**, 36–48.

Goudsmit, E. M. (1976). Galactogen catabolism by embryos of the freshwater snails, *Bulimnaea megasoma* and *Lymnaea stagnalis*. *Comp. Biochem. Physiol. B* **53B**, 439–442.

Hieter, P. A., Hendricks, M. B., Hemminki, K., and Weinberg, E. S. (1979). Histone gene switch in the sea urchin embryo. Identification of late embryonic messenger ribonucleic acids and the control of their synthesis. *Biochemistry* **18**, 2707–2716.

Hinegardner, R. (1974). Cellular DNA content of the mollusca. *Comp. Biochem. Physiol. A* **47A**, 447–460.

Horstmann, H. J. (1956). Der Galactogengehalt der Eier von *Lymnaea stagnalis* während der Embryonal Entwicklung. *Biochem. Z.* **328**, 342–347.

Humphreys, T. (1973). RNA and protein synthesis during early animal embryogenesis. *In* "Developmental Regulation: Aspects of Cell Differentiation" (S. J. Coward, ed.), p. 1. Academic Press, New York.

Jeffery, W. R. (1983). Maternal RNA and the embryonic localization problem. *In* "Gene Control of Early Embryonic Development" (A. Siddiqui, ed.). CRC Press, New York, N.Y. (in press).

Jockusch, B. (1968). Protein synthesis during the first three cleavages in pond snail eggs (*Lymnaea stagnalis*) *Z. Naturforsch., B: Anorg. Chem., Org. Chem., Biochem. Biophys., Biol.* **23B**, 1512–1516.

Karp, G. C. (1973). Autoradiographic patterns of ^{3}H-uridine incorporation during the development of the mollusc, *Acmaea scutum*. *J. Embryol. Exp. Morphol.* **29**, 15–25.

Karp, G. C., and Whiteley, A. H. (1973). DNA-RNA hybridization studies of gene activity during the development of the gastropod, *Acmaea scutum*. *Exp. Cell Res.* **78**, 236–241.

Kedes, L. H., and Gross, P. R. (1969). Synthesis and function of messenger RNA during early embryonic development. *J. Mol. Biol.* **42**, 559–576.

Kidder, G. (1972a). Gene transcription in mosaic embryos. I. The pattern of RNA synthesis in early development of the coot clam, *Mulinia lateralis*. *J. Exp. Zool.* **180**, 55–74.

Kidder, G. (1972b). Gene transcription in mosaic embryos. II. Polyribosomes and messenger RNA in early develop of the coot clam, *Mulinia lateralis*. *J. Exp. Zool.* **180**, 75–84.

Kidder, G. (1976a). The ribosomal RNA cistrons in clam gametes. *Dev. Biol.* **49**, 132–142.

Kidder, G. (1976b). RNA synthesis and the ribosomal cistrons in early molluscan development. *Am. Zool.* **16**, 501–520.

Kidder, G. M., Clark, A. J., and Gerdes, P. A. (1977). Polyadenylic acid in *Ilyanassa*. Localization and size distribution of newly-synthesized poly(A) in embryonic and larval stages. *Differentiation* **9**, 77–84.

Koser, R. B., and Collier, J. R. (1971). The molecular weight and thermolability of *Ilyanassa* ribosomal RNA. *Biochim. Biophys. Acta* **254**, 272–277.

Koser, R. B., and Collier, J. R. (1972). Characterization of the ribosomal RNA precursor in *Ilyanassa*. *Exp. Cell Res.* **70**, 124–128.

Koser, R. B., and Collier, J. R. (1976). An electrophoretic analysis of RNA synthesis in the normal and lobeless *Ilyanassa* embryo. *Differentiation* **6**, 47–52.

Lankester, E. R. (1877). Notes on the embryology and classification of the animal kingdom: Comprising a revision of speculation relative to the origin and significance of the germ-layers. I. The planula theory. *Q. J. Microsc. Sci.* **17**, 399–454.

MacKay, S., and Newrock, K. M. (1982). Histone subtypes and switches in synthesis of histone subtypes during *Ilyanassa* development. *Dev. Biol.* **93**, 430–437.

McLean, K. W., and Whiteley, A. H. (1973). Characteristics of DNA from the oyster, *Crassostrea gigas*. *Biochim. Biophys. Acta* **335**, 35–41.

McLean, K. W., and Whiteley, A. H. (1974). RNA synthesis during the early development of the pacific oyster, *Crassostera gigas*. *Exp. Cell Res.* **87**, 132–138.

Manning, J. E., Schmidt, C. W., and Davidson, N. (1975). Interspersion of repetitive and nonrepetitive DNA sequences in the *Drosophila melanogaster* genome. *Cell* **2**, 141–155.

Maul, G. G., and Avdalovic, N. (1980). Nuclear envelope proteins from *Spisula solidissima* germinal vesicles. *Exp. Cell Res.* **130**, 229–240.

Mirkes, P. E. (1970). Protein synthesis before and after fertilization in the egg of *Ilyanassa obsoleta*. *Exp. Cell Res.* **60**, 115–118.

Mirkes, P. E. (1972). Polysomes and protein synthesis during development of *Ilyanassa obsoleta*. *Exp. Cell Res.* **74**, 503–508.

Moon, R. T., and Morrill, J. B. (1979). Further studies on the electrophoretically mobile acid phosphatases of the developing embryo of *Lymnaea palustris* (Gastropoda, Pulmonata). *Acta Embryol. Exp.* **1**, 3–15.

Morrill, J. B. (1963). Bound amino acids of egg albumen and free amino acids in the larval and adult *Limnaea palustris*. *Acta Embryol. Morphol. Exp.* **6**, 339–343.

Morrill, J. B. (1964). Protein content and dipeptidase activity of normal and cobalt-treated embryos of *Limnaea palustris*. *Acta Embryol. Morphol. Exp.* **7**, 131–142.

Morrill, J. B. (1973). Biochemical and electrophoretic analysis of acid and alkaline phosphatase activity in the developing embryo of *Physa fontinalis* (Gastropoda, Pulmonata). *Acta Embryol. Exp.* **6**, 61–82.

Morrill, J. B., and Norris, E. (1965). Electrophoretic analysis of hydrolytic enzymes in the *Ilyanassa* embryo. *Acta Embryol. Morphol. Exp.* **8**, 232–238.

Morrill, J. B., Norris, E., and Smith, S. D. (1964). Electro- and immunoelectrophoretic patterns of egg albumen of the pond snail *Limnaea stagnalis*. *Acta Embryol. Morphol. Exp.* **7**, 155–166.

Morrill, J. B., Rubin, R. W., and Grandi, M. (1976). Protein synthesis and differentiation during pulmonate development. *Am. Zool.* **16**, 547–562.

Naus, C. G., and Kidder, G. M. (1982). Regulation of expression of the ribosomal RNA cistrons in *Ilyanassa* embryos: Nucleolus-like bodies and nucleologenesis. *J. Exp. Zool.* **219**, 51–66.

Nelson, M. S., and Scandalios, J. G. (1977). Developmental expression and biochemical characterization of catalase and aminopeptidase of *Nassarius obsoleta*. *J. Exp. Zool.* **199**, 257–268.

Newrock, K. M., and Raff, R. A. (1975). Polar lobe specific regulation of translation in embryos of *Ilyanassa obsoleta*. *Dev. Biol.* **42**, 242–261.

Newrock, K. M., Alfageme, C. R., Nardi, R. V., and Cohen, L. H. (1977). Histone changes during chromatin remodeling in embryogenesis. *Cold Spring Harbor Symp. Quant. Biol.* **42**, 421–431.

Norris, E., and Morrill, J. B. (1964). An electrophoretic analysis of hydrolytic enzymes in adult organs and developing embryo of *Limnaea palustris*. *Acta Embryol. Morphol. Exp.* **7**, 29–41.

Raff, R. A., Newrock, K. M., and Secrist, R. D. (1975). Regulation of microtubule protein synthesis in embryos of the marine snail, *Ilyanassa obsoleta*. *Dev. Biol.* **44**, 369–374.

Raven, C. P. (1972). Chemical embryology of mollusca. *In* "Chemical Zoology" (M. Florkin and B. T. Scheer, eds.), Vol. 7, pp. 155–185. Academic Press, New York.

Rebhun, L. I., White, D., Sander, G., and Nettie, I. (1973). Cleavage inhibition in marine eggs by puromycin and 6-dimethylaminopurine. *Exp. Cell Res.* **77,** 312–318.

Risley, M. S., and Eckhardt, R. A. (1981). H1 histone variants in *Xenopus laevis. Dev. Biol.* **84,** 79–87.

Rosenthal, E. T., Hunt, T., and Ruderman, J. V. (1980). Selective translation of mRNA controls the pattern of protein synthesis during early development of the surf clam, *Spisula solidissima. Cell* **20,** 487–494.

Ruderman, J. V., and Gross, P. R. (1974). Histones and histone synthesis in sea urchin development. *Dev. Biol.* **36,** 286–298.

Schmekel, L., and Fioroni, P. (1976). Cell differentiation during early development of *Nassarius reticulatus* L. (Gastropoda, Prosobranchia). II. Morphological changes of nuclei and nucleoli. *Cell Tissue Res.* **168,** 361–371.

Sconzo, G., Pirrone, A. M., Mutolo, V., and Giudice, G. (1970). Synthesis of ribosomal RNA during sea urchin development. III. Evidence for an activation of transcription. *Biochim. Biophys. Acta* **199,** 435–440.

Steele, R. E., Merrifield, P. A., and Ruderman, J. V. (1978). Reiteration frequency and expression of histone genes in the early embryo of the clam, *Spisula solidissima. J. Cell Biol.* **79,** 358a.

Surrey, S., Ginzburg, I., and Nemer, M. (1979). Ribosomal RNA synthesis in pre- and post-gastrula-stage sea urchin embryos. *Dev. Biol.* **71,** 83–99.

Tapaswi, P. K. (1972). RNA synthesis during "oogenesis to the onset of fertilization" in *Limnaea* (Mollusc). *Z. Naturforsch, B: Anorg. Chem., Org. Chem., Biochem., Biophys., Biol.* **27B,** 581–582.

Tapaswi, P. K. (1974). Further investigations on transcription during oogenesis and immediately after activation by sperm in *Limnaea* (Mollusca) eggs. *Acta Embryol. Exp.* **2,** 191–195.

Teitelman, G. (1973). Protein synthesis during *Ilyanassa* development: Effect of the polar lobe. *J. Embryol. Exp. Morphol.* **29,** 267–281.

van den Biggelaar, J. A. M. (1971a). RNA synthesis during cleavage of the *Lymnaea* egg. *Exp. Cell Res.* **67,** 207–210.

van den Biggelaar, J. A. M. (1971b). Timing of the phases of the cell cycle with tritiated thymidine and Feulgen cytophotometry during the period of synchronous division in *Lymnaea. J. Embryol. Exp. Morphol.* **26,** 353–366.

van den Biggelaar, J. A. M. (1971c). Timing of the phase of the cell cycle during the period of asynchronous division up to the 49-cell stage in *Lymnaea. J. Embryol. Exp. Morphol.* **26,** 367–391.

van den Biggelaar, J. A. M., Dorresteijn, A. W. C., de Laat, S. W., and Bluemink, J. G. (1981). The role of topographical factors in cell interaction and determination of cell lines in molluscan development. *In* "Internation Cell Biology 1980–1981" (H. G. Schweiger, ed.), pp. 526–538. Springer-Verlag, Berlin and New York.

van der Wal, U. P. (1976). The mobilization of the yolk of *Lymnaea stagnalis.* II. The localization and function of the newly synthesized proteins in the yolk granules during early embryogenesis. *Proc. K. Ned. Akad. Wet., Ser. C* **79,** 405–420.

van Dongen, C. A. M., Mikkers, F. E. P., de Bruyn, C., and Verheggen, T. P. E. M. (1981). Molecular composition of the polar lobe of first cleavage stage embryos in comparison with the lobeless embryo in *Nassarius reticulatus* (Mollusca) as analyzed by isotachophoresis. *In* "Analytical Isotachophoresis" (F. M. Everaerts, ed.), pp. 207–216. Elsevier, Amsterdam.

Wahli, W., Dawid, I. G., Ryffel, G. U., and Weber, R. (1981). Vitellogenesis and the vitellogenin gene family. *Science* **212,** 298–304.

Wegnez, M., and Denis, H. (1979). Biochemical research on oogenesis. Transfer RNA is fully charged in the 42-S storage particles of *Xenopus laevis* oocytes. *Eur. J. Biochem.* **98,** 67–75.

Weintraub, H., Flint, S. J., Lefrak, I. M., and Grainger, R. M. (1977). The generation and propagation of variegated chromosome structures. *Cold Spring Harbor Symp. Quant. Biol.* **42,** 401–407.

Wilson, E. B. (1928). "The Cell in Development and Heredity." Macmillan, New York.

Woodland, H. R., and Adamson, E. D. (1977). The synthesis and storage of histones during the oogenesis of *Xenopus laevis*. *Dev. Biol.* **57,** 118–135.

Woodland, H. R., Flynn, J. M., and Wyllie, A. J. (1979). Utilization of stored mRNA in *Xenopus* embryos and its replacement by newly synthesized transcripts; Histone H1 synthesis using interspecies hybrids. *Cell* **18,** 165–171.

8

Physiological Ecology of Marine Molluscan Larvae

B. L. BAYNE

Natural Environmental Research Council
Institute for Marine Environmental Research
Plymouth, United Kingdom

I. Introduction

Physiological ecology encompasses the responses of an animal to its environment, viewed primarily in terms of the acquisition and utilization of energy and nutrients. There is no clear distinction here between the biochemistry of energy metabolism, the expression of these biochemical processes as energy flow in the individual, and aspects of behavior. On the contrary, the physiological ecologist must draw on knowledge of both the biochemical and the behavioral processes involved. Nevertheless, the physiologist must provide the understanding of the means by which the animal (in this case the larva) gains energy and nutrients

THE MOLLUSCA, VOL. 3
Development

from the environment, how this gain is partitioned among the various metabolic requirements of maintenance, movement, and growth, and then to interpret these processes in the context of the environmental variables that comprise the habitat of the animal. A further dimension is provided by considering the evolutionary challenges that underline the life-history patterns observed, and the extent to which an understanding of the physiological energetics of the individual can help in interpreting evolutionary events.

The questions posed are ecological ones, therefore, but they require a physiological approach to the functioning of the whole organism. Much of the available relevant information on molluscan larvae derives from laboratory studies originally designed to facilitate the cultivation of particular species, usually for commercial rather than scientific objectives. Furthermore, the aquacultural interest has largely been confined to lamellibranch molluscs, primarily oysters, with the result that many molluscan larval types have not been researched, at least to the point where the detailed physiological energetics can be resolved. This bias shows recent signs of change, however, particularly through a growing interest in fundamental questions of life-history strategies and of the survival of populations.

This chapter necessarily reflects the current bias in the literature and concentrates on a few examples from work on oysters, clams, mussels, and some gastropod species, notably *Nassarius* and *Crepidula*. Many of the studies available for review were not designed to understand aspects of energy balance, but rather to describe rates of growth under different environmental conditions. Such work has been dramatically successful, in some cases, in optimizing hatchery conditions for growth and survival of larvae (e.g., the advances made in the commercial cultivation of oysters in the last decade). However, it is only drawn on for this chapter when it illuminates aspects of feeding behavior and metabolic expenditure relevant to problems of energy flow.

This chapter is also primarily concerned with planktotrophic molluscan larvae, that is those that rely on planktonic organisms (and possibly suspended organic detritus) as food. In some species (notably oysters in the genus *Ostrea*), the larvae may first be brooded within the mantle cavity of the adult, to be released later as shelled veligers. In other species (e.g., *Nassarius*) the eggs first develop within egg capsules; shelled veligers escape from the capsules at the start of their planktonic phase. In yet other species fertilization occurs within the water mass and the larvae that develop are planktonic from the start. All of these developmental types nevertheless share a dependence on planktonic food for at least part of their larval life. This is in contrast to nonplanktotrophic, or lecithotrophic, larvae which are nourished by yolk deposits in the egg and which, although they may be pelagic for some period of time prior to metamorphosis, are not dependent on a planktonic food source for successful development. Jablonski and Lutz

(1980) provide a useful review of the various molluscan larval types, with reference to the main taxonomic and ecological groupings.

The pelagic phase of planktotrophic larval life can be divided into a phase of growth and development followed by a delay period in which development to a morphogenic condition at which metamorphosis is possible (i.e., a condition of metamorphic competence) is complete, but settlement out of the plankton, and metamorphosis to a benthic way of life, is delayed until a suitable substratum is found (Scheltema, 1967). This is a useful distinction to make, not least in terms of the behavior and ecology of the larvae. From the point of view of physiological energetics, the gradual transition from a free-swimming veliger to a swimming/crawling pediveliger may or may not be marked by changes in the rate of metabolism and feeding efficiency (Pechenik, 1980). These questions will be discussed briefly, but the morphogenic aspects of the delay of metamorphosis and the processes within metamorphosis are beyond the scope of the chapter.

The aim of this chapter is, therefore, to provide a summary of informatin on the main components of energy balance in planktotrophic larvae (feeding, metabolism, and growth) set in an ecological context and using, in the main, illustrations from recent studies on a few species of bivalve and gastropod molluscs.

II. Feeding

A. Methods of Feeding

Molluscan veliger larvae feed by means of ciliary currents on the velum. Descriptions of the feeding currents are given by Yonge (1926; oyster larvae) Werner (1955; *Crepidula* larvae), Thompson (1959; nudibranch larvae), Fretter (1967) and Fretter and Montgomery (1968; prosobranch larvae). Strathmann et al. (1972) and Strathmann and Leise (1979) have studied suspension feeding in various marine invertebrate larvae including molluscs. In the veliger the cilia are arranged as two bands on the velum edge, enclosing the ciliated food groove between them. The pre-oral band consists of long compound cilia that produce both the swimming and the feeding currents. The post-oral band is composed of shorter cilia that beat towards the pre-oral band and the mouth. The combined effort of these two ciliary bands is to collect suspended particles in the food groove, from which they are taken to the mouth, either to be ingested or to be rejected over the oral palp or foot.

Strathmann and Leise (1979) observed that the pre-oral cilia move faster than the particles in the feeding current and they suggest that the cilia overtake the particles and push them into the food groove, with the post-oral cilia helping to retain particles in the groove and possibly slowing the current from the pre-oral

TABLE I

Volume of Water Passed in the Pre-Oral Ciliary Current, and Clearance Rates in Three Molluscan Veligers[a]

Species	Pre-oral cilium length (μm)	Volume through the preoral band[b] (μm³/sec μm)	Fraction caught from n particles	Clearance rate[b] (μm³/sec μm)
Crassostrea gigas	30	8,200	0.44	3,600
Tritonia diomedea	40	21,000	0.25	4,900
Nassarius obsoletus	70	110,000	0.15	17,000

[a] From Strathman and Leise, 1979.

[b] Per μm velar edge.

cilia to facilitate particle capture. Direct evidence for this mechanism, and indeed for any other proposed mechanism, is lacking. However, feeding by means of a simple sieve, comprised of adjacent cilia, seems unlikely. The metachronal nature of the pre-oral ciliary beat suggests that the effective sieve diameter would be the metachronal wavelength, which may exceed 20 μm in some veligers, and thus would exceed the diameter of captured particles (Strathmann and Leise, 1979).

Jørgensen (1981) has postulated a principle for particle retention in ciliary suspension feeders based on velocity gradients between neighboring ciliary currents. By this mechanism a particle moving in the pre-oral ciliary current would experience a velocity gradient between this and the surface current in the food groove. Thus, the particle would be exposed to transverse forces which Jørgensen supposes will cause the particle to migrate perpendicularly to the through-current, resulting in its capture. The post-oral cilia might contribute to the surface current in the same way as that proposed by Jørgensen (1981) for the laterofrontal cilia on the adult eulamellibranch gill; particle retention is then a complex function of particle size and the magnitude of the velocity gradients. However, the theory of particle behavior in complex patterns of water flow is yet to be developed, and Jørgensen's ideas have yet to be tested.

Strathmann and Leise (1979) were able to estimate clearance rate, or the volume of water swept clear of particles per unit time, per micron of velar edge, by measuring the pre-oral cilium length and calculating the volume of water passed through the pre-oral band per second (Table I). In these three species, the longer the pre-oral cilia the greater the volume of water passed but the smaller the efficiency of particle capture; nevertheless, the species with long pre-oral cilia had higher clearance rates per micron of velar edge than those with shorter cilia. Taking estimates of the total length of the velar edge for *Nassarius* (Thorson, 1946; Scheltema, 1962) and for *Crassostrea* (Galstoff, 1964) the values in Table

I calculate to ~7.0 μl/h/larva for *Crassostrea* and ~80 μl/h/larva for *Nassarius* and are in good agreement with observed clearance rates at particle concentrations >20 cells/μl (see next section).

B. Particle Selection and Clearance

Rates of clearance of particles from suspension are dependent on particle size, concentration of particles, larval size, and temperature. For two bivalve species (Fig. 1) maximum retention efficiencies were recorded for particle diameters 2–6 μm (Sprung, 1982); particles larger than 9 μm were not retained by *Mytilus* larvae (Riisgard et al., 1980) and Wilson (1980) reported reduced retention by *Ostrea* larvae of particles smaller than 3 μm. Clearance rates are highest at low particle concentrations (Fig. 2) declining at concentrations greater than 5–10 cells/μl in two bivalve veligers (Walne, 1965; Sprung, 1982) and at concentrations greater than 50–100 cells μl in *Nassarius* (Pechenik and Fisher, 1979). Gallager and Mann (1980) used a fluorometer to measure the rates of particle clearance by the larvae of *Teredo navalis* and *Aplysia californica* at constant food

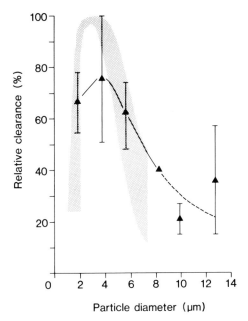

Fig. 1. Relative clearance of particles in suspension by bivalve veligers. ▲, *Ostrea edulis*, 203 μm shell length (from Walne, 1965, calculated with clearance rate of cells of *Phaeodactylum* set at 100%; means ± 95% c.l. The stippled curve is for *Mytilus edulis*, 5–13 days old (from Riisgard et al., 1980 and Sprung, 1982).

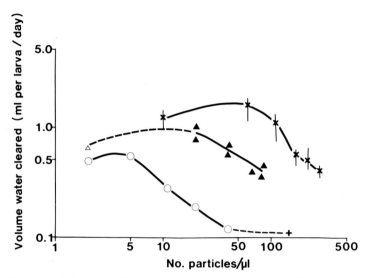

Fig. 2. Rates of clearance of *Isochrysis galbana* in suspension by veliger larvae, plotted on logarithmic scales. △, *Ostrea edulis*, shell length 200 μm (from Bruce et al., 1939); ▲, *O. edulis*, shell length 219 μm (From Walne, 1965), ○, *Mytilus edulis*, shell length 187 μm (from M. Sprung, personal communication); +, *M. edulis*, shell length 170 μm (from Bayne, 1965); x, *Nassarius obsoletus*, shell length 295 μm (from Pechenik and Fisher, 1979).

concentrations. At concentrations from 10^3–10^6 cells/ml (*Isochrysis galbana*), the grazing rate by *Teredo* was quite uniform at $\sim 10^4$ cells/larva/day; *Aplysia* increased grazing rate from $\sim 5 \times 10^3$ to $\sim 10^5$ cells/larva/day. These values are in fair agreement with the results of Wilson (1979; *Ostrea edulis*) and Pechenik and Fisher (1979; *N. obsoletus*).

Clearance rates increase with the size of the larva at any one particle concentration (Wilson, 1980). Data from Walne (1956, 1965) yield an allometric relationship for clearance rate (CR) by *Ostrea* larvae (CR: μm/larva/day) against dry flesh weight (W; μg) of: CR = 0.67 $W^{0.53}$. More recent data for *Ostrea* by M. Helm (personal communication) yield a similar intercept but higher proportionality constant: CR = 0.73 $W^{1.11}$, which is consistent with exponential rates of growth achieved in Helm's experiments. *Nassarius* veligers show higher rates of clearance with maxima at higher particle concentrations than the smaller bivalve veligers. However, in such experiments it may be difficult to identify the numbers of cells removed from suspension but not ingested by the larvae, that is, the pseudofeces. At high cell concentrations, as many as 75% of the cells may be rejected in this way (Walne, 1965). The reported effects of temperature on clearance rates are very varied; Wilson (1980) estimated Q_{10} values >2.0 for various temperature increments between 9.2 and 35.0°C (*Ostrea* larvae) but higher Q_{10} values have also been observed (Bayne, 1976). Sprung (1982) estimated Q_{10} values of 4 to 5 between 6 and 12°C and <2 between 12 and 18°C for clearance rates by *Mytilus edulis*.

Mollusc veligers possess ciliary rejection currents at the mouth, either in the oral palp (bivalves) or foot (prosobranchs, opisthobranchs). Rejection appears to be based on particle concentration and size, rather than any property of quality of the item as food. Rejected particles may be bound in mucus (Yonge, 1926) but the role of mucus in feeding by veligers needs to be reassessed. When starved larvae are placed in a suspension of food the gut is filled rapidly and further clearance of particles is accompanied by rejection and a reduced rate of ingestion (Fretter and Montgomery, 1968; Babinchak and Ukeles, 1979). At low particle concentrations, feeding may be continuous, but at much higher concentrations high rates of rejection appear to interfere with feeding and, ultimately, with growth (Walne, 1965; Fretter and Montgomery, 1968; Mapstone, 1970; Sprung, 1982). The relationship between natural particle concentrations (phytoplankton, organic detritus, and silt) and larval feeding requires further study.

It seems doubtful that veligers can distinguish between particles of similar size, shape, and density. Paulson and Scheltema (1968) found that *Nassarius obsoletus* veligers distinguished between three phytoplankton species in a mixture, but given a narrow size range for maximal retention efficiencies (Fig. 1) selection could have been based on size. Different prosobranch larvae ingested many species of phytoplankton equally (Fretter and Montgomery, 1968). However, more readily digestible plant cells were retained for a longer time in the stomach than undigestible cells and inorganic particles, which were both passed rapidly to the intestine for defecation. In these circumstances, and with resuspension of fecal particles occuring, "selection" for the more digestible cells would appear in particle counts over time. Babinchak and Ukeles (1979) used epifluorescence microscopy to observe feeding by *Crassostrea virginica* veligers and confirmed these observations; cells of *Monochrysis lutheri* were lysed and digested rapidly, whereas cells of *Chlorella autotrophica* were ingested but not lysed and were rapidly voided.

Millar (1955), Thompson (1959), Fretter and Montgomery (1968), Pilkington and Fretter (1970), Williams (1980), and Bickell et al. (1981) all describe the movement and possible mechanisms for sorting of particles in the veliger digestive system. Particles not digested can be isolated in the stomach and moved directly to the intestine whereas particles to be digested may be concentrated separately, subjected to shearing forces and parturition and eventually taken into the digestive gland for intracellular digestion. Shearing occurs by rotation of the food bolus against the gastric shield in the stomach, effected by transverse beating of the style-sac cilia. The result is the mechanical disruption of some cells and one distinction between digestible and undigestible particles (quite apart from any enzymatic deficiencies in the larva) may be differences in the resistance to these mechanical shearing forces (cf. *Monochrysis* and *Chlorella*).

Control of particle selection may therefore occur at various stages in feeding, including the ciliary feeding current itself (size selection, largely passive), the rejection tracts around the mouth (control over the quantity of food ingested) and the stomach and digestive gland (affecting attributes of food quality/digestibility). There is also a possible behavioral component of food selection, which may be significant in the natural environment. Veligers normally swim in a cyclical pattern of upward movement (more or less spiral) followed by passive sinking. This may be varied both according to the presence or absence of food particles and in respect to stimuli that may signify the quality of food present (Mapstone, 1970). There is uncertainty regarding the efficiency with which veligers may feed during passive sinking, when the pre-oral cilia are not beating, but the larvae do seem capable of altering their vertical position in the water column in order to optimize their feeding conditions.

C. Ingestion

The form of the clearance rate curves in Fig. 2 implies an increase in the rate of ingestion, with increased particle concentration, to a maximum value above which the ingestion rate should remain independent of particle concentration, assuming that all particles cleared from suspension are ingested. This is the *functional response* to food, for which various mathematical relationships have been derived (Mullin et al., 1975). Parsons et al. (1967) used a modified empirical function:

$$r = R_{max}(1 - e^{-kp}e^{kp_0})$$

where r is the ingestion rate, R_{max} is the maximum or limiting value for ingestion rate, k is a constant, p the particle concentration, and p_0 is the threshold concentration necessary to initiate feeding. The existence (and ecological relevance of) a feeding threshold for planktonic copepods is controversial (Frost, 1975; Conover, 1981); suitable data for molluscan veligers do not exist. The form of the functional response may also be affected at high particle concentrations as discussed earlier.

Pechenik and Fisher (1979) recorded an increase in ingestion rate for *Nassarius* larvae feeding on *Isochrysis galbana* over the concentration range 10–64 cells/μl with no significant variation in ingestion rate from 64 to 282 cells/μl. Wilson (1979), who used a flow-through culturing apparatus, observed a reduction in ingestion rate for *Ostrea* larvae at concentrations greater than 200–300 cells μl. Gallager and Mann (1980) present data for *Teredo* and *Aplysia* larvae.

Figure 3 is compiled from data in Walne (1965) and Wilson (1980) to illustrate the relationship between particle diameter and ingestion rate in *Ostrea edulis* larvae at constant concentration (by volume) of particles. Maximum ingestion rate (R_{max}), when calculated as numbers of cells per day, declines with an

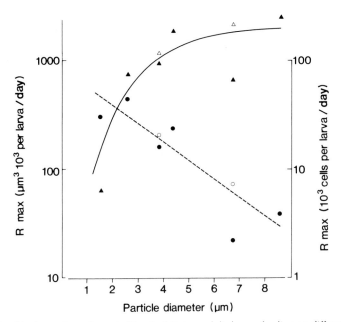

Fig. 3. Maximum ingestion rate (R_{max}) of *Ostrea edulis* larvae feeding on different species of phytoplankton, calculated in terms of volume (triangles and the continuous line) and in terms of number (circles and the dashed line). Values calculated from Walne (1965) (▲, ●) and from Wilson (1980) (△, ○).

increase in particle diameter. The limiting particle concentration, at which $r = R_{max}$, also decreases with increasing particle size, for example from ~350 cells/μl for *Thalassomonas caeca* (diameter 1.5–2.00 μm) to ~20 cells/μl for *Dunaliella tertiolecta* (diameter 6.0–7.0 μm). When R_{max} is calculated as the volume of cells ingested per day, values increase asymptotically with increase in particle size, with maximal values for particles >4 um in diameter, consistent with observations on relative clearance (Fig. 1). The asymptotic value in Fig. 3 represents a maximum daily intake of ~0.45 μg dry matter/larva/day, equivalent to 50% of larval body weight per day.

The rate at which cells are ingested is a function not only of clearance rate, particle concentration, and particle size, but also of the rate at which the gut is emptied (i.e., the gut retention time); this may, in turn, be directly proportional to particle size. Pechenik and Fisher (1979) found that gut retention time for *Nassarius* was longest (>180 min) for *Dunaliella* (cell volume ~300 μm³), shortest (<90 min) for *Isochrysis galbana* (57 μm³) and intermediate (~120 min) for *Thalassiosira pseudonana* (120–160 μm³). Bearing in mind the earlier discussion of rapid gut clearance of inorganic particles or even of algal cells with cell walls, the relationship between gut retention time and ingestion rate is likely to be complex. Smaller phytoplankton species appear to possess larger amounts of organic carbon per cell volume than larger species (Strathmann, 1967) so differences in R_{max} in volume units may not be reflected as differences in ingested carbon (Pechenik and Fisher, 1979).

Rates of ingestion also increase with the size of the larva. Pechenik (1980) records values for *Nassarius obsoletus* and *Crepidula fornicata* feeding on *Isochrysis*, which, when related to body weight, can be described by an allometric equation of the form:

$$\text{Ingestion rate} = aW^b$$

where b takes values from 0.5 to 0.6. Data from Walne (1965) give an exponent value $b = 0.54$ for *Ostrea edulis* larvae and (from Sprung, 1982) of $b = 0.87$ for *Mytilus edulis* larvae at 12°C. Sprung calculated the following equation for *Mytilus* larvae feeding at 12°C:

$$\text{Number of } \textit{Isochrysis} \text{ cells ingested/h} = 769 \, W^{0.87}$$

where W is the ash-free dry-tissue weight in μg. Unpublished data from M. Helm (personal communication) are plotted with data from Pechenik (1980) in Fig. 4. Ingestion rates are also very dependent on the temperature. For example, for

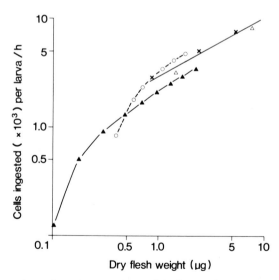

Fig. 4. The maximum ingestion rate (R_{max}) of the larvae of *Nassarius obsoletus* (x), *Crepidula fornicata* (△) (both from Pechenik, 1980), *Ostrea edulis* (○), and *Crassostrea gigas* (▲) [both from M. Helm, personal communication; 24–25°C, 32°/oo (*O. edulis*) and 25°/oo (*C. gigas*], related to dry flesh weight.

Crassostrea larvae feeding on *Monochrysis*, Q_{10} (16–24°C) = 3.4 (Ukeles and Sweeney, 1969); *Ostrea* feeding on *Isochrysis*, Q_{10} (17–25°C) = 2.4 (Walne, 1965); *Mytilus* larvae feeding on *Isochrysis*, Q_{10} = 3.2 (Bayne, 1965).

Lehman (1976) and Lam and Frost (1976) included some of these relationships in their models designed to elucidate general rules for optimal foraging by filter feeders. Although developed with planktonic crustacea in mind, aspects of these models also apply to filter-feeding planktonic veligers. Lehman's model sets the net rate of energy gain equal to the rate of energy assimilation from the diet minus the sum of the energy costs of clearance and of the rejection of unwanted particles. Assimilated energy is considered a function of gut passage time; the cost of clearance is assumed to be a function of drag in the feeding apparatus under conditions of viscous flow; and the cost of rejection is assumed to vary linearly with the number of particles rejected. The model of Lam and Frost (1976) is also based on the optimization of the net rate of energy intake, but includes animal size as a variable and assumes assimilation efficiency is constant.

Some of these terms have not been determined by experimentation with veligers, and analytical solutions are not possible. However, the qualitative behavior and conclusions of the two models are similar and agree with some observations on molluscan larvae. For example, in both cases clearance rate should increase with particle concentration to a maximum that corresponds with the concentration at which the gut first becomes filled; clearance rate should then decline, because a full gut may be maintained at the higher cell concentrations by a lower rate of clearance. The rate of ingestion should rise sigmoidally with increased particle concentration, the inflexion coinciding with the point at which the gut becomes fully packed with cells. If an upper limit on clearance rate is imposed by morphological or physiological constraints (Lam and Frost, 1976), the clearance rate may be truncated at an intermediate particle concentration, with implications for ingestion rate also. Results in Fig. 2 suggest a qualitative agreement with model predictions for clearance rates; sigmoidal ingestion rates, with inflexion at very low particle concentrations, are indicated in the data of Walne (1965) for *Ostrea* larvae.

Lehman (1976) also predicts that when the rejection costs are high, all particles filtered, including undigestible particles, should be ingested. Lam and Frost (1976) predict that an animal should maintain a balanced net rate of energy gain at lower food-cell concentrations as the particle size increases, and this is also consistent with observation on *Ostrea* larvae (Walne, 1965). As Hughes (1980) points out, experiments should be specifically designed to test the predictions of these models, rather than relying on *a posteriori* validation. The generality of the models, their reliance on reasonable physiological premises, and their apparent concurrence with some recorded data suggest that such experiments would be worthwhile.

D. Food Value

1. Dissolved Organic Matter

Stephens (1981) has reviewed studies on the trophic role of dissolved organic matter (particularly dissolved free amino acids, or DFAA) in marine animals and emphasized the need for quantitative studies of net flux under conditions that do not result in an unstirred layer between the medium and the animal (Wright, 1979). Meeting all the criteria in experiments with larvae is difficult (Fankboner and De Burgh, 1978; Sorokin and Wyshkwartzer, 1973) due to the need for concentration of the larvae in a small volume of water, with resultant inhibition of ciliary activity. Rice et al. (1980) measured a maximum rate of uptake of DFAA by veligers of *Ostrea edulis* as 8.9×10^{-5} μmol/h/larva, with a K_t (concentration of DFAA at which net flux is 50% of the maximum) of 35 μ*M*, a value the authors considered artificially high due to crowding of the larvae; at 10 μ*M* (glycine) the recorded rate of ~1.0×10^{-5} μmol/h could account for 20% of the total oxidative requirements of the larva (calculated from values for oxygen consumption discussed next).

Whatever the true values for net flux of amino acids in mollusc larvae, there is some evidence of a significant trophic function. Davis and Chanley (1956) recorded increased growth rates of oyster larvae when cultures were supplemented with water-soluble vitamins and Courtwright et al. (1971) enhanced development of *Mytilus edulis* larvae by adding lysine and aspartic acid to the medium. Gustafson (1980) experimented with *Mya arenaria* larvae at low (5–8 μ*M*) concentrations of a mixture of DFAA and found a significant improvement in survival over fully starved individuals; larvae exposed to DFAA maintained higher protein and lipid reserves than starved larvae. These experiments suggest that if larvae experience periods of starvation due to a diffuse or overdispersed food supply, dissolved organic matter may have a saving function allowing longer survival.

2. Bacteria and Detritus

Single cells of bacteria may be too small for retention by veliger larvae. However, according to Sorokin (1981) 20–30% of the total microbial population in seawater may exist within aggregates exceeding 5 μm in size, making them available to veligers. The abundance of these bacterial aggregates in coastal waters (1.0–1.5 g m³) suggests that they may serve as food for larvae. Ukeles and Sweeney (1969) observed a depression in the rate of ingestion of *Monochrysis* cells by *Crassostrea* larvae in bacterial concentrations of 2.7×10^5 ml, which is within the range for bacterial numbers in temperate coastal waters, but their counts were done by plating techniques and may be underestimates.

Rigorous experimental work on the food value of bacteria to mollusc larvae is lacking. High bacterial numbers are usually avoided in larval cultures due to their role in disease (Tubiash et al., 1965) but the results of experiments have been conflicting, often because of the difficulty of discriminating between harmful and potentially useful strains. Davis (1953) found that nine species of bacteria did not support the growth of *Crassostrea virginica* larvae. Hidu and Tubiash (1963) concluded that antibiotics enhanced the growth of *Mercenaria* larvae by limiting harmful bacteria and encouraging useful types. Helm and Millican (1977) found that the growth of *Cassostrea gigas* larvae was consistently better in the absence of penicillin–streptomycin mixtures. Millar and Scott (1967) grew *Ostrea edulis* larvae on *Monochrysis* in bacteria-free medium and concluded that growth rates were not impaired, but Martin and Mengus (1977) reported that living bacteria, when offered with algal cells, improved larval growth rate (*Mytilus galloprovincialis*).

There is no sharp distinction between bacterial aggregates and "detritus" as potential food particles in the sea, but experiments to determine the food value of detritus are beset by the difficulties of preparing a reproducible and natural source of detritus. Chanley and Normandin (1967) were able to grow *Mercenaria* larvae on dried, pulverized particles of *Ulva* and brown algae but further experiments, preferably under natural conditions of particle concentration and quality, are needed. At high concentrations of any particle, from inorganic silt (Davis, 1960) to dinoflagellate trichocysts (Ukeles and Sweeney, 1969) and living algal cells (see earlier discussion), the feeding of veligers may be disrupted, with reduced growth rates as a result. Experiments at sea are needed to establish whether these events occur naturally.

3. Algae

The food value of different algae, once ingested, may be a function of the digestibility of the cells, of the production of toxins by the algae, and of any fundamental difference in nutrient content. Factors affecting digestibility have been discussed, but the relationship with food value is not necessarily simple. Mapstone (1970) found that *Nassarius reticulatus* veligers took between 24 and 38 mins from first ingestion to first defecation with three algal species, but the algae differed in their support of larval growth. An unknown factor in such experiments is the production of toxic metabolites.

Algal cells produce extracellular metabolites (reviewed by Wangersky, 1978) which may be growth inhibiting or stimulating to larvae. Bayne (1965) fed *Mytilus edulis* larvae with *Monochrysis lutheri* cells (which supported larval growth) suspended in the medium from *Nannochloris atomus* cultures (a species which did not support growth) and recorded a depression of growth (see also Loosanoff, 1954). Wilson (1979) concluded that cultures of *Isochrysis galbana* at the late exponential and early stationary growth phases contained substances that stimulated the feeding of *Ostrea* larvae, whereas older "collapsed" cultures inhibited feeding. Calabrese and Davis (1970) consider that toxins produced by algal cells may be internal or external to the cell. Larvae cultured with internally toxic cells feel normally but die; larvae with externally toxic cells do not feed.

Pechenik and Fisher (1979) point out that the nutritional value of different phytoplankton species should be considered in the context of the different nutritional requirements of the larvae; this has not been attempted for any molluscan veliger. Gross differences in total protein, carbohydrate, or lipid content are unlikely to explain differences in food value, but more subtle differences in micronutrients may be significant. Walne (1970) concluded that differences in food value are unlikely to be explained by differences in amino acid composition and he emphasized the importance of culture conditions (light intensity, nutrient concentrations, age) on the gross biochemical composition of the algal cells. However, recent studies on fatty acids suggest significant dietary requirements. Waldock and Nasciomento (1979) found that the fatty acid composition of the triacylglycerols of the larvae of *Crassostrea gigas* resembled the composition of the algal lipids in the diet. Using three algal species, all of which supported some growth of the larvae, they were unable to correlate differences in fatty acid composition with growth-rate differences. However, Langdon and Waldock (1981), working with *C. gigas* spat, found poor growth with *Dunaliella tertiolecta* that correlated with the absence in this species of the polyunsaturated fatty acids 20:5 ω3 and 22:6 ω3. The rate of growth of the spat did not correlate with total lipid content of the algae, but by adding 22:6 ω3 in microcapsules growth was increased. *Dunaliella tertiolecta* is also a poor food for *Nassarius* larvae (Pechenik and Fisher, 1979). Studies with natural food complemented with encapsulated diets should help in further understanding algal food quality.

E. Absorption and Assimilation of Food

Absorption is defined as the uptake of ingested food across the gut wall. The term has frequently been used synonymously with the term assimilation. Absorption is to be preferred, however, and assimilation should be used to refer to the absorbed ration minus the energy lost as excreta.

Conover (1978) reviews methods used to measure absorption efficiency that is, the efficiency with which the ingested ration is digested. The assumption is often made that all nonabsorbed material (or energy) from the ingested ration is lost as particulate fecal matter; this ignores losses of dissolved material which may be released from the food during digestion but not subsequently absorbed by the animal (Johannes and Satomi, 1967). In practice it is difficult to discriminate between dissolved losses of this kind and true excreta (i.e., material absorbed across the gut wall and then excreted in dissolved form). In some studies, therefore, particularly those involving radiolabeled food particles, fecal and excretory losses are compounded and the resulting measure of organic matter retained by the animal actually represents assimilated ration, or ingested ration minus the sum of fecal and excretory losses.

Pechenik and Fisher (1979; see also Pechenik, 1980) considered the magnitude of these losses to *Nassarius* larvae feeding on three species of radiolabeled phytoplankton. Approximately 28% of the radioactivity lost during the first hour from larvae fed with *Isochrysis* was particulate (i.e., retained by a 0.8 µm Millipore filter); equivalent values were 15% for *Thalassiosira* and 60% for *Dunaliella*. These values amount to a very considerable loss of dissolved organic matter from the diet. The calculated assimilation efficiencies were 25–39%, with a mean of 32%.

The use of radiolabeled food to measure absorption and assimilation poses considerable problems of interpretation (Conover and Francis, 1973) which, if ignored, lead to underestimates of true values by an unknown amount. Walne (1965) used ^{32}P-labeled algal cells and deduced an efficiency of 29% (range 15–45%) for *Ostrea* larvae; this is probably an underestimate of the true assimilation efficiency. From considerations of the energy balance equation (see later) the absorbed ration may also be indirectly calculated. Gabbott and Holland (1973) used this approach to calculate absorption efficiencies of 28–52% for *Ostrea* larvae between 178 and 219 µm shell length. (Original values recalculated using a more reliable estimate of the energy value of *Isochrysis* cells). By a similar approach M. Helm (personal communication) estimates an assimilation efficiency for *Ostrea* of 35.0 ± 6.2%. These results will be discussed later.

Although Walne's (1965) experiments probably underestimate the true efficiencies, they were designed to be comparable between experiments and therefore throw some light on the effects of temperature, ration, and larval size on food absorption by *Ostrea* veligers. Over a temperature range from 15 to 30°C and ration level from 25 to 300 cells µl, larvae of all sizes assimilated a similar fraction of the ingested ration of *Isochrysis*. Some values are plotted in Fig. 5 to show the decline in the weight-related assimilated ration (as percentage of body weight per day) during larval growth, from 25% at a shell length of 170 µm to 13% at 260 µm. Using other species of algae, Walne (1965) showed that the number of cells retained by the larvae decreased with increase in algal cell size (discussed earlier) but the dry weight of cells assimilated per day was less variable across diets. Data from Pechenik (1980), recalculated to approximate true assimilation, suggest a decline in assimilated ration (*Nassarius* larvae) from 60 to 20% of body weight per day.

Some recent interest has concerned the role of symbiotic algae in the nutrition of the larvae of tridacnid clams. La Barbera (1975) reared the larvae of *Tridacna maxima* and *T. squamosa* but found no evidence of zooxanthellae in the eggs or in the larvae; zooxanthellae first appeared in juveniles 28 days after metamorphosis. Jameson (1976) confirmed, for *T. crocea*, *T. maxima*, and *Hippopus*

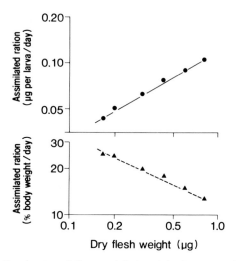

Fig. 5. The assimilated ration of *Ostrea edulis* larvae feeding on *Isochrysis galbana* (21°C; 100 cells μl), related to dry flesh weight. Data from Walne (1965).

hippopus, that zooxanthellae were first acquired 19–40 days after metamorphosis, after which the rates of growth of the juveniles increased sharply. Gwyther and Munro (1981), however, reared *T. maxima* larvae with and without algal food supplement. The unfed and fed larvae grew from 100 μm to 204 μm at similar rates, although the unfed larvae did not metamorphose successfully. The authors conclude that zooxanthellae were acquired by the eggs and provided nutrient to the free-swimming larvae, but that the larvae also fed on phytoplankton and this provided an energy store necessary for metamorphosis.

III. Respiration

Heat loss is usually estimated indirectly as the caloric (or Joule) equivalents of oxygen consumption. Direct calorimetric measurements of heat output by molluscan veligers have not been reported. Alternatively, catabolic losses can be measured as the decline in biochemical components (protein, lipid, carbohydrate) during periods of starvation. Commonly, however, it is the rate of oxygen consumption that has been measured, either using Cartesian divers as in the classical work of Zeuthen (1947) or various types of differential microrespirometers (e.g., equipment due to Grunbaum et al., 1955). Most techniques rely on crowding the larvae into small-volume experimental vessels, but this is likely to result in various experimental artifacts (Walne, 1966; Millar and Scott, 1967).

Some values for rates of oxygen consumption (μl O_2/h) by molluscan veligers are plotted against the size of the larva (as μg dry flesh weight) in Fig. 6. Values range from <0.001 μl/h for small bivalve larvae to ~0.1 μl/h for larger pros­obranch veligers, or from 2.5 to 10.0 ml O_2/h/g dry weight. The exponent relating the rate of oxygen consumption to body weight varies between 0.7 and >1.0; in a careful study using single larvae, Zeuthen (1947, 1953) established an exponent of 0.8 for the larval stages of *Mytilus edulis*. Sprung (1982) calculated the following equation for *Mytilus edulis* larvae at 12°C:

$$\mu l\ O_2\ \text{respired/h} = 1.57 \times 10^{-3}\ W^{0.72}$$

where W is the ash-free dry-tissue weight in μg. Zeuthen (1947) concluded that rates for many larvae approximately doubled between resting and swimming

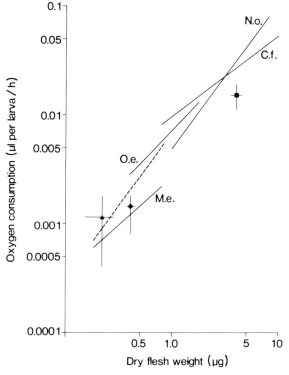

Fig. 6. Rates of oxygen consumption by veliger larvae, related to dry flesh weight. N.o., *Nassarius obsoletus* (Pechenik, 1980); C.f., *Crepidula fornicata* (Pechenik, 1980); M.e., *Mytilus edulis* (Zeuthen, 1947); O.e., *Ostrea edulis* (M. Helm, personal communication); dashed line, *O. edulis* (Gabbott and Holland, 1973, calculated as oxygen equivalents of biochemical losses); ■, *Nassarius reticulatus* (Zeuthen, 1947); ●, *Littorina* sp. (Holland, 1978—calculated from biochemical losses); ▲, combined value for bivalves (Holland, 1978). All values taken at temperatures and ration levels at or close to the growth optima.

rates of activity. Vernberg (1972) and Pechenik (1980) recorded a reduction in oxygen consumption by *Nassarius* larvae as they progressed from an actively swimming stage to the swimming–crawling behavior typical of the pediveliger larva just prior to metamorphosis.

The effects of temperature and salinity on oxygen consumption rates vary between species and with the pretreatment conditions of the adults, gametes, and larvae in any one experimental series. Vernberg and Vernberg (1975) recorded Q_{10} values for 7-day-old *Nassarius* veligers at 30‰ salinity of 3.0 (from 15 to 20°C) and 1.7 (from 20 to 30°C). The Q_{10} values varied with the age of the larva and with the salinity of the medium, being generally higher in the older larvae at lower salinities. Lough and Gonor (1973b) recorded the oxygen consumption by the larvae of the bivalve *Adula californianus* at five temperatures and four salinities. Rates were highest at 18°C, declining at 21°C and there was evidence of higher Q_{10} values at the lower salinities (13 and 33‰). As is to be expected, no simple rules for predicting rates of oxygen consumption under different conditions emerge from these and other studies (see also Sprung, 1982), although the data in Fig. 6 suggest no major quantitative differences in the intensity of respiration between the few species of bivalves and gastropods represented.

Measurements of oxygen consumption by larvae are technically difficult and time consuming. Biochemical measurements of catabolic losses during starvation offer an alternative although, even during short periods without food, the larvae may reduce their level of swimming activity, rendering the biochemical estimates of energy losses as minimal values. Holland (1978) has reviewed the literature on the biochemical composition of the larvae of benthic marine invertebrates, including changes occurring during starvation. Molluscan veligers have small amounts of carbohydrate in their tissues, and large amounts of protein and lipid. [Larvae of the shipworm, *Teredo pedicellata,* may be an exception, with large stores of glycogen available during their short pelagic lives (Lane et al., 1952); more estimates of the biochemical composition of lecithotrophic larvae would be interesting.] During starvation, protein and lipid are both lost from bivalve larvae, in proportions that vary with the age of the larva and the duration of starvation. Over a two-day starvation period, *Ostrea edulis* larvae lost approximately equal amounts of protein and lipid, although because of the greater calorific yield per unit weight of lipid, the larvae obtained more energy from the lipid than from the protein loss (Millar and Scott, 1967; Helm et al., 1973; Gabbott and Holland, 1973). During 7 d of starvation the ratio of protein : lipid : carbohydrate losses for *Mytilus edulis* larvae were 1.0 : 0.24 : 0.11 (Bayne, 1976) and for *Mya arenaria* larvae were 1.0 : 0.30 : 0.04 (Gustafson, 1980). Gustafson (1980) records an oxygen to nitrogen ratio (O:N ratio, indicating the catabolic balance between protein, lipid, and carbohydrate; Bayne, 1976) for *M. arenaria* larvae that declined during starvation from ~40 at two days to ~10 at 14 days;

this suggests a much increased reliance on protein as the energy substrate during prolonged starvation.

When these biochemical losses are converted to caloric equivalents, they agree reasonably well with published values for direct oxygen consumption measurements (Crisp, 1974; Holland, 1978). Weight-related values for *Ostrea edulis* larvae, calculated by Gabbott and Holland (1973), have been plotted in Fig. 6 as a dashed line.

Crisp (1974) used an average value for oxygen consumption of 5 ml O_2/h/g dry weight to predict survival times for starving larvae relying on protein and lipid as energy reserves. A larva that utilizes between 25% and 50% of its dry tissue as lipid may survive between four and eight days; for an equivalent protein loss, the survival time may be two to four days. However, in laboratory experiments veligers have been shown to survive for very much longer periods without food, for example, 20–30 d for *Mytilus edulis* larvae at 15–16°C (Bayne, 1976) and much longer (up to 150 days in sterilized seawater at 12°C) for older *Mytilus* larvae (Sprung, 1982). This apparent discrepancy may be due in part to a reduction in metabolic rate during starvation, so prolonging the period of energy supply from the reserves. It may also be due to a sparing effect of dissolved organic substances: Gustafson (1980) has demonstrated this in *Mya arenaria* larvae cultured in an amino acid mixture (5.1 μM) without particulate food. These larvae survived longer than comprehensively starved larvae and maintained their reserves of protein and lipid at higher levels.

IV. Growth

The growth of molluscan veligers has usually been described as increase in the length (occasionally length and height or length × height) of the shell; seldom has growth in weight been recorded. Shell length is a convenient routine measurement to make, and knowledge of the relationship between shell length and height may be of use in the identification of larvae (Loosanoff et al., 1966; Chanley and Andrews, 1971). But shell length may not provide a reliable estimate of growth in tissue weight. In bivalve veligers, the weight of the inorganic shell declines, as a proportion of total dry weight, during larval development (Holland and Spencer, 1973). M. Helm (personal communication) has observed an increase in dry tissue (organic) weight of *Ostrea* larvae from 21 to 30% (shell length 180–190 μm) to 37–40% (300–310 μm); an unknown proportion of this organic weight is due to the organic component of the shell. A similarly variable relationship between shell length and tissue weight is likely in gastropods (Scheltema, 1967; Pechenik, 1980).

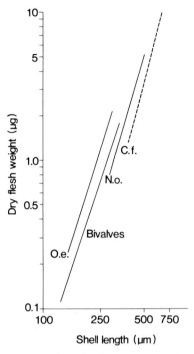

Fig. 7. Dry flesh (organic) weight related to shell length. C.f., *Crepidula fornicata* (Pechenik, 1980; data recalculated to be simply allometric); N.o., *Nassarius obsoletus* (Pechenik, 1980); O.e., *Ostrea edulis* (M. Helm, personal communication); Bivalves, a general relationship derived from Walne (1965), Holland and Spencer (1973), Malouf and Breese, (1977), Waldock and Nasciomento (1979), and B. L. Bayne (unpublished).

The shell length/body weight relationship is made more complex by the capacity for veligers to add new shell during starvation, in spite of the loss of body tissue. For example, *Crepidula fornicata* larvae grew ~50 μm in shell length when subjected to a temperature cycle between 30 and 35°C but declined in carbon content from 1.2 μg C to <1.0 μg/larva (Lucas and Costlow, 1979). A more extreme example is recorded by Gustafson (1980), who observed that the antibiotic chloramphenicol inhibited the growth of body tissue in *Mya arenaria* larvae, but had little effect on shell growth. The larvae grew from ~75 μm to 220 μm shell length in 12 days, but failed to increase body mass.

In the discussion that follows, conversions of length to organic (tissue plus organic matrix of the shell) weight used are taken either from the original publications or from parallel studies made on similar material (Fig. 7). These relationships can only be used with any confidence for growth under optimum or near-optimum conditions.

A. Growth Curves

When growth in length has been recorded from the earliest shelled stage (prodissoconch 1 in bivalves, protoconch 1 in gastropods) to the larva fully competent to metamorphose, the resulting curve of growth may be linear or asymptotic and is occasionally sigmoidal. There is evidence of greater sigmoidicity in the growth curve when growth rate is reduced by low temperatures or reduced ration (Loosanoff et al., 1951; Sprung, 1982). When conditions for rearing the larvae in the laboratory are fully optimal, growth may be linear; improvements in hatchery practices, including manipulations of water quality and improved larval diets (M. Helm, personal communication) have even resulted in exponential rates of growth in shell length by larval oysters. If the larvae enter a period of delayed metamorphosis, due to the absence of a suitable settlement substrate, the rate of growth declines and becomes asymptotic (Bayne, 1965). However, data on rates of growth under natural conditions in the field are rare. Some recent data due to Sprung (1982) are plotted in Fig. 8 with the growth curves fitted by eye.

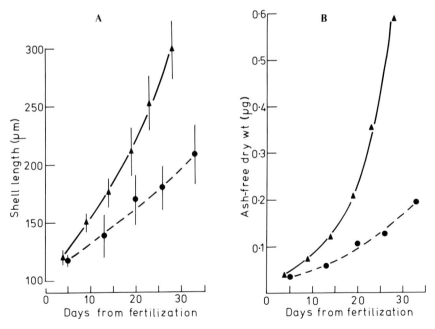

Fig. 8. Curves of growth in shell length (**A**) and in weight (**B**) for *Mytilus edulis* larvae cultured at 12°C with either 2,000 (●) or 10,000 (▲) cells of *Isochrysis galbana* per milliliter as food; from Sprung (1982). In (**A**) values are means ± s.d.; in (**B**) mean length measurements were calculated to ash-free dry weight using the equation: ash-free dry wt (μg) = 1.85×10^{-8}. Length (μm)$^{3.03}$.

It is not possible, from the available information, to deduce a form of growth curve that would be generally applicable to molluscan larvae. Some studies assume linear growth over time (i.e., rate = $(l_{t+n} - l_t)/n$) or assume an exponential rate of growth [i.e., rate = $(\ln{}_{t+n} - \ln l_t)/n]$ where $(t + n) - t$ is the increment of time under consideration. More complex, and more general, equations describing growth may be used; R. Clarke (unpublished) has described such a generalized equation, employing four parameters, which encompasses a family of curves from the hyperbolic to the sigmoidal:

$$L_t = l_\infty \left[1 - e^{-kt}(1 - \left(\frac{l_0}{l_\infty} \right)^\delta) \right]^{1/\delta}$$

where L_t is length at time t, l_0 is length at time 0, l_∞ is maximum length, k is the rate parameter, and δ a parameter describing the shape of the curve ($\delta = 1$ for a hyperbolic curve; $\delta = 0$ for a sigmoidal curve). When fitted by means of weighted, nonlinear least squares interation, comparisons between rates of growth in different experiments may then be possible in terms of $k \pm$ S.D.

B. Endogenous Factors and Growth

A common observation in experiments with veligers is the considerable variability in the lengths of larvae from the same parents grown under the same conditions of temperature, salinity, and ration. For example, Loosanoff et al. (1951) report shell lengths of *Mercenaria mercenaria* larvae ranging from 107 μm to 226 μm after eight days at 30°C. Bayne (1965) recorded an increase in the variance : mean ratio of the shell lengths of *Mytilus edulis* with increased rates of growth. There is large genetic contribution to this variability (Losee, 1979), amounting to 25% in *Mytilus* larvae from a single population (Innes and Haley, 1977) and between 25% and 50% in *Crassostrea virginica* larvae from different populations (Newkirk et al., 1977). The genetic component may be both additive and nonadditive. These findings are consistent with the large amount of genetic variability, measured electrophoretically, in populations of adult bivalves (Koehn, 1982). Sing and Zouros (1978) have reported a positive correlation between the growth rate and genetic heterozygosity among juvenile *C. virginica*.

In experiments with the larvae of *Mytilus edulis* (Innes and Haley, 1977) and *Crassostrea virginica* (Newkirk, 1978), the expression of family and population differences in growth rate depended upon the salinity of the medium in which the larvae were reared. That is, there was significant genotype–environment interaction and the relative importance of genes having an effect on growth rate was different at different salinities; this interaction applied both to the additive and the nonadditive genetic effects. Innes and Haley (1977) follow Levins (1968) in suggesting that bivalve populations that are exposed to fluctuating salinities in coastal waters may have adapted by expressing a range of genotypes within the population. More populations need to be studied, however, to discriminate between adaptation of this type and the extent of genetic isolation between populations (Newkirk, 1978).

In a series of experiments with *Crassostrea gigas*, Lannan (1980a,b,c) confirmed a genetic and a nongenetic component in the survival of larvae under unstressed rearing conditions. At least a part of the genetic variance was related to the state of the gonadal development of the parent oysters; the rate and the timing of gonadal development is, in part, under genetic control. Larval survival was increased when fertilization took place between gametes spawned at an optimal time in the adults' gametogenic cycle. When mating occurred before or after this optimal period, gamete viability was reduced. These conclusions are particularly important for hatchery practice (Lannon, 1980c) but may also be significant in natural populations. Environmental factors that disturb the normal pattern of gamete development and spawning could result in less viable offspring. In addition, Bayne (1972) and Bayne et al. (1975) have demonstrated that the impacts of stressful conditions on adult mussels are reflected in reduced viability of the eggs. The full significance of these effects for the survival and growth of larvae in nature remains to be established.

Another endogenous factor to be considered is the size of the larva itself. As discussed earlier, the rate of growth normally declines with increase in size and in age. Some values for the rates of growth in length over three-day periods [$K_3^l = (\ln l_{t+3} - \ln l_t)/3$] are plotted against shell length in Fig. 9; the decline in the

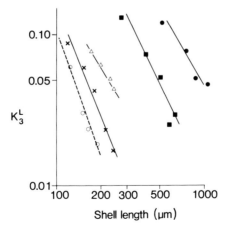

Fig. 9. Rates of growth of veliger larvae, calculated as the length growth coefficient over three days (K_3^l), related to shell length. $\bigcirc$, *Mercenaria mercenaria* (Loosanoff et al., 1951; 24°C); X, *Mytilus edulis* (Bayne, 1965, 17°C); $\triangle$, *Ostrea edulis* (Walne, 1965, 23°C): $\blacksquare$, *Nassarius obsoletus* (Scheltema, 1962, 25°C) $\bullet$, *Conus textile*, (Perron, 1980, 25°C).

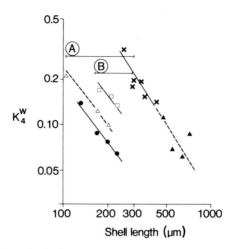

Fig. 10. Rates of growth of veliger larvae, calculated as the weight growth coefficient over four days (K_4^W), related to shell length. All values at 20°C. ●, *Mytilus edulis* (Bayne, 1965); △, *Crassostrea gigas* (Helm and Millican, 1977); □, *Ostrea edulis* (Walne, 1965); x, *Nassarius obsoletus* (Pechenik, 1980); ▲, *Crepidula fornicata* (Pechenik, 1980); ○, *C. gigas*, and ○, *O. edulis* (both from M. Helm, personal communication).

rate of growth with increase in size, together with the fundamental similarities in K for different species, are apparent. In Fig. 10 some values for rates of growth in tissue (organic) weight are plotted for four-day periods. As a result of recent improvements in hatchery practices, oyster larvae may now be reared to metamorphosis along an exponential growth curve (M. Helm, personal communication). Some values are plotted in Fig. 10 to illustrate the comparison for *Crassostrea gigas* (Helm and Millican, 1977) and *Ostrea edulis* (Walne, 1965). Results for *Nassarius* and *Crepidula* (Pechenik, 1980) suggest a similar rate of growth when converted for differences in size.

C. Exogenous Factors and Growth

Of the various physicochemical factors that are likely to affect the growth of larvae temperature, salinity and food have received the most attention. Calabrese and Davis (1966, 1970) and Calabrese (1969) studied the effects of pH on bivalve embryos and larvae and Davis and Hidu (1969), Loosanoff (1962) and Walne (1970) have considered the effects of suspended silt; various toxicants have also been evaluated as affecting the survival and growth of young bivalve larvae (Walne, 1970; Lehnberg and Theede, 1979; MacInnes and Calabrese, 1979).

1. Temperature

Rates of growth of larvae have been shown to increase with a rise in temperature to an optimum and then to decline with further temperature increase. Ursin (1963) described the relationship between temperature and the time to complete a specified amount of growth as a symmetrical catenary curve:

$$y = y_0 \cosh p(x - x_0)$$

where y is time, x is temperature, x_0 is the temperature at which development is most rapid, y_0 is the development time at x_0 and p is a temperature coefficient. The reciprocal expression is:

$$\frac{1}{y} = \frac{1}{y_0} [\cosh p(x - x_0)]^{-1}$$

where the reciprocal of y is the rate of development. Ursin (1963) illustrated the use of this relationship with data from Ansell (1961) on the growth of the larvae of *Venus striatula*. The optimal temperature for growth (x_0) for these larvae was 18°C, the maximum rate of development from fertilization to a shell length of 155 μm was 8.25 days (equivalent to ~12.5 days for development to metamorphosis) and the value for the temperature coefficient p was 0.20.

The result of a similar analysis for three other bivalve species and a gastropod are plotted in Fig. 11, with the goodness of fit illustrated for *Mytilus edulis*. The temperature coefficients for these species lie between 0.18 and 0.24; the predicted temperature optima and the rates of development from fertilization to metamophosis are apparent from the graphs.

The assumption of symmetry around the temperature optimum for growth may not always hold; for some species there is an abrupt transition from the optimum to an upper lethal temperature (Lucas and Costlow, 1979). Loosanoff (1959) calculated a linear relationship between temperature and the number of days from fertilization to settlement in *Mercenaria mercenaria*:

$$\text{Days to settling} = 37.9 - T \, (°C).$$

However, at 15 and at 33°C, growth and development were abnormal and mortality was high. Loosenoff's (1959) data have been replotted in Fig. 12 to show the considerable variability that is usually encountered in such experiments, together with a fitted catenary curve, which should be truncated above 30°C.

Ansell (1968) summarized data on the rate of growth of adult *Mercenaria mercenaria* throughout its geographical range and arrived at a mean temperature optimum of 20°C; this is in contrast to growth of the larvae, with its optimum ~30°C (Fig. 12). Similarly for *Ostrea edulis*, adults grow maximally at 15–18°C (Walne, 1958; Mann, 1979; Newell et al., 1977) whereas larval growth has an

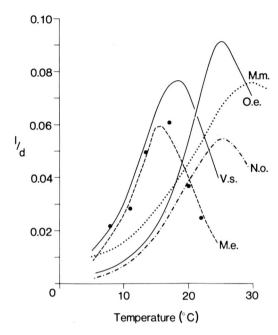

Fig. 11. Rates of growth of veliger larvae, calculated as the reciprocal of time (in days) from fertilization to the pediveliger stage, related to temperature. Responses are plotted as symmetrical catenary curves (see text), with mean values for observations shown only for *Mytilus edulis* (M.e.) (from Bayne, 1965). V.s., *Venus striatula* (Ansell, 1961); N.o., *Nassarius obsoletus* (Scheltema, 1962); O.e., *Ostrea edulis* (Walne, 1965); M.m., *Mercenaria mercenaria* (Loosanoff, 1959).

optimum at ~26°C (Fig. 11). The explanation for these observations, and their ecological significance, remains obscure. Other bivalve species show greater similarity between larvae and adult (e.g., *Mytilus edulis*, Bayne, 1965; Seed, 1976). Dehnel (1955) compared rates of growth of four gastropod larvae from populations in California and Alaska. Rates of growth from the northern populations were from 2 to 9 times faster than from the southern populations, when compared at similar temperatures. Dehnel calculated Q_{10} values for growth rate and shows temperature ranges over which the values are relatively low; for the southern populations the range was 10–22°C, for the northern populations 8–16°C. Similar studies comparing larval and adult growth rates would be of considerable interest.

An evaluation of the effects of temperature on larval growth is not complete without considering the period during which the larva is competent to metamorphose but is unable to do so for lack of a suitable substrate to stimulate settlement. During such a delay period (Scheltema, 1961) most larvae seem able to continue growth while retaining metamorphic competency, at least for a period (Bayne, 1965; Pechenik, 1980). At lower temperatures the larvae may delay metamorphosis for longer than at higher temperatures and may grow to a greater size.

Lutz and Jablonski (1978) have suggested that a negative correlation between temperature and shell length at metamorphosis may serve as a useful tool in paleoclimatology. They document the relationship for *Mya arenaria* larvae with values from the literature. Loosanoff (1959) examined this relationship for *Mercenaria mercenaria* larvae reared at different temperatures in the laboratory and concluded that there was no consistent relationship between temperature and either larval size or shape; however, the extent to which larvae in these cultures delayed metamorphosis is not clear. There is evidence from samples taken in the field (Sullivan, 1948; Scheltema, 1961; Bayne, 1965; Jablonski and Lutz, 1980) that a period of delayed metamorphosis may be common, in some environments at least, and a negative correlation between temperature and larval-shell length might be expected in these circumstances.

In the natural environment larvae may be exposed to short-term changes in

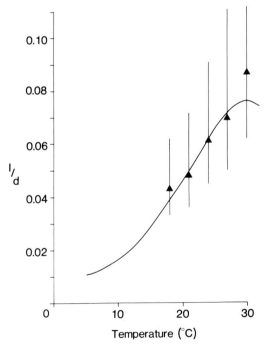

Fig. 12.   The rate of growth of *Mercenaria mercenaria* larvae, calculated as the reciprocal of time (in days) from fertilization to the pediveliger stage, related to temperature. Values are means ± range, from Loosanoff (1959). The curve is fitted as a symmetrical catenary curve (see text).

temperature, but experiments on the influence of temperature changes on rates of growth are very rare. Kennedy et al. (1974) found that *Mulinia lateralis* larvae survived brief exposures to otherwise lethal temperatures. Lucas and Costlow (1979) carried out a thorough study of the growth of *Crepidula fornicata* larvae in response to cycling temperatures. Daily cycles of ~5°C amplitude had a predictable effect on growth, with rates of growth taking values typical for the mean temperature at cycles between 15 and 30°C. At higher temperatures growth was retarded in proportion to the time spent at unfavorable temperature (32.5 or 35°C).

2. Salinity

The growth rate of mollusc larvae responds to changes in salinity. Tolerance varies between species, but the range of salinities tolerated is normally wide and encompasses the salinity variation that may be experienced in the field. For example *Nassarius obsoletus* larvae grew at >60% of the maximum rate of salinities from 21 to 33‰, but did not complete larval development at 17‰ (Scheltema, 1965). Some values for the salinity range over which growth is >50% of the maximum observed are listed for various bivalve species in Table II. Generally, larvae continue to grow at near normal rates at salinities close to the lethal maxima and minima; development of the eggs, however, often requires a narrower salinty range (Fig. 13).

TABLE II

Salinity Range over Which Various Bivalve Species Can Grow at >50% of Their Maximum Observed Growth Rates

Species	Salinity range (‰)	Temperature (°C)	Authority
Ostrea edulis	25–35	25	Davis and Ansell (1962)
Mya arenaria	10–32	18–20	Stickney (1964)
Mercenaria mercenaria	17.5–>28	27.5	Calabrese and Davis (1970)
Mulinia lateralis	17.5–35	25	Calabrese (1969)
Mytilus edulis			
Helsingor, Denmark	14.5–30	16	Bayne (1965)
Millford, Connecticut	15–40	17.5	Hrs-Brenko and Calabrese (1969)
Ostrea Lake, Nova Scotia	16–30	16	Innes and Haley (1977)
Crassostrea virginica			
Millford, Connecticut	12.5–>27	30	Calabrese and Davis (1970)
New Brunswick	12–30	23	Newkirk (1978)
Crassostrea gigas	15–34	28	Helm and Millican (1977)

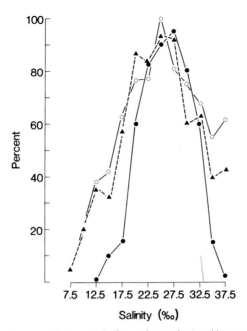

Fig. 13. Egg development (●), survival (○), and growth (▲) of larvae of *Mulinia lateralis,* calculated as percentage of maximum values, related to salinity. (From Calabrese, 1969.)

Most interest currently centers on the variation between populations of the same species as to the effects of salinity on larval growth. Stickney (1964) observed that the larvae of *Mya arenaria* from Chesapeake Bay grew faster at a lower salinity than larvae from New England. Bayne (1965) showed that larvae from an adult population of mussels at high salinity had reduced rates of growth at salinities <24‰, whereas larvae from the Oresund, where salinities are lower, grew normally down to 14‰. The salinity at which the adults are maintained prior to spawning may influence the salinity tolerance of the larvae. Helm and Millican (1977) held adult *Crassostrea gigas* for 40 days at 31 and 26‰, then induced spawning. Larvae from the adults at high salinity grew more rapidly than those from low salinity, at both 27 and 30‰; at lower salinities (15–22.5‰) there were no differences between growth rates. Newkirk (1978) has demonstrated genetic differences between populations of *Crassostrea virginica* expressed as larval rates of growth at different salinities; genetic differences between populations from the same estuary depended upon the salinity at which the larvae were reared (see also Newkirk, 1980).

For most of the species studied, a wide salinity tolerance in growth rate is to be expected, both on ecological and evolutionary grounds (Innes and Haley, 1977), because the species are mostly estuarine in their distribution. But it is the combined effects of temperature and salinity on growth that is expected to provide a better indication of abiotic environmental effects than either factor in isolation. For many species larval tolerance of high and low temperatures declines as the salinity is reduced. Lough (1975) reevaluated earlier data on the combined effects of temperature and salinity on the survival and growth of various bivalve larvae, using multiple regression analysis and response surface techniques. For most species there was a marked interaction between temperature and salinity affecting growth; maximum growth generally required a narrower range of temperature–salinity conditions than maximum survival (Lough and Gonor, 1971, 1973a; Lough 1975). The larvae of *Mytilus edulis,* however, grew normally within a wide salinity range (Table II) within which growth was dependent only on temperature (Hrs-Brenko and Calabrese, 1969; Lough, 1974). In experiments such as these it is important to consider the survival and quality of the food species as well as the larvae themselves at the salinity and temperature extremes (Davis and Calabrese, 1964; Cain, 1973; Helm and Millican, 1977).

Combined experiments of this type describe, in part at least, the fundamental niche of the species. The species so far considered, however, show considerable overlap in their tolerance limits and a proper understanding of possible competitive interactions between larvae would require information on the potential food resources of the larvae.

3. Food

Planktotrophic veliger larvae, unlike lecithotrophic larvae, feed upon the cells of phytoplankton (and possibly other organic material in suspension) and are dependent on a net energy gain from such food for successful growth and development. Considerable research has been carried out to identify good and indifferent phytoplankton food species, the main criterion to be measured being the rate of growth in length of the larvae when fed on particular species (Davis, 1953; Davis and Guillard, 1958; Loosanoff and Davis, 1963; Walne, 1963, 1970, 1974). These studies identified some algal species that could be relied upon to support larval growth in most cases (e.g., *Isochrysis galbana, Monochrysis lutheri, Tetraselmis suecica, Thalassiosira pseudonana, Chaetoceros calcitrans;* see Walne, 1974; Helm and Millican, 1977; Helm, 1977; various papers in Smith and Chanley, 1975; Pechenik and Fisher, 1979; Ewart and Epifanio, 1981), other species that were less reliable (e.g., *Chlorella* sp., *Phaeodactylum tricoruntum;* see Davis and Calabrese, 1964; Calabrese and Davis, 1970; Wilson, 1978; Perron, 1980), and a few species that have proved, in the majority of trials, not to be capable of supporting growth (e.g., *Olisthodiscus* sp, and in some cases *Dunaliella tetiolecta* see Pilkington and Fretter, 1970; Kempf and Willows, 1977; Pechenik and Fisher, 1979; Langdon and Waldock, 1981; but cf. Russell-Hunter et al., 1972, Lucas and Costlow, 1979).

There are several possible explanations for the low food quality of some algae.

Some species are probably too small, making capture by the larvae difficult. Davis and Calabrese (1964; see also Loosanoff and Davis, 1963) found *Chlorella* sp. (an alga with a cell wall) to increase in value to clam and oyster larvae as the temperature increased above 20°C. They suggested that the enzymes needed to digest the cell wall may only be fully active at high temperatures, in contrast with other digestive enzymes that are effective against naked flagellates such as *Isochrysis*. It may also be that the physical rupturing of species with cell walls that seems a necessary precursor to digestion may be more effective at higher temperature through more vigorous ciliary activity in the stomach. Fretter and Montgomery (1968) suggested that the food value of an alga to prosobranch larvae depended on the ease with which the cell wall could be damaged by mechanical treatment in the stomach. Variability in essential nutrients, and the possible production of toxins by some algae, which have been discussed earlier, also affect food quality.

Several authors have shown that by feeding larvae on a mixed algal diet, growth may be more rapid than when the individual algal species are fed singly (Davis and Guillard, 1958; Bayne, 1965; Walne and Spencer, 1968; Pilkington and Fretter, 1970; Calabrese and Davis, 1970; Helm, 1977; Kempf and Willows, 1977; Chia and Koss, 1978). For example, Helm (1977) quotes the following four-day growth coefficients $[K_4^L = (\ln l_{t+x} - \ln l_t)/4]$ for *Ostrea* larvae:

Isochrysis galbana (100 cells/μl)	:	0.057
Tetraselmis suecica (10 cells/μl)	:	0.051
Mixture (*Iso.* 50 + *Tetra.* 5 cells/μl)	:	0.062

But mixtures need not always result in faster growth. Nasciomento (1980) recorded the following four-day coefficients for *Crassostrea gigas* larvae:

Isochrysis galbana (100 cells/μl)	:	0.052
Chaetoceros calcitrans (100 cells/μl)	:	0.091
Mixture (*Iso.* 50 + *Chaet.* 50 cells/μl)	:	0.088

Results of this kind suggest that some mixtures may enhance growth by complementing deficiencies of essential nutrients in the individual algal species, or possibly masking potentially toxic factors. Whatever the explanation (and the problem clearly merits further research), larvae in the natural environment will normally experience very complex mixtures of potential food items. Many studies have suggested an equivalence between rates of growth in the laboratory and in the field (Bayne, 1965; Walne, 1965, 1970; Pechenik, 1980), but the laboratory experiments are usually carried out at food-cell concentrations higher than normally encountered in the field.

Evidence reviewed earlier has indicated that the feeding rate of larvae in a suspension of food depends on the concentration and size of the algal cells.

Walne (1965) analyzed these relationships for *Ostrea edulis* larvae feeding and growing on *Isochrysis galbana* and demonstrated linear relationships between growth increments of the larvae and the numbers of cells captured and assimilated (Walne, 1965; Fig. 25 and 31). Some values for increments of growth in weight of two species of bivalve larvae feeding at different cell concentrations are plotted in Fig. 14. In most cases (and see also Pilkington and Fretter, 1970) growth rates rise to maximal values between 10 and 100 cells μl. These findings are consistent with determinations of the cell concentrations at which ingestion and assimilation rates are maximal (Pechenik and Fisher, 1979), although at very high concentrations feeding and growth may be impaired.

The concentrations of phytoplankton that are optimal for growth in the laboratory represent approximately 100–500 μg C/liter and are similar to those typical of maximum ration for a variety of crustacean zooplankton species (reviewed by Parsons et al., 1977). They are nevertheless high relative to mean concentrations recorded in the field and this apparent anomaly between the food concentration necessary for maximal growth in the laboratory and the observed concentrations and growth rates in nature warrants further research. The explanation may lie, in part, in the patchy nature of phytoplankton distributions. In such a patchy environment the ability to survive long periods without food, and to grow rapidly

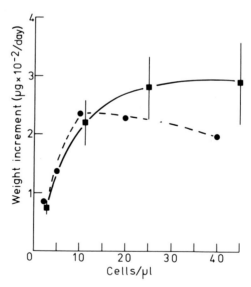

Fig. 14. The growth of *Mytilus edulis* and *Crassostrea gigas* larvae as increments in dry flesh weight, related to the concentration of food. ●, *M. edulis:* daily increments, initial shell length 200 μm(ash-free dry weight ~0.180 μg), fed cells of *Isochrysis galbana.* (From Sprung, 1982.) ■, *C. gigas:* mean ± s.d. Four-day increments, initial dry weight 0.027 μg, fed *Chaetoceros calcitrans.* (From Nasciomento, 1980.)

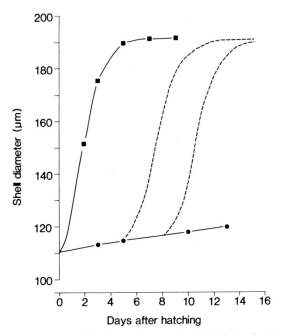

Fig. 15. The growth of *Doridella obscura* larvae when fed (■; *Isochrysis* at 100 cells µl) and when starved (●). The dashed lines illustrate growth when fed *Isochrysis* after two different periods of starvation. All at 25°C. (Redrawn from Perron and Turner, 1977.)

should food become available, would have adaptive value. This has indeed been observed for bivalves (Loosanoff and Davis, 1963; Bayne, 1965; Millar and Scott, 1967), prosobranchs (Scheltema, 1961; Pilkington and Fretter, 1970), and an opistobranch (Perron and Turner, 1977; see Fig. 15).

An ecologically significant physiological measurement that relates the available food level to the growth potential of the individual is the maintenance ration, that is, the ration, usually expressed as percentage of body weight per day, at which the animal's rate of growth (and hence growth efficiency—see Section IV,D) is zero. Sprung (1982) has calculated the maintenance ration for *Mytilus edulis* larvae to be 6.5% body wt/day at 6°C and 17.5% body wt/day at 12°C. These values are equivalent to a maintenance requirement of ~20 µg C/liter within the size spectrum of particles taken as food by the larvae.

D. The Energy Budget

The balance in the individual larva between the gains from the diet and the losses due to respiration and excretion can be represented by the simple expression:

$$G = C - F - U - R$$

where G is growth, C is the ingested ration, F the fecal losses, U the losses of dissolved material (the excreta), and R the metabolic losses assessed as respiration. The absorbed ration $A = C - F$. Three measures of efficiency are of primary interest:

$$\text{Absorption efficiency } (e) \quad = \quad 100(A/C)$$
$$\text{Gross growth efficiency } (K_1) \quad = \quad 100(G/C)$$
$$\text{Net growth efficiency } (K_2) \quad = \quad 100(G/A)$$

It has become commonplace to calculate these efficiencies in order to draw conclusions about growth and production in different animal types. But efficiencies of absorption and growth are not constants for a particular species, and they change as a direct consequence of the animal/environment interaction. Absorption efficiency is expected to alter with food quality and quantity, with larval size and possibly with temperature. Growth efficiency might vary with larval size, ration level and with temperature. It is our knowledge of the variability in these efficiencies that provides ecological insight and an understanding of growth. Conover (1978) provides an excellent review of these and related topics concerning marine species in general. There has been little relevant work with molluscan larvae.

Absorption efficiencies have been measured using algal cells labelled with ^{32}P (Walne, 1965) or ^{14}C (Pechenik and Fisher, 1979; Pechenik, 1980) or estimated indirectly from knowledge of other components of the energy budget (Gabbott and Holland, 1973; M. Helm, personal communication). Pechenik and Fisher (1979) correctly refer to their measurements as "retention efficiency" and point out that they probably underestimate true absorption efficiency (see earlier discussion p. 312). The published values generally agree in suggesting rather low absorption efficiency in veligers:

Ostrea edulis	15–45% (Walne, 1965)
	28–52% (Gabbott and Holland, 1973)
	29–46% (M. Helm, personal communication)
Mytilus edulis	18–44% (Sprung, 1982)
Nassarius obsoletus	17–58% (Pechenik, 1980)
Crepidula fornicata	69% (Pechenik, 1980)
Bittium alternatum	35–45% (Pechenik, 1980)

These values are all somewhat lower than those for planktonic copepods as summarized by Conover (1978), although, in the absence of more direct measurements, the significance of this is difficult to assess. Pechenik and Fisher (1979) found no consistent relationship between ingestion rate and retention (absorption) efficiency in *Nassarius obsoletus* larvae (see also Walne, 1965, in

Ostrea). However, retention efficiency declined with more rapid movement of food through the gut. Fretter and Montgomery (1968) had earlier commented that the digestibility of phytoplankton cells to gastropod larvae was proportional to the length of time the cells were retained in the stomach. All of the values for absorption efficiency recorded above derive from experiments at high food-cell concentrations. Under such circumstances in adult bivalves many cells are passed directly to the intestine without entry into the digestive gland and absorption efficiency is reduced as a result; a similar situation may occur in the larvae. Walne (1965) found that *Ostrea* larvae absorbed a fixed proportion of the food caught over a wide range of temperatures.

Welch (1968) suggested an inverse correlation between absorption efficiency and net growth efficiency (K_2) for a variety of species. Conover (1978) recalculated and replotted values from the literature to support Welch's assertion. This relationship would predict high K_2 values for veliger larvae, therefore, given the low absorption efficiencies. There is some support for this although there have been few measurements made. Jørgensen (1952) calculated K_2 values of 73% for *Mytilus edulis* larvae (see also Jørgensen, 1976) and 62 and 63% for two gastropod veligers, *Littorina littorea* and *Nassarius reticulatus*. Walne (1965) estimated that young *Ostrea edulis* larvae grew with K_2 efficiencies of 68–80% and Gabbott and Holland (1973) calculated, for the same species, K_2 equal to 79% for newly released, and 55% for 10-day-old larvae.

Values of K_1 must, of course, be less than K_2. Pechenik (1980) quotes K_1 equal to 11 and 18% for *Nassarius* and *Crepidula* larvae, respectively. Table III lists measurements made by M. Helm (personal communication) on *Ostrea edulis* larvae at 22–24°C, fed with *Isochrysis galbana* and *Chaetoceros calcitrans* each at 50 cells µl; K_1 was 35.0 ± 6.2%. These experiments were carried out under the best possible conditions for larval growth, based on many years experience with the species, and the efficiency values are probably maximal for the particular regime of temperature and ration; the ration levels were high, at 115 ± 18% of body weight cleared from suspension per day. Table IV presents data from Sprung (1982) to show the range of values calculated for assimilation efficiency and gross and net growth efficiencies for *Mytilus* larvae at 12°C, fed *Isochrysis* at different cell concentrations. There was no consistent relationship between assimilation efficiency [estimated as (respiration + growth) / consumption] and either larval size or cell concentration. Both growth efficiencies were highest at 10 cells/µl, however, and declined with increase in shell length. Figure 16 shows growth by *Mytilus edulis* larvae as a function of ration, taken from two independent studies. This treatment of the data predicts a maintenance ration of ~10% of body weight per day, maximum growth efficiency occurring at a ration of 40–50% body weight per day and a maximal growth rate at a ration of >80% body weight per day. More studies, under a variety of experimental conditions and with natural diets at levels normally to be found in nature, are needed.

TABLE III

Energy Balance in *Ostrea edulis* Larvae[a]

Day	Mean shell length (μm)	Ash-free dry weight (μg)	Dry weight algal cells removed from suspension (μg per larva/d) (C)	Oxygen consumed (μl/larva d)	Dry weight equivalent of oxygen consumption (R)	Growth (μg/larva d) (G)	$(G + R)/C$ =assim. efficiency	G/C =K_1	$G/(G + R)$ =K_2
0	171	0.40	0.31	0.067	0.06	0.07	0.42	0.22	0.54
1	184	0.47	0.51	0.079	0.07	0.10	0.33	0.20	0.59
2	197	0.57	0.70	0.097	0.08	0.12	0.29	0.17	0.60
3	211	0.69	0.90	0.119	0.10	0.17	0.30	0.19	0.63
4	227	0.86	1.13	0.148	0.12	0.22	0.30	0.19	0.65
5	243	1.08	1.37	0.187	0.16	0.31	0.34	0.23	0.66
6	261	1.39	1.63	0.243	0.20	0.40	0.37	0.24	0.67
7	279	1.79	1.89	0.316	0.26	0.61	0.46	0.32	0.70
8	300	2.40	—	—	—	—	—	—	—
Mean ± S.D.							0.35 ± .06	0.22 ± .05	0.63 ± .05

[a] Results from M. Helm (personal communication). Larvae were grown at ~10/ml in 125 liter rearing vessels at 22–24°C, 29–33o. Cell concentrations were automatically controlled to 50 cells/μl *Isochrysis galbana* (Tahiti variety) and 50 cells/μl *Chaetoceros calcitrans*. Oxygen consumption values converted to weight equivalents assuming 1.2 ml $O_2 \equiv 7$ mg organic material oxidized.

TABLE IV

Assimilation and Growth Efficiencies in *Mytilus edulis* larvae[a]

Shell length (μm)	Concentration *Isochrysis* (cells/μl)	Assimilation efficiency (e)	Gross growth efficiency (K_1)	Net growth efficiency (K_2)
120	2	0.44	0.22	0.50
120	5	0.30	0.19	0.61
120	10	0.41	0.30	0.73
120	20	0.39	0.28	0.72
120	40	0.38	0.26	0.70
200	2	0.35	0.16	0.46
200	5	0.21	0.12	0.58
200	10	0.28	0.20	0.71
200	20	0.28	0.20	0.70
200	40	0.29	0.20	0.67
250	2	0.32	0.14	0.45
250	5	0.18	0.10	0.57
250	10	0.24	0.17	0.69
250	20	0.25	0.17	0.68
250	40	0.27	0.17	0.65

[a] Results from Sprung (1982). Larvae were grown at 12°C and fed on various concentrations of *Isochrysis galbana*. Assimilation efficiency = (Respiration + growth)/consumption; Gross growth efficiency = growth/consumption; Net growth efficiency = growth/(growth + respiration).

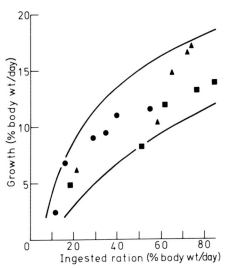

Fig. 16. The growth of *Mytilus edulis* larvae, as a percentage of body weight per day, related to ingested ration, also calculated as percent of body weight per day. ●, 17°C; *Isochrysis* and *Monochrysis* food cell mixture; larvae 180–230 μm shell length. (From Bayne, 1965.) ▲, 18°C, larvae 200 μm, ■, 18°C, larvae 250 μm; both fed *Isochrysis* only. (From Sprung, 1982.)

V. Conclusion

Mollusc veligers undoubtedly play an important part in the pelagic coastal ecosystem, yet their contribution to trophic relationships, to competitive interactions, and to energy flow remain almost entirely unstudied. In contrast, a wealth of data exists from studies of oyster and other bivalve larvae undertaken with objectives related to aquaculture, and much is known, as a result, of feeding, growth, and the rate of development of some veligers. It is now relatively easy to handle these larvae in the laboratory and to rear them successfully through to metamorphosis. It should now be possible, therefore, both to conduct meaningful experiments at sea and to examine in detail aspects of larval bioenergetics in the laboratory, in order to understand the ecology of these animals and to evaluate questions concerning adaptation, such as the costs and the benefits of different reproductive tactics, from the point of view of the physiology of the larvae. The information reviewed in this chapter points to considerable similarities of feeding mechanisms and rates of metabolism, biochemical composition, and growth among different molluscan groups suggesting a common set of consequences of the veliger body plan. Such a consistency of functional attributes, if confirmed by more comparative experiments, should be of particular benefit to more ecologically oriented research on marine molluscs in general and perhaps allow general statements concerning the energetic consequences of planktotrophy for the phylum.

Note Added in Proof

Manahan and Crisp (1982) have recently reported experiments with oyster larvae (*Crassostrea gigas*) that demonstrate the efficient uptake of dissolved amino acids (see p. 310) at naturally occurring substrate concentrations ($0.6–6 \times 10^{-6} M$ glycine). The observed rates of uptake of this one amino acid could account for approximately 10% of the protein synthesis in the growing larva. The oyster larvae were shown to compete efficiently with bacteria for dissolved amino acids, and the research indicates a potential nutritional role for dissolved organic molecules in the physiology of veliger larvae.

Acknowledgments

I am extremely grateful to Mr. M. Helm (MAFF Fisheries Experimental Station, Conwy, North Wales) and Dr. M. Sprung (Biologische Anstalt Helgoland, W. Germany) for permission to quote from their unpublished results and for helpful discussions. R. Strathmann commented critically on the manuscript. I have liberally used results from the late Peter Walne's illuminating experiments on oysters. My thanks also to my host laboratory, the Institute for Marine Environmental Research, for library and secretarial assistance.

References

Ansel, A. D. (1961). Reproduction, growth and mortality of *Venus striatula* (da Costa) in Kames Bay, Millport. *J. Mar. Biol. Assoc. U.K.* **41**, 191–215.

Ansell, A. D. (1968). The rate of growth of the hard clam *Mercenaria mercenaria* (L) throughout the geographical range. *J. Cons., Cons. Int. Explor. Mer* **31**, 364–409.

Babinchak, J., and Ukeles, R. (1979). Epifluoresence microscopy, a technique for the study of feeding in *Crassostrea virginica* veliger larvae. *Mar. Biol. (Berlin)* **51**, 69–76.

Bayne, B. L. (1965). Growth and the delay of metamorphosis of the larvae of *Mytilus edulis* (L). *Ophelia* **2**, 1–47.

Bayne, B. L. (1972). Some effects of stress in the adult on the larval development of *Mytilus edulis*. *Nature (London)* **237**, 459.

Bayne, B. L., ed. (1976). "Marine Mussels, Their Ecology and Physiology." Cambridge Univ. Press, London and New York.

Bayne, B. L., Gabbott, P. A., and Widdows, J. (1975). Some effects of stress in the adult on the eggs and larvae of *Mytilus edulis* L. *J. Mar. Biol. Assoc. U.K.* **55**, 675–689.

Bickell, L. R., Chia, F. S., and Crawford, B. J. (1981). Morphogenesis of the digestive system during metamorphosis of the nudibranch *Doridella steinbergae* (Gastropoda): Conversion from phytoplanktovore to carnivore. *Mar. Biol. (Berlin)* **62**, 1–16.

Bruce, J. R., Knight, M., and Parke, M. W. (1939). The rearing of oyster larvae on an algal diet. *J. Mar. Biol. Assoc. U.K.* **24**, 337–374.

Cain, T. D. (1973). The combined effects of temperature and salinity on embryos and larvae of the clam *Rangia cuneata*. *Mar. Biol. (Berlin)* **21**, 1–6.

Calabrese, A. (1969). Individual and combined effects of salinity and temperature on embryos and larvae of the coot clam, *Mulinia lateralis* (Say). *Biol. Bull. (Woods Hole, Mass.)* **137**, 417–428.

Calabrese, A., and Davis, H. C. (1966). The pH tolerance of embryos and larvae of *Mercenaria mercenaria* and *Crassostrea virginica*. *Biol. Bull. (Woods Hole, Mass.)* **131**, 427–436.

Calabrese, A., and Davis, H. C. (1970). Tolerances and requirements of embryos and larvae of bivalve moluscs. *Helgol. Wiss. Meeresunters.* **20**, 553–564.

Chanley, P., and Andrews, J. D. (1971). Aids for identification of bivalve larvae of Virginia. *Malacologia* **10**, 45–120.

Chanley, P., and Normandin, R. F. (1967). Use of artificial foods for larvae of the hard clam *Mercenaria mercenaria* (L). *Proc. Natl. Shellfish Assoc.* **57**, 31–37.

Chia, F., and Koss, R. (1978). Development and metamorphosis of the long planktotrophic larvae of *Rostanga pulchra* MacFarland (Mollusca: Nudibranchia). *Mar. Biol. (Berlin)* **46**, 109–119.

Conover, R. J. (1978). Transformation of organic matter. *In* "Marine Ecology" (O. Kinne, ed.), Vol. 4, pp. 221–499. Wiley, New York.

Conover, R. J. (1981). Nutrional strategies for feeding on small suspended particles. *In* "Analysis of Marine Ecosystems" (A. R. Longhurst, ed.), pp. 363–395. Academic Press, New York.

Conover, R. J., and Francis, V. (1973). The use of radioactive isotopes to measure the transfer of materials in aquatic food chains. *Mar. Biol. (Berlin)* **18**, 272–283.

Courtright, R. C., Breese, W. P., and Krueger, H. (1971). Formulation of a synthetic seawater for bioassays with *Mytilus edulis* embryos. *Water Res.* **5**, 877–888.

Crisp, D. J. (1974). Energy relations of marine invertebrate larvae. *Thallasia Jugosl.* **10**, 103–120.

Davis, H. C. (1953). On food and feeding of larvae of the American oyster *Crassostrea virginica*. *Biol. Bull. (Woods Hole, Mass.)* **104**, 334–350.

Davis, H. C. (1960). Effects of turbidity producing materials in seawater on eggs and larvae of the clam *Venus (Mercenaria) mercenaria*. *Biol. Bull. (Woods Hole, Mass.)* **118**, 48–54.

Davis, H. C., and Ansell, A. D. (1962). Survival and growth of larvae of the European oyster *Ostrea edulis* at lowered salinities. *Biol. Bull. (Woods Hole, Mass.)* **122**, 33–39.

Davis, H. C., and Calabrese, A. (1964). Combined effects of temperature and salinity on development of eggs and growth of larvae of *M. mercenaria* and *C. virginica*. *Fish. Bull.* **63**, 643–655.

Davis, H. C., and Chanley, P. E. (1956). Effects of some dissolved substances on bivalve larvae. *Proc. Natl. Shellfish. Assoc.* **46**, 59–74.

Davis, H. C., and Guillard, R. R. (1958). Relative value of ten genera of micro-organisms as foods for oyster and clam larvae. *Fish. Bull.* **58**, 293–304.

Davis, H. C., and Hidu, H. (1969). Effects of turbidity-producing substances in sea water on eggs and larvae of three genera of bivalve molluscs. *Veliger* **11**, 316–323.

Dehnel, P. A. (1955). Rates of growth of gastropods as a function of latitude. *Physiol. Zool.* **28**, 115–144.

Ewart, J. W., and Epifanio, C. E. (1981). A tropical flagellate food for larval and juvenile cysters, *Crassostrea virginica*. *Aquaculture* **22**, 297–300.

Frankboner, P. V., and De Burgh, M. E. (1978). Comparative rates of dissolved organic carbon accumulation of juveniles and pediveligers of the Japanese oyster *Crassostrea gigas* Thunberg. *Aquaculture* **13**, 205–212.

Fretter, V. (1967). The prosobranch veliger. *Proc. Malacol. Soc. London* **37**, 357–366.

Fretter, V., and Montgomery, M. C. (1968). The treatment of food by prosobranch veligers. *J. Mar. Biol. Assoc. U.K.* **48**, 499–520.

Frost, B. W. (1975). A threshold feeding behaviour in *Calanus pacificus*. *Limnol. Oceanogr.* **17**, 805–815.

Gabbott, P. A., and Holland, D. L. (1973). Growth and metabolism of *Ostrea edulis* larvae. *Nature (London)* **241**, 475–476.

Gallager, S. M., and Mann, R. (1980). An apparatus for the measurement of grazing activity of filter feeders at constant food concentrations. *Mar. Biol. Lett.* **1**, 341–349.

Galstoff, P. S. (1964). The American oyster. *Fish Bull.* **64**, 1–480.

Grunbaum, B. W., Siegal, B. V., Schulz, A. R., and Kirk, P. L. (1955). Determination of oxygen uptake by tissue growth in an all glass differential microrespirometer. *Mikrochim. Acta* **6**, 1069–1075.

Gustafson, R. G. (1980). Dissolved free amino acids in the nutrition of larvae of the soft-shell clam *Mya arenaria*. University of Maine, Orono (unpublished M.Sc. thesis).

Gwyther, J., and Munro, J. L. (1981). Spawning induction and rearing of larvae of tridacnid clams (Bivalvia: Tridacnidae). *Aquaculture* **24**, 197–217.

Helm, M. M. (1977). Mixed algal feeding of *Ostrea edulis* larvae with *Isochrysis galbana* and *Tetraselmis suecica*. *J. Mar. Biol. Assoc. U.K.* **57**, 1019–1029.

Helm, M. M., and Millican, P. F. (1977). Experiments in the hatchery rearing of Pacific oyster larvae (*Crassostrea gigas* Thunberg). *Aquaculture* **11**, 1–12.

Helm, M. M., Holland, D. L., and Stephenson, R. R. (1973). The effect of supplemental algal feeding of a hatchery breeding stock of *Ostrea edulis* L. on larval vigour. *J. Mar. Biol Assoc. U.K.* **53**, 673–684.

Hidu, H., and Tubiash, H. S. (1963). A bacterial basis for the growth of antibiotic treated bivalve larvae. *Proc. Natl. Shellfish. Assoc.* **54**, 25–39.

Holland, D. L. (1978). Lipid reserves and energy metabolism in the larvae of benthic marine invertebrates. *Biochem. Biophys. Perspect. Mar. Biol.* **4**, 85–123.

Holland, D. L., and Spencer, B. E. (1973). Biochemical changes in fed and starved oysters, *Ostrea edulis*. L. during larval development, metamorphosis and early spat growth. *J. Mar. Biol. Assoc. U.K.* **53**, 287–298.

Hrs-Brenko, M., and Calabrase, A. (1969). The combined effects of salinity and temperature on larvae of the mussel *Mytilus edulis*. *Mar. Biol. (Berlin)* **4**, 224–226.

Hughes, R. N. (1980). Optimal foraging theory in the marine context. *Oceanogr. Mar. Biol.* **18**, 423–481.

Innes, D. J., and Haley, L. E. (1977). Genetic aspects of larval growth under reduced salinity in *Mytilus edulis*. *Biol. Bull. (Woods Hole, Mass.)* **153**, 312–321.

Jablonski, D., and Lutz, R. A. (1980). Molluscan larval shell morphology. *In* "Skeletal Growth of Aquatic Organisms" (D. C. Rhoads and R. A. Lutz, eds.), pp. 323–377. Plenum, New York.

Jameson, S. C. (1976). Early life history of the giant clams *Tridacna crocea, Tridacna maxima* and *Hippopus hippopus*. *Pac. Sic.* **30**, 219–233.

Johannes, R. E., and Satomi, M. (1967). Measuring organic matter retained by aquatic invertebrates. *J. Fish. Res. Board Can.* **24**, 2467–2471.

Jørgensen, C. B. (1952). Efficiency of growth in *Mytilus edulis* and two gastropod veligers. *Nature (London)* **170**, 714.

Jørgensen, C. B. (1976). Growth efficiencies and factors controlling size in some mytilid bivalves, especially *Mytilus edulis* L: Review and interpretation. *Ophelia* **15**, 175–192.

Jørgensen, C. B. (1981). A hydromechanical principle for particle retention in *Mytilus edulis* and other ciliary suspension feeders. *Mar. Biol. (Berlin)* **61**, 277–282.

Kempf, S. C., and Willows, A. O. D. (1977). Laboratory culture of the nudibranch *Tritonia diomedea* Bergh (Tritoniidae, Opisthobranchia) and some aspects of its behavioural development. *J. Exp. Mar. Biol. Ecol.* **30**, 261–276.

Kennedy, V. S., Roosenberg, W. H., Zion, H. H., and Castagna, M. (1974). Temperature–time relationships for survival of embryos and larvae of *Mulinia lateralis* (Mollusca: Bivalvia). *Mar. Biol. (Berlin)* **24**, 137–145.

La Barbera, M. (1975). Larval and post larval development of the giant clams *Tridacna maxima* and *Tridacna squamosa* (Bivalvia. Tridacnidae). *Malacologia* **15**, 69–79.

Lam, R. K., and Frost, B. W. (1976). Model of copepod filtering response to changes in size and concentration of food. *Limnol. Oceanogr.* **21**, 490–500.

Lane, C. E., Posner, G. S., and Greenfield, L. J. (1952). Distribution of glycogen in the shipworm. *Bull. Mar. Sci.* **2**, 385–392.

Langdon, C. J., and Waldock, M. J. (1981). The effect of algal and artificial diets on the growth and fatty acid composition of *Crassostrea gigas* spat. *J. Mar. Biol. Assoc. U.K.* **61**, 431–448.

Lannan, J. E. (1980a). Broodstock management of *Crassostrea gigas*. I. Genetic and environmental variation in survival in the larval rearing system. *Aquaculture* **21**, 323–336.

Lannan, J. E. (1980b). Broodstock management of *Crassostrea gigas*. II. Broodstock conditioning to maximize larval survival. *Aquaculture* **21**, 337–345.

Lannan, J. E. (1980c). Broodstock management of *Crassostrea gigas*. III. Selective breeding for improved larval survival. *Aquaculture* **21**, 347–351.

Lehman, J. T. (1976). The filter-feeder as an optimal forager, and the predicted shapes of feeding curves. *Limnol. Oceanogr.* **21**, 501–516.

Lehnberg, W., and Theede, H. (1979). Kombinierte Wirkungen von Temperatur, Salzgehalt und Cadmium anf Entwicklung, Wachstum und Mortalifat der Larven von *Mytilus edulis* aus der westlichen ostree. *Helgol. Wiss. Meeresunters.* **32**, 179–199.

Levins, R. (1968). "Evolution in Changing Environments." Princeton Univ. Press, Princeton, New Jersey.

Loosanoff, V. L. (1954). New advances in the study of bivalve larvae. *Am. Sci.* **42**, 607–624.

Loosanoff, V. L. (1959). The size and shape of metamorphosing larvae of *Venus (Mercenaria) mercenaria* grown at different temperatures *Biol. Bull. (Woods Hole, Mass.)* **117**, 308–318.

Loosanoff, V. L. (1962). Effect of turbidity on some larval and adult bivalves. *Proc. Annu. Gulf Caribb. Fish. Inst.* **14**, 80–94.

Loosanoff, V. L., and Davis, H. C. (1963). Rearing of bivalve molluscs. *Adv. Mar. Biol.* **1**, 1–136.

Loosanoff, V. L., Miller, W. S., and Smith, P. B. (1951). Growth and setting of larvae of *Venus mercenaria* in relation to temperature. *J. Mar. Res.* **10**, 59–81.

Loosanoff, V. L., Davis, H. C., and Chanley, P. E. (1966). Dimensions and shapes of larvae of some marine bivalve molluscs. *Malacologia* **4**, 351–435.

Losee, E. (1979). Relationship between larval and spat growth rates in the oyster *Crassostrea virginica*. *Aquaculture* **16**, 123–126.

Lough, R. G. (1974). A re-evaluation of the combined effects of temperature and salinity on survival and growth of *Mytilus edulis* larvae using response surface techniques. *Proc. Natl. Shellfish. Assoc.* **64**, 73–76.

Lough, R. G. (1975). A re-evaluation of the combined effects of temperature and salinity on survival and growth of bivalve larvae using response surface techniques. *Fish. Bull.* **73**, 86–94.

Lough, R. G., and Gonor, J. J. (1971). Early embryonic stages of *Adula californiensis* (Pelecypoda: Mytilidae) and the effect of temperature and salinity on development rate. *Mar. Biol. (Berlin)* **8**, 118–125.

Lough, R. G., and Gonor, J. J. (1973a). A response–surface approach to the combined effects of temperature and salinity on the larval development of *Adula californiensis* (Pelecypoda: Mytilidae). I. Survival and growth of three and fifteen day old larvae. *Mar. Biol. (Berlin)* **22**, 241–250.

Lough, R. G., and Gonor, J. J. (1973b). A response-surface approach to the combined effects of temperature and salinity in the larval development of *Adula californiensis* (Pelecypoda: Mytilidae). II. Long-term larval survival and growth in relation to respiration. *Mar. Biol. (Berlin)* **22**, 295–305.

Lucas, J. S., and Costlow, J. D., Jr. (1979). Effects of various temperature cycles on the larval development of the gastropod mollusc *Crepidula fornicata*. *Mar. Biol. (Berlin)* **51**, 111–117.

Lutz, R. A., and Jablonski, D. (1978). Larval bivalve shell morphometry: A new palaeoclimatic tool? *Science* **202**, 51–53.

MacInnes, J. R., and Calabrese, A. (1979). Combined effects of salinity, temperature and copper on embryos and early larvae of the American oyster, *Crassostrea virginica*. *Arch. Environ. Contam. Toxicol.* **8**, 553–562.

Malouf, R. E., and Breese, W. P. (1977). Food consumption and growth of larvae of the Pacific oyster, *Crassostrea gigas* (Thunberg), in a constant flow rearing system. *Proc. Natl. Shellfish. Assoc.* **67**, 7–16.

Manahan, D. T., and Crisp, D. J. (1982). The role of dissolved organic material in the nutrition of pelagic larvae; amino acid uptake by bivalve veligers. *Amer. Zool.* **22**, 635–646.

Mann, R. (1979). Some biochemical and physiological aspects of growth and gametogenesis in *Crassostrea gigas* and *Ostrea edulis* grown at sustained elevated temperatures. *J. Mar. Biol. Assoc. U.K.* **59**, 95–110.

Mapstone, G. M. (1970). Feeding activities of veligers of *Nassarius reticulatus* and *Crepidula fornicata* and the use of artificial foods in maintaining cultures of these larvae. *Helgol. Wiss. Meeresunters.* **20**, 505–575.

Martin, Y. R., and Mengus, B. M. (1977). Utilisation de souches bacteriennes selectionees dans l'alimentation des larves de *Mytilus galloprovincialis* Limk (Mollusque bivalve) en elevages experimentaux. *Aquaculture* **10**, 253–262.

Millar, R. H. (1955). Notes on the mechanism of food movement in the gut of the larval oyster *Ostrea edulis*. *Q. J. Microsc. Sci.* **96**, 539–544.

Millar, R. H., and Scott, J. M. (1967). The larvae of the oyster *Ostrea edulis* during starvation. *J. Mar. Biol. Assoc. U.K.* **47**, 475–484.

Mullin, M. M., Stewart, E. F., and Fuglister, F. J. (1975). Ingestion by planktonic grazers as a function of concentration of food. *Limnol. Oceanogr.* **20**, 259–262.

Nasciomento, I. A. (1980). Growth of the larvae of *Crassostrea gigas* Thunberg, fed with different algal species at high cell concentrations. *J. Cons., Cons. Int. Explor. Mer.* **89**, 134–139.

Newell, R. C., Johnson, L. G., and Kofoed, L. H. (1977). Adjustment of the components of energy balance in response to temperature change in *Ostrea edulis*. *Oecologia* **30**, 97–110.

Newkirk, G. F. (1978). Interaction of genotype and salinity in larvae of the oyster *Crassostrea virginica*. *Mar. Biol. (Berlin)* **48**, 227–234.

Newkirk, G. F. (1980). Review of the genetics and the potential for selective breeding of commercially important bivalves. *Aquaculture* **19**, 209–228.

Newkirk, G. F., Haley, L. E., Waugh, D. L., and Doyle, R. (1977). Genetics of larvae and spat growth rate with the oyster *Crassostrea virginica*. *Mar. Biol. (Berlin)* **41**, 49–52.

Parsons, T. R., Takahashi, M., and Hargrave, B. (1977). "Biological Oceanographic Processes," 2nd ed. Pergamon, Oxford.

Parsons, T. R., Le Brasseur, R. J., and Fulton, J. D. (1967). Some observations on the dependence of zooplankton grazing on the cell size and concentration of phytoplankton blooms. *J. Oceanogr. Soc. Jpn.* **23**, 10–17.

Paulson, T. C., and Scheltema, R. S. (1968). Selective feeding on algal cells by the veliger larvae of *Nassarius obsoletus* (Gastropoda, Prosbranchia). *Biol. Bull. (Woods Hole, Mass.)* **134**, 481–489.

Pechenik, J. A. (1980). Growth and energy balance during the larval lives of three prosobranch gastropods. *J. Exp. Mar. Biol. Ecol.* **44**, 1–28.

Pechenik, J. A., and Fisher, N. S. (1979). Feeding, assimilation and growth of mud snail larvae, *Nassarius obsoletus* (Say), on three different algal diets. *J. Exp. Mar. Biol. Ecol.* **38**, 57–80.

Perron, F. E. (1980). Laboratory culture of the larvae of *Conus textile* L. (Gastropoda: Toxoglossa). *J. Exp. Mar. Biol. Ecol.* **42**, 27–38.

Perron, F. E., and Turner, R. D. (1977). Development, metamorphosis and natural history of the nudibranch *Doridella obscura* Verrill (Coramibidae: Opisthobranchia). *J. Exp. Mar. Biol. Ecol.* **27**, 171–185.

Pilkington, M. C., and Fretter, V. (1970). Some factors affecting the growth of prosobranch veligers. *Helgol. Wiss. Meeresunters.* **20**, 576–593.

Rice, M. A., Wallis, K., and Stephens, G. C. (1980). Influx and netflux of amino acids into larval and juvenile European flat oysters, *Ostrea edulis* L. *J. Exp. Mar. Biol. Ecol.* **48**, 51–59.

Riisgard, H. U., Randlov, A., and Kristensen, P. S. (1980). Rates of water processing, oxygen consumption and efficiency of particle retention in veligers and young post-metamorphic *Mytilus edulis*. *Ophelia* **19**, 37–47.

Russell-Hunter, W. D., Apley, M. L., and Hunter, R. D. (1972). Early life history of *Melampus* and the significance of semilunar synchrony. *Biol. Bull. (Woods Hole, Mass.)* **143**, 623–656.

Scheltema, R. S. (1961). Metamorphosis of the veliger larvae of *Nassarius obsoletus* (Gastropoda) in response to bottom sediment. *Biol. Bull. (Woods Hole, Mass.)* **120**, 92–109.

Scheltema, R. S. (1962). Pelagic larvae of New England intertidal gastropods I *Nassarius obsoletus* Say and *Nassarius vibex* Say. *Trans. Am. Microsc. Soc.* **81**, 1–11.

Scheltema, R. S. (1965). The relationship of salinity to larval survival and development in *Nassarius obsoletus* (Gastropoda). *Biol. Bull. (Woods Hole, Mass.)* **129**, 340–354.

Scheltema, R. S. (1967). The relationship of temperature to the larval development of *Nassarius obsoletus* (Gastropoda). *Biol. Bull. (Woods Hole, Mass.)* **132**, 253–265.

Seed, R. (1976). Ecology. *In* "Marine Mussels, Their Ecology and Physiology" (B. L. Bayne, ed.), pp. 13–65. Cambridge Univ. Press, London and New York.

Sing, S. M., and Zouros, E. (1978). Genetic variation associated with growth rate in the American oyster (*Crassostrea virginica*). *Evolution* **32**, 342–353.

Smith, W. L., and Chanley, M. H., eds. (1975). "Culture of Marine Invertebrate Animals." Plenum, New York.

Sorokin, Y. I. (1981). Microheterotrophic organisms in marine ecosystems. *In* "Analysis of Marine Ecosystems" (A. R. Longhurst, ed.), pp. 293–342. Academic Press, New York.

Sorokin, Y. I., and Wyshkwartzev, D. I. (1973). Feeding on dissolved organic matter by some marine animals. *Aquaculture* **2**, 141–148.

Sprung, M. (1982). Untersuchungen zum Energiebudget der Larven der Miesmuschel, *Mytilus edulis* L. Dissertation Universität Kiel (unpublished doctorate thesis).

Stephens, G. C. (1981). The trophic role of dissolved organic material *In* ''Analysis of Marine Ecosystems'' (A. R. Longhurst, ed.), pp. 271–291. Academic Press, New York.

Stickney, A. P. (1964). Salinity, temperature and food requirements of soft-shell clam larvae in laboratory culture. *Ecology* **45**, 283–291.

Strathmann, R. R. (1967). Estimating the organic carbon content of phytoplankton from cell volume or plasma volume. *Limnol. Oceanogr.* **12**, 411–418.

Strathmann, R. R., and Leise, E. (1979). On feeding mechanisms and clearance rates of molluscan veligers. *Biol. Bull. (Woods Hole, Mass.)* **157**, 524–535.

Strathmann, R. R., Jahn, T. L., and Fonseca, J. R. C. (1972). Suspension feeding by marine invertebrate larvae: Clearance of particles by ciliated bands of a rotifer, pluteus and trochophore. *Biol. Bull. (Woods Hole, Mass.)* **142**, 505–519.

Sullivan, C. M. (1948). Bivalve larvae of Malpique Bay, P. E. I. *Bull. Fish Res. Board Can.* **77**.

Thompson, T. E. (1959). Feeding in nudibranch larvae. *J. Mar. Biol. Assoc. U.K.* **38**, 239–248.

Thorson, G. (1946). Reproduction and larval development of Danish marine bottom invertebrates. *Medd. Dan. Fisk.- Havunders., Ser. Plankton* **4**, 1–523.

Tubiash, H. S. Chanley, P. E., and Leifson, E. (1965). Bacterial necrosis, a disease of larval and juvenile bivalve molluscs. *J. Bacteriol.* **90**, 1036–1044.

Ukeles, R., and Sweeney, B. M. (1969). Influence of dinoflagellate trichocysts and other factors on the feeding of *Crassostrea virginica* larvae on *Monochrysis lutheri. Limnol. Oceanogr.* **14**, 403–410.

Ursin, E. (1963). On the incorporation of temperature in the von Bertalanffy growth equation. *Medd. Dan. Fisk.- Havunders.* **4**, 1–16.

Vernberg, W. B. (1972). Metabolic–environmental interaction in the marine plankton. *Proc. Eur. Mar. Biol. Symp., 5th, 1970* pp. 189–196.

Vernberg, W. B., and Vernberg, F. J. (1975). The physiological ecology of larval *Nassarius obsoletus* Say. *Proc. Eur. Mar. Biol. Symp., 9th, 1974* pp. 179–190.

Waldock, M. J., and Nasciomento, I. A. (1979). The triacylglycerol composition of *Crassostrea gigas* larvae fed on different algal diets. *Mar. Biol. Lett.* **1**, 77–86.

Walne, P. R. (1956). Experimental rearing of the larvae of *Ostrea edulis* L. in the laboratory. *Fish. Invest., London, Ser.* 2 **20** (9), 1–23.

Walne, P. R. (1958). Growth of oysters (*Ostrea edulis* L.). *J. Mar. Biol. Assoc. U.K.* **37**, 591–602.

Walne, P. R. (1963). Observations on the food value of seven species of algae to the larvae of *Ostrea edulis* L. Feeding experiments. *J. Mar. Biol. Assoc. U.K.* **43**, 767–784.

Walne, P. R. (1965). Observations on the influence of food supply and temperature on the feeding and growth of the larvae of *Ostrea edulis* L. *Fish. Invest., London, Ser.* 2 **24** (1), 1–45.

Walne, P. R. (1966). Experiments in the large-scale culture of the larvae of *Ostrea edulis* L. *Fish. Invest., London, Ser.* 2 **25** (4), 1–53.

Walne, P. R. (1970). Present problems in the culture of the larvae of *Ostrea edulis. Helgol. Wiss. Meeresunters.* **20**, 514–525.

Walne, P. R. (1974). ''Culture of Bivalve Molluscs.'' Fishing News (Books) Ltd., Surrey, England.

Walne, P. R., and Spencer, B. E. (1968). Some factors affecting the growth of the larvae of *Ostrea edulis. Int. Counc. Explor. Sea, C. M. Pap. Rep.* **K15**, 1–7 (mimeo.).

Wangersky, P. J. (1978). Production of dissolved organic matter. *In* ''Marine Ecology'' (O. Kinne, ed.), Vol. 4, pp. 115–220. Wiley, New York.

Welch, H. E. (1968). Relationships between assimilation efficiencies and growth efficiencies for aquatic consumers. *Ecology* **49**, 755–759.

Werner, B. (1955). Uber die Anatomie, die Entwicklung und Biologie des Veligers and der Veliconcha von *Crepidula fornicata* L. (Gastropoda Prosobranchia). *Helgol. Wiss. Meeresunters.* **5**, 169–217.

Williams, L. G. (1980). Development and feeding of larvae of the nudibranch *Hermissenda crassicornis* and *Aeolida papillosa. Malacologia* **20**, 99–116.

Wilson, J. H. (1978). The food value of *Phaeodactylum tricornutum* Bohlin to the larvae of *Ostrea edulis* L and *Crassostrea gigas* Thunberg. *Aquaculture* **13**, 313–323.

Wilson, J. H. (1979). Observations on the grazing rates and growth of *Ostrea edulis* L. larvae when fed algal cultures of different ages. *J. Exp. Mar. Biol. Ecol.* **38**, 187–199.

Wilson, J. H. (1980). Particle retention and selection by larvae and spat of *Ostrea edulis* in algal suspensions *Mar. Biol. (Berlin)* **57**, 135–145.

Wright, S. H. (1979). Effect of activity of lateral cilia on transport of amino acids in gills of *Mytilus californianus*. *J. Exp. Zool.* **209**, 209–220.

Yonge, C. M. (1926). Structure and physiology of the organs of feeding and digestion in *Ostrea edulis*. *J. Mar. Biol. Assoc. U.K.* **14**, 295–386.

Zeuthen, E. (1947). Body size and metabolic rate in the animal kingdom with special regard to the marine microfauna. *C. R. Tray. Lab. Carlsberg, Ser. Chim.* **26**, 17–161.

Zeuthen, E. (1953). Oxygen uptake as related to body size in organisms. *Q. Rev. Biol.* **28**, 1–12.

Index